BIOCHEMISTRY RESEARCH TRENDS

LEUCINE

BIOLOGY, CONSUMPTION AND BENEFITS

BIOCHEMISTRY RESEARCH TRENDS

Additional books in this series can be found on Nova's website under the Series tab.

Additional e-books in this series can be found on Nova's website under the e-book tab.

BIOCHEMISTRY RESEARCH TRENDS

LEUCINE

BIOLOGY, CONSUMPTION AND BENEFITS

SOPHIA R. NEWMAN
EDITOR

nova publishers
New York

Library of Congress Cataloging-in-Publication Data

ISBN: 978-1-63482-593-1

Published by Nova Science Publishers, Inc. † New York

Contents

PREFACE

Leucine is a branched-chain essential amino acid. It accounts for approximately one-fifth of the requirements for essential amino acids. Leucine facilitates insulin-stimulated glucose uptake and insulin signaling in skeletal muscle cells. This book includes chapters on the physiological roles of leucine and the effects of leucine supplementation; the syntheses, properties and applications of the main leucine containing polymers and materials made of them that up to now have been studied; and the effects of hydroxy- beta methylbutyrate (HMB) and alpha-ketoisocaproate (KIC), which are metabolites of leucine on muscle function and sports performance based on recent findings.

Chapter 1 – Leucine facilitates insulin-stimulated glucose uptake and insulin signaling in skeletal muscle cells with a mechanism that involves mammalian target of rapamycin complex 1 and complex 2(mTORC1 and mTORC2). Ursolic acid inhibits leucine-stimulated mTORC1 signaling by suppressing mTOR localization to lysosomes. Whole body protein mass is preserved and protein anabolism is enhanced by furnishing a leucine-rich diet during caloric restriction. Enhanced leucine intake may be of utility in inhibiting too much protein waste when there is an excessive weight loss. The leucine metabolite beta-hydroxy-beta-methylbutyrate is effective in inhibiting exercise-related muscle injury in healthy trained and untrained subjects and muscle wastage in chronic diseases. The recommended dose to maintain or improve muscle mass and function in health and disease is 3 g/day. Leucine treatment enhances oxidative capacity through complete carbohydrate oxidation and increased mitochondrial density in skeletal muscle cells. A significant increase of serum insulin-like growth factor 1 and protein/RNA ratio in the liver was observed in leucine-supplemented rats. Chronic leucine supplementation enhances insulin sensitivity and body weight by minimizing oxidative stress and improving insulin signaling pathway in animals receiving a high-fat diet. Leucine directly facilitates insulin signaling via a Gαi protein-dependent intracellular signaling pathway in hepatocytes. Dietary leucine upregulates whole-body insulin sensitivity independent of body fat in diet-induced obese rats. Leucine exerts a preventive action on high-fat diet-induced obesity, hyperglycemia, and dyslipidemia in animals. The upregulated expression of SIRT1 following leucine supplementation may bring about diminished acetylation of PGC1α and FoxO1, which is associated with attenuation of high-fat diet-induced mitochondrial dysfunction, insulin resistance, and obesity. Immunity to gastrointestinal parasites during lactation is compromised if there is leucine and methionine deficiency. Plasma lipid profile has been evaluated In adult rats treated with dexamethasone, an experimental model of dyslipidemia, Plasma levels of triglyceride and very low density lipoprotein-cholesterol underwent a significant decline and high density lipoprotein-cholesterol level was elevated following one week of leucine supplementation combined with resistance exercise, despite no improvements in plasma total cholesterol and low density lipoprotein–cholesterol levels. Following exercise-induced muscle damage, high-dose leucine supplementation may help maintain force output during isometric contractions.

Chapter 2 – During last decades great efforts have been made to get polymers constituted by α-amino acid units since they form smart materials, show catalytic activity, are promising as micro and nanocontainers for drug delivery purposes, as biodegradable and biocompatible materials, to name a few. Unique physical-chemical properties of amino acid based polymers are of great interest as well. Among these properties self-assembling mechanisms of peptides and proteins nowadays represent an object of extensive studies. Specifically, leucine rich sequences are usually involved in "coiled-coil" motifs in which α-helices with a short repeat sequence are coiled intertwiningly. Leucine zippers are involved in the dimerization of a number of transcription factors, playing the hydrophobic interactions provided by leucine residues a remarkable stabilizing effect. Properties and functions of biomacromolecules and synthetic polymers can be integrated in a single material. Hence, hybrid block copolymers based on peptides are receiving great attention due to the capability to self-assemble into highly organized nanoscale structures and their wide applications in the biomedical field that covers neural tissue engineering; scaffolds for bone regeneration; dental implants; drug, gene and protein delivery and biosensors for detection of viruses and other pathogens. These molecular chimeras might have advantageous properties provided by the synthetic block such as economy, solubility, melt processability, versatility, or rubber elasticity as well as those given by the peptide block (e.g., structure, functionality, high mechanical performance or biodegradability). Hydrophilic–hydrophobic characteristics of the considered polymers can be easily controlled by changing their specific α-amino acid units and even by the incorporation of hydrophilic polyethylene glycol segments. These copolymers are highly interesting to get suitable polymeric nanoparticles for drug delivery uses as well as new temperature responsive polymers. Finally, the development of hybrid materials based on conducting and biodegradable polymers (e.g., those containing leucine units) merits attention since these hybrids may be a promising solution to cover aspects like stimulation of either the proliferation or differentiation of various cell types. Amino acid based polymers are especially promising as biodegradable biomaterials since can give rise to non-toxic and nutritious (physiological) degradation products. New biodegradable polymers such as poly(ester amide)s and poly(ester urea)s have been developed and even their interest as new biomedical materials have been clearly demonstrated. It was shown that leucine-based polymers possess much better solubility in suitable organic solvents like ethanol. Besides, leucine-based biodegradable polymers are safer for patients as compared with e.g., phenylalanine-based analogues. The present review deals with syntheses, properties and applications of the main leucine containing polymers and materials made of them that up to now have been studied.

Chapter 3 – Alpha-ketoisocaproate (KIC) and Beta hydroxy- beta methylbutyrate (HMB) are metabolites of the amino acid leucine. The first step in HMB metabolism is the reversible transamination of leucine to KIC, which occurs mainly extrahepatically. Following this enzymatic reaction, KIC is converted to HMB by two routes: through the cytosolic enzyme KIC dioxygenase; or through formation of isovaleryl-CoA by branched-chain ketoacid dehydrogenase (BCKD) in liver, which after some steps results in HMB formation. Under normal conditions the majority of KIC is converted into isovaleryl-CoA, with approximately 5% of leucine consumed metabolized into HMB. HMB and KIC have been proposed to enhance physical performance and muscle hypertrophy via their effects on muscle proteolysis and protein synthesis. Studies investigating the mechanisms by which HMB promotes increases in strength and muscle hypertrophy are still in preliminary stages. Based on the results of these studies, it is postulated that enhancements from leucine metabolite supplementation involve the following mechanisms: improvements in sarcolemal integrity; upregulation of IGF-1 gene expression in skeletal muscle and IGF-1 production by liver; stimulation of protein synthesis by activating the mTOR signaling pathway; suppression of skeletal muscle proteolysis by the inhibition of the ubiquitin–proteasome system; and modulation of lipopolysaccharides, tumor necrosis factor alpha (TNF-alpha) and angiotensin 2, attenuating protein synthesis depression in catabolic conditions. In turn, KIC has been used in conjunction with glycine and arginine (marketed as GAKIC). Acute GAKIC ingestion has been demonstrated to significantly enhance human performance during intense muscle contractions and repeated cycling sprints. The proposed mechanism of action of GAKIC includes: removal of deaminated nitrogen waste produced during exercise; regeneration of L-leucine via leucine dehydrogenase enzyme; promoting HMB synthesis via KIC dioxygenase enzyme; reducing ammonia levels during intense exercise; and increasing ATP levels through Isovaleryl-CoA and Acetyl-CoA formation. Interestingly, KIC ingestion does not result in the same effects of GAKIC, suggesting that glycine and arginine are both necessary for the enhanced performance. In this chapter the effects of HMB and KIC on muscle function and sports performance will be discussed, based on recent research findings.

In: Leucine
Editor: Sophia R. Newman

ISBN: 978-1-63482-593-1

Chapter 1

PHYSIOLOGICAL ROLES OF LEUCINE AND EFFECTS OF LEUCINE SUPPLEMENTATION

Tzi Bun Ng* and Charlene Cheuk Wing Ng
School of Biomedical Sciences,Faculty of Medicine,The Chinese University of Hong Kong, New Territories, Hong Kong, China

ABSTRACT

Leucine facilitates insulin-stimulated glucose uptake and insulin signaling in skeletal muscle cells with a mechanism that involves mammalian target of rapamycin complex 1 and complex 2(mTORC1 and mTORC2). Ursolic acid inhibits leucine-stimulated mTORC1 signaling by suppressing mTOR localization to lysosomes. Whole body protein mass is preserved and protein anabolism is enhanced by furnishing a leucine-rich diet during caloric restriction. Enhanced leucine intake may be of utility in inhibiting too much protein waste when there is an excessive weight loss. The leucine metabolite beta-hydroxy-beta-methylbutyrate is effective in inhibiting exercise-related muscle injury in healthy trained and untrained subjects and muscle wastage in chronic diseases. The recommended dose to maintain or improve muscle mass and function in health and disease is 3 g/day. Leucine treatment enhances oxidative capacity through complete carbohydrate oxidation and increased mitochondrial density in skeletal muscle cells. A significant increase of serum insulin-like growth factor 1 and protein/RNA ratio in

* Corresponding author, Email: b021770@mailserv.cuhk.edu.hk, fax: 852 26035123, phone: 852 39436872.

the liver was observed in leucine-supplemented rats. Chronic leucine supplementation enhances insulin sensitivity and body weight by minimizing oxidative stress and improving insulin signaling pathway in animals receiving a high-fat diet. Leucine directly facilitates insulin signaling via a Gαi protein-dependent intracellular signaling pathway in hepatocytes. Dietary leucine upregulates whole-body insulin sensitivity independent of body fat in diet-induced obese rats. Leucine exerts a preventive action on high-fat diet-induced obesity, hyperglycemia, and dyslipidemia in animals. The upregulated expression of SIRT1 following leucine supplementation may bring about diminished acetylation of PGC1α and FoxO1, which is associated with attenuation of high-fat diet-induced mitochondrial dysfunction, insulin resistance, and obesity. Immunity to gastrointestinal parasites during lactation is compromised if there is leucine and methionine deficiency. Plasma lipid profile has been evaluated In adult rats treated with dexamethasone, an experimental model of dyslipidemia, Plasma levels of triglyceride and very low density lipoprotein-cholesterol underwent a significant decline and high density lipoprotein-cholesterol level was elevated following one week of leucine supplementation combined with resistance exercise, despite no improvements in plasma total cholesterol and low density lipoprotein–cholesterol levels. Following exercise-induced muscle damage, high-dose leucine supplementation may help maintain force output during isometric contractions

Keywords: Leucine, physiological roles, supplementation

INTRODUCTION

Leucine is a branched-chain essential amino acid. It accounts for approximately one-fifth of the requirements for essential amino acids. The diverse metabolic effects of leucine have drawn the attention of many researchers. Since the amino acid has as many as six 6 codons, it is found in abundance in some dairy and cereal proteins. Individual dietary consumption may reach 250 mg/kg/day. Leucine affects a multitude of cellular functions with different mechanisms, which comprise enzyme activation allosterically, induction of ATP generation and insulin secretion from the beta cells, and activation of signaling pathways. It enhances the secretion of anabolic hormones and increases protein deposition and growth through the augmentation of protein synthesis and inhibition of proteolysis. Its signaling pathway may involve the mammalian target of rapamycin complex and rapamycin-insensitive pathways responding to a leucine "transceptor," which combines leucine cellular transport, energized by the intracellular-extracellular glutamine gradient, and a signaling response to alterations and water balance and cell volume. In animal studies, dietary leucine supplementation reversed deleterious consequences of a high-fat diet (Millward, 2012). The account below aims at discussing the physiological roles of leucine and the effects of leucine supplementation.

Leucine Facilitates Insulin-Stimulated Glucose Uptake and Insulin Signaling in Skeletal Muscle Cells through Pathways Involving mTORC1 and mTORC2

Leucine enhanced uptake of glucose and heightened insulin sensitivity in skeletal muscle. The amino acid did not affect basal but upregulated insulin-induced glucose uptake or protein kinase B AKT (phosphorylation). Its effect on the insulin-induced glucose uptake and AKT phosphorylation was attenuated by an inhibitor of mammalian target of rapamycin complex 1 (mTORC1). Leucine partially reversed the inhibition of insulin-stimulated glucose uptake and AKT phosphorylation by the phosphatidylinositol 3-kinase inhibitor, wortmannin. The use of specific siRNA to knock down the core component of the mammalian target of rapamycin complex 2 (mTORC2) abrogated leucine-facilitated insulin-induced AKT phosphorylation. Thus, the mechanism involves both mTORC1and mTORC2 (Liu et al., 2014).

Leucine-Stimulated mTORC1 Signaling by Suppressing mTOR Localization to Lysosome

Ursolic acid, a pentacyclic triterpenoid from medicinal herbs and fruits, suppressed activation of the mTORC1 signaling pathway brought about by leucine in C2C12 myotubes. The target of ursolic acid lies downstream of the mTOR regulators Akt, tuberous sclerosis complex 1/2, and Ras homolog enriched in brain. Ursolic acid suppressed the binding of RagB to the mTOR activator Raptor and also suppressed leucine-induced mTOR lysosomal localization (Ou et al., 2014).

Chronic Leucine Supplementation Increases Body Weight and Insulin Sensitivity in Rats on High-Fat Diet Likely by Promoting Insulin Signaling in Insulin-Target Tissues

Leucine supplementation increased insulin sensitivity and prevented insulin resistance in rats on a high-fat diet but not in rats fed a normal chow diet.

The augmented insulin sensitivity brought about by leucine supplementation mitigated oxidative stress and enhanced insulin signaling, but did not bring about changes in ectopic fat accumulation in liver and muscle, and circulating concentrations of cytokines and lipids. Chronic leucine supplementation did not increase phosphorylation of insulin receptor substrate-1 (IRS-1) on serine 302, but upregulated basal IRS-1 phosphorylation on tyrosine 632 as well as insulin-stimulated phosphorylation of protein kinase B and mTOR in insulin-target tissues (adipose tissue liver, and skeletal muscle) of rats fed a high-fat diet (Li et al., 2013).

Leucine Facilitates Insulin Signaling through a Gαi Protein-Dependent Signaling Pathway in Hepatocytes

In cultured hepatocytes, insulin-induced phosphorylation of Akt at residue 473 facilitated by leucine was not inhibited by knocking down the key component of mTORC1 or mTORC 2 complexes. However, the phosphorylation was adversely affected by suppression of c-Src (PP2), PI3K, Gαi protein (pertussis toxin or siRNA against Gαi1 gene, or β-arrestin 2 (siRNA)).

Leucine also facilitated insulin-mediated inhibition of glucose production and expression of key gluconeogenic genes through a Gαi1 protein-dependent signaling pathway (Yang et al., 2013).

Dietary Leucine Improves Whole-Body Insulin Sensitivity Independent of Body Fat in Diet-Induced Obese Sprague-Dawley Rats

Differential gene expression of insulin receptor substrate (IRS) and protein kinase B (Akt) may account for alterations in insulin sensitivity in obese rats. Changes in gene expression of IRS and Akt were observed in liver tissue following dietary administration of leucine. The body weight and fat mass of rats receiving leucine and skim milk in the diet were reduced compared to those receiving casein. Enhanced insulin sensitivity was discernible only after administration of leucine (Eller et al., 2013)

Concurrent Intake of Leucine, Phenylalanine (1 mmol/kg Fat-Free Body Mass/Amino Acid) and Glucose (25 g) Has an Additive Stimulatory Effect on Circulating Insulin Level but a Less Than Additive Effect on Circulating Glucose Level

In healthy subjects, the insulin response to the combination of leucine and phenylalanine was essentially attributed to leucine, while the glucose response was mainly ascribed to phenylalanine. Consumption of only leucine, only phenylalanine or leucine and phenylalanine together, did not bring about any changes in plasma glucose level (Iverson et al., 2013).

Leucine Improves Protein Nutritional Status and Regulates Hepatic Lipid Metabolism in Calorie-Restricted Rats

In rats, leucine supplementation during calorie restriction resulted in preservation of body lean mass and protein mass and improvement of protein anabolism since there was a rise in the plasma levels and hepatic mRNA expression of albumin and insulin-like growth factor-1, as well as an elevation in the plasma concentrations of leptin and interleukin-6. Ketone bodies were increased. The hepatic expression of 3-hydroxy-3-methyl-glutaryl-CoA reductase (the key cholesterol biosynthesis enzyme) and sterol regulatory element-binding transcription factor 1 underwent an increase despite a lack of an effect on cholesterol and triglyceride concentrations. Hence leucine supplementation may alleviate excessive protein waste during considerable weight reduction (Pedroso et al., 2014).

Leucine Supplementation Combined with Resistance Exercise Improves the Plasma Lipid Profile of Dexamethasone-Treated Rats

Seven days of leucine supplementation and resistance exercise did not lower plasma total cholesterol and low density lipoprotein-cholesterol concentrations in dexamethasone-treated adult rats, an experimental model of dyslipidemia. However, plasma levels of triglyceride and very low density lipoprotein-cholesterol underwent a decline and high density lipoprotein-cholesterol levels were elevated (Nicastro et al., 2012).

Excessive Leucine-mTORC1-Signalling of Bovine Milk-Based Infant Formula: Missing Link to Understand Obesity in Early Childhood

Leucine serves as a maternal-neonatal relay for neonatal β-cell proliferation dependent on mTORC1 and insulin secretion. The mTORC1 target S6K1 stimulates the differentiation of mesenchymal stem cells into adipocytes and induces insulin resistance. Infant formulas have a higher leucine content than human milk. Enhanced leucine-mediated mTORC1-S6K1 signalling induced by infant formulas may account for the enhanced adipogenesis and production of lifelong increased adipocyte numbers. Attenuation of mTORC1 signalling of infant formulas by leucine restriction to physiological lower levels of human milk is likely to prevent childhood obesity and obesity-related metabolic diseases (Melnik, 2012).

Leucine Signaling in the Pathogenesis of Type 2 Diabetes and Obesity

Leucine-mediated mTORC1-S6K1-signaling contributes to adipogenesis, elevating the risk of obesity-mediated insulin resistance. Overconsumption of leucine-abundant proteins accounts for escalated mTORC1-dependent insulin secretion, accelerated β-cell growth and β-cell proliferation enhancing early replicative β-cell senescence leading to β-cell apoptosis. Disruption of β-cell mass regulation with accelerated β-cell proliferation, apoptosis, and insulin resistance are features of type 2 diabetes, all related to mTORC1 hyperactivation. Metformin used for anti-diabetic therapy inhibits leucine-mediated mTORC1 signaling. Flavonoids and polyphenols from plants with anti-obesity and antidiabetic activities are natural mTORC1 inhibitors. Bariatric surgery for treatment of obesity causes the elevated plasma concentrations of leucine and other branched-chain amino acids to fall. Reduction of leucine-mediated mTORC1 signaling by setting the maximum daily consumption of leucine-rich animal and dairy proteins may prevent T type 2 diabetes, obesity, cancer and neurodegeneration (Melnik,2012).

Leucine Supplementation Increases Serum Insulin-Like Growth Factor 1 Concentration and Liver Protein/RNA Ratio in Rats after a Period of Nutritional Recovery

Leucine supplementation in rats subjected to 50% food restriction for seven days, followed by fourteen days of nutritional recovery, brought about a rise in serum insulin-like growth factor 1 and hepatic protein/RNA ratio but no alterations in carcass, liver, or gastrocnemius protein content in comparison with control animals. Thus leucine supplementation did not lead to a better protein status, though some indicators of an anabolic state showed an improvement (Pedrosa et al., 2013).

Effects of Leucine and Its Metabolite B-Hydroxy-B-Methylbutyrate on Human Skeletal Muscle Protein Metabolism

Leucine increased anabolic signalling (mTOR) to a larger extent than its metabolite. The metabolite also slowed muscle protein breakdown in an insulin-independent manner, suggesting additional or distinct mechanism(s) compared with those of leucine (Wilkinson et al., 2013).

Acute Uremia Suppresses Leucine-Induced Signal Transduction in Skeletal Muscle

Gross abnormalities were observed in bilateral ureteral ligated rats (an acute uremia model) in comparison with sham-operated pair-fed control rats. Levels of several signaling proteins were elevated. Leucine-induced phosphorylation of mTOR and the downstream proteins, 4e-BP1 and S6K1, was totally inhibited. LC3B-II, a specific autophagosomal membrane-associated protein and autophagy marker, was present at heightened concentrations in uremia. Leucine lowered LC3B-II levels in controls, but not in uremic rats. In uremia, severe resistance to leucine-induced activation of the mTOR anabolic signaling pathway was observed. There were elevated levels of muscle IL-6 mRNA, and lowered levels of IGF-1 mRNA leading to reduction of the anabolic effect of nutritional therapy (McIntire et al., 2014)

Leucine Treatment Enhances Oxidative Capacity through Complete Carbohydrate Oxidation and Increased Mitochondrial Density in Skeletal Muscle Cells

Basal as well as peak oxidative metabolism in murine myoblast cells (C2C12) and human rhabdomyosarcoma cells (RD) were elevated following exposure to leucine (100 or 500 μM) for 24 or 48 h. Leucine treatment produced a greater mitochondrial proton leak associated with heightened energy expenditure, and curtailed basal extracellular acidification rate. Upregulated expression of peroxisome proliferator-activated receptor coactivator 1 alpha which enhances mitochondrial biosynthesis and cytochrome C content were observed after exposure to leucine for 48 h (Vaughan et al., 2013).

Leucine Supplementation Increases SIRT1 Expression and Prevents Mitochondrial Dysfunction and Metabolic Disorders in High-Fat Diet-Induced Obese Mice

It has been demonstrated leucine supplementation minimizes high-fat-diet-induced obesity, hyperglycemia, and dyslipidemia in animal models. Leucine supplementation to male C57BL/6J mice fed a high-fat diet for two months resulted in augmented expression of Sirtuin 1 (SIRT1) and nicotinamide phosphoribosyltransferase and also elevated intracellular NAD(+) levels, and hence diminished acetylation of peroxisome proliferator-activated receptor-γ coactivator 1α (PGC1α) and forkhead box O1 (FoxO1). PGC1α deacetylation activated genes regulating mitochondrial biogenesis and fatty acid oxidation, thus enhancing mitochondrial function and forestalling obesity caused by the high-fat diet. The attenuated FoxO1 acetylation was accompanied by reduced expression of pseudokinase tribble 3 (TRB3) and undermined the association between TRB3 and Akt, which enhanced insulin sensitivity and enhanced glucose metabolism (Li et al., 2012).

Effect of Post-Exercise Leucine-Protein Feeding on Neutrophil Function, Immunomodulatory Plasma Metabolites and Cortisol during a 6-Day Block of Intense Cycling

Twelve male cyclists participated in a randomized double-blind crossover, in which they consumed protein/leucine/carbohydrate/fat beverage (leucine-protein 20/7.5/89/22 g per h, respectively) or isocaloric carbohydrate/fat control beverage (control 119/22 g per h) for one to three hours post-exercise, during six days of high-intensity training. On day six, cyclists who had ingested leucine-protein beverage demonstrated curtailed pre-exercise cortisol and palmitoylcarnitine during exercise, and increased post-exercise neutrophil superoxide, compared with control. The modified plasma amino acid and acylcarnitine concentrations after consumption of leucine-protein beverage may at least partially account for the acute post-exercise attenuation in neutrophil function and elevated exercise-induced neutrophil oxidative burst on day 6, which will affect neutrophil-dependent processes in the period of recovery from intense training(Pedrosa et al., 2013).

Leucine and Methionine Deficiency Impairs Immunity to Gastrointestinal Parasites during Lactation

Lactating rats reinfected with the parasite *Nippostrongylus brasiliensis* and receiving fed low-crude protein food demonstrated poor lactational performance and decreased resistance to parasites compared with rats fed high-crude protein food. Compared to rats fed high-protein food (250 g crude protein /kg), rats receiving low-protein food (150 g crude protein /kg), high-protein Met-deficient food and high-protein Leu-deficient food exhibited a lower dam weight gain and litter weight gain, and diminished resistance to parasites reflected by the number of worm eggs found in the colon (Sakkas et al., 2013).

Similar Effects of Leucine- Rich and Regular Dairy Products on Muscle Mass and Functions of Older Polymyalgia Rheumatica Patients: A Randomized Crossover Trial

A 16-week low intensity home (Helsinki rheumatoid association)-based exercise program (conducting the maximum number of stand- ups twice a day) combined with post-exercise milk protein supplementation (a high-leucine whey protein enriched dairy product) for older polymyalgia rheumatica patients yielded satisfactory results seen in enhanced muscle mass and muscle functions as reflected by by hand grip strength, force platform countermovement jump performance, chair stand test, and walking speed (Björkman et al., 2011).

Leucine-Protein Supplemented High-Carbohydrate Postexercise Feeding Enhances Subsequent Cycling Performance in Well-Trained Men

Consumption of a leucine-protein supplement and also high-carbohydrate food after intense training on successive days improves high-intensity endurance performance and may minimize muscle membrane disruption in well-trained male cyclists (Thomson et al., 2011).

Effect of Leucine Supplementation on Indices of Muscle Damage Following Drop Jumps and Resistance Exercise

After muscle damage brought about by exercise, high-dose leucine supplementation (250 mg/kg body mass taken half an hour before, during and after exercise and the morning of each recovery day) following exercise may be useful for keeping the force output during isometric contractions, but not force output needed for complex physical tasks thus probably restricting its ergogenic efficacy (Kirby., 2012).

Tolerable Upper Intake Level of Leucine in Adult Men

Five healthy men aged 20-35 each received graded stepwise escalation in leucine consumption of 50, 150, 250, 500, 750, 1000, and 1250 mg/kg/day corresponding to the estimated average requirement, and estimated average requirement×3, ×5, ×10, ×15, ×20, and ×25 to a total of 29 studies. The strategy graded stepwise escalation was adopted to reduce possible severe adverse consequences. The men were provided a maintenance diet for 2 days before each leucine level containing 1 g/kg/day of protein and 1.7× measured the resting metabolic rate. A battery of markers comprising alanine aminotransferase, insulin, glucose, and ammonia was employed for monitoring adverse conseqences. Plasma leucine levels was elevated beyond consumption of 500 mg/kg/day. The metabolic limit to oxidize leucine lied between 550 and

700 mg/kg/day. A rise in blood ammonia concentration occurred when leucine consumption exceeded 500 mg/kg/day. No alterations in hepatic alanine aminotransferase activity were detected. Plasma glucose level underwent a decline but stayed within the normal range and no alterations in insulin occurred. It was concluded from the acute investigation that consumption exceeding 500 mg/kg/day or about 39 g/day may constitute a health risk (Pencharz et al., 2012).

Supplementation with Leucine Metabolite Beta-Hydroxy-Beta-Methylbutyrate in Health and Disease

A systematic review of randomized trials revealed that the leucine metabolite can prevent muscle injury associated with exercise in healthy trained and untrained individuals and also muscle loss during AIDS, cancer, and chronic obstructive pulmonary disease. The recommended intake of this safe leucine metabolite to maintain or improve muscle mass and function in healthy and diseased states is 3 g/day (Molfino et al., 2013).

Leucine-Sensitive Hyperinsulinemic Hypoglycemia in Patients with Loss of Function Mutations in Gene Coding for 3-Hydroxyacyl-CoA Dehydrogenase

Loss-of-function mutations in HADH, gene encoding an enzyme catalyzing mitochondrial fatty acid β-oxidation, short chain 3-hydroxyacyl-CoA dehydrogenase, lead to protein-sensitive hyperinsulinemic hypoglycemia. Mutations in GLUD1 encoding glutamate dehydrogenase also bring about the same disease (caused by leucine sensitivity). 3-Hydroxyacyl-CoA dehydrogenase (HADH) has a direct protein-protein interaction, which is absent in patients with HADH mutations giving rise to leucine-induced hyperinsulinemic hypoglycemia. This is not accompanied with loss of the suppressive action of GTP on glutamate dehydrogenase (as in patients with GLUD1 mutations) (Heslegrave et al., 2012).

Autophagy and Leucine Promote Chronological Longevity and Respiration Proficiency during Calorie Restriction in Yeast

In the budding yeast *Saccharomyces cerevisiae,* calorie restriction induced autophagy. Autophagy is necessary for full extension of chronological life span by calorie restriction in most cases (depending on atg allele, strain, and leucine availability) and promotes mitochondrial respiration proficiency during aging under calorie restriction conditions (Aris et al., 2013).

Conclusion

Amino acids, and especially leucine, can play a role as potent insulin secretagogues when given in conjunction with carbohydrate. Leucine can be employed to improve postprandial glycemic control, and to enhance translation, muscle protein synthesis and net protein accretion in an insulin-independent fashion. The use of leucine as a pharmaconutrient in health and disease has been suggested, for instance, for the prevention and therapy of sarcopenia and/or type 2 diabetes. The pharmaconutrient value of leucine in investigations of a longer duration remains to be established (van Loon, 2012).

References

Aris JP, Alvers AL, Ferraiuolo RA, Fishwick LK, Hanvivatpong A, Hu D, Kirlew C, Leonard MT, Losin KJ, Marraffini M, Seo AY, Swanberg V, Westcott JL, Wood MS, Leeuwenburgh C, Dunn WA Jr. Autophagy and leucine promote chronological longevity and respiration proficiency during calorie restriction in yeast. *Exp Gerontol.* 2013;48(10):1107-19.

Björkman MP, Pilvi TK, Kekkonen RA, Korpela R, Tilvis RS. Similar effects of leucine rich and regular dairy products on muscle mass and functions of older polymyalgia rheumatica patients: a randomized crossover trial. *J Nutr Health Aging.* 2011;15(6):462-7.

Eller LK, Saha DC, Shearer J, Reimer RA. Dietary leucine improves whole-body insulin sensitivity independent of body fat in diet-induced obese Sprague-Dawley rats. *J Nutr Biochem.* 2013;24(7):1285-94.

Heslegrave AJ, Kapoor RR, Eaton S, Chadefaux B, Akcay T, Simsek E, Flanagan SE, Ellard S, Hussain K. Leucine-sensitive hyperinsulinaemic hypoglycaemia in patients with loss of function mutations in 3-Hydroxyacyl-CoA Dehydrogenase. *Orphanet J Rare Dis.* 2012;7:25.

Iverson JF, Gannon MC, Nuttall FQ. Ingestion of leucine + phenylalanine with glucose produces an additive effect on serum insulin but less than additive effect on plasma glucose. *J Amino Acids.* 2013;2013:964637.

Kirby TJ Effect of leucine supplementation on indices of muscle damage following drop jumps and resistance exercise. *Amino Acids.* 2012;42(5):1987-96.

Li H, Xu M, Lee J, He C, Xie Z. Leucine supplementation increases SIRT1 expression and prevents mitochondrial dysfunction and metabolic disorders in high-fat diet-induced obese mice. *Am J Physiol Endocrinol Metab.* 2012;303(10):E1234-44.

Li X, Wang X, Liu R, Ma Y, Guo H, Hao L, Yao P, Liu L, Sun X, He K, Cao W,Yang X. Chronic leucine supplementation increases body weight and insulin sensitivity in rats on high- fat diet likely by promoting insulin signaling in insulin-target tissues. *Mol Nutr Food Res.* 213;57(6):1067-79.

Liu H, Liu R, Xiong Y, Li X, Wang X, Ma Y, Guo H, Hao L, Yao P, Liu L, Wang D,Yang X. Leucine facilitates the insulin-stimulated glucose uptake and insulin signaling in skeletal muscle cells: involving mTORC1 and mTORC2. *Amino Acids.* 2014;46(8):1971-9.

McIntire KL, Chen Y, Sood S, Rabkin R. Acute uremia suppresses leucine-induced signal transduction in skeletal muscle. *Kidney Int.* 2014;85(2):374-82.

Melnik BC. Leucine signaling in the pathogenesis of type 2 diabetes and obesity. *World J Diabetes.* 2012;3(3):38-53.

Melnik BC. Excessive Leucine-mTORC1-Signalling of Cow Milk-Based Infant Formula: The Missing Link to Understand Early Childhood Obesity. *J Obes.* 2012;2012:197653.

Millward DJ. Knowledge gained from studies of leucine consumption in animals and humans. *J Nutr.* 2012;142(12):2212S-2219S.

Molfino A, Gioia G, Rossi Fanelli F, Muscaritoli M. Beta-hydroxy-beta-methylbutyrate supplementation in health and disease: a systematic review of randomized trials. Amino Acids. 2013;45(6):1273-92.

Nicastro H, da Luz CR, Chaves DF, das Neves W, Valente KS, Lancha AH Jr. Leucine supplementation combined with resistance exercise improves the plasma lipid profile of dexamethasone-treated rats. *Lipids Health Dis.* 2012;11:7.

Ou X, Liu M, Luo H, Dong LQ, Liu F. Ursolic acid inhibits leucine-stimulated mTORC1 signaling by suppressing mTOR localization to lysosome. *PLoS One*. 2014;9(4):e95393.

Pedroso JA, Nishimura LS, de Matos-Neto EM, Donato J Jr, Tirapegui J. Leucine improves protein nutritional status and regulates hepatic lipid metabolism in calorie-restricted rats. *Cell Biochem Funct.* 2014 Jun;32(4):326-32.

Pencharz PB, Elango R, Ball RO. Determination of the tolerable upper intake level of leucine in adult men. *J Nutr.* 2012;142(12):2220S-2224S.

Sakkas P, Jones LA, Houdijk JG, Athanasiadou S, Knox DP, Kyriazakis I. n. Leucine and methionine deficiency impairs immunity to gastrointestinal parasites Nippostrongylus brasiliensis during lactation. *Br J Nutr.* 2013;109(2):273-82.

Thomson JS, Ali A, Rowlands DS. Leucine-protein supplemented recovery feeding enhances subsequent cycling performance in well-trained men. *Appl Physiol Nutr Metab.* 2011;36(2):242-53

van Loon LJ. Leucine as a pharmaconutrient in health and disease. *Curr Opin Clin Nutr Metab Care.* 2012;15(1):71-7.

Vaughan RA, Garcia-Smith R, Gannon NP, Bisoffi M, Trujillo KA, Conn CA. Leucine treatment enhances oxidative capacity through complete carbohydrate oxidation and increased mitochondrial density in skeletal muscle cells. *Amino Acids.* 2013;45(4):901-11.

Yang X, Mei S, Wang X, Li X, Liu R, Ma Y, Hao L, Yao P, Liu L, Sun X, Gu H,Liu Z, Cao W. Leucine facilitates insulin signaling through a Gαi protein-dependent signaling pathway in hepatocytes. *J Biol Chem.* 2013;288(13):9313-20.

In: Leucine
Editor: Sophia R. Newman

ISBN: 978-1-63482-593-1

Chapter 2

LEUCINE BASED POLYMERS: SYNTHESIS AND APPLICATIONS

Ramaz Katsarava[1] and Jordi Puiggalí[2]
[1]Institute of Chemistry and Molecular Engineering, Agricultural University of Georgia, Tblisi, Georgia
[2]Departament d'Enginyeria Química, Universitat Politècnica de Catalunya, Barcelona, Spain

ABSTRACT

During last decades great efforts have been made to get polymers constituted by α-amino acid units since they form smart materials, show catalytic activity, are promising as micro and nanocontainers for drug delivery purposes, as biodegradable and biocompatible materials, to name a few.

Unique physical-chemical properties of amino acid based polymers are of great interest as well. Among these properties self-assembling mechanisms of peptides and proteins nowadays represent an object of extensive studies. Specifically, leucine rich sequences are usually involved in "coiled-coil" motifs in which α-helices with a short repeat sequence are coiled intertwiningly. Leucine zippers are involved in the dimerization of a number of transcription factors, playing the hydrophobic interactions provided by leucine residues a remarkable stabilizing effect. Properties and functions of biomacromolecules and synthetic polymers can be integrated in a single material. Hence, hybrid block copolymers based on peptides are receiving great attention due to the capability to self-assemble into highly organized nanoscale structures and their wide applications in the biomedical field that covers neural tissue engineering; scaffolds for bone regeneration; dental implants; drug, gene and protein delivery and biosensors for detection of viruses and other pathogens. These molecular chimeras might have advantageous properties provided by the synthetic block such as economy, solubility, melt processability, versatility, or rubber elasticity as well as those given by the peptide block (e.g., structure, functionality, high mechanical performance or biodegradability). Hydrophilic–hydrophobic characteristics of the considered polymers can be easily controlled by changing their specific α-amino acid units and even by the incorporation of hydrophilic polyethylene glycol segments. These copolymers are highly interesting to get suitable polymeric nanoparticles for drug delivery uses as well as new temperature responsive polymers. Finally, the development of hybrid materials based on conducting and biodegradable polymers (e.g., those containing leucine units) merits attention since these hybrids may be a promising solution to cover aspects like stimulation of either the proliferation or differentiation of various cell types. Amino acid based polymers are especially promising as biodegradable biomaterials since can give rise to non-toxic and nutritious (physiological) degradation products. New biodegradable polymers such as poly(ester amide)s and poly(ester urea)s have been developed and even their interest as new biomedical materials have been clearly demonstrated. It was shown that leucine-based polymers possess much better solubility in suitable organic solvents like ethanol. Besides, leucine-based biodegradable polymers are safer for patients as compared with e.g., phenylalanine-based analogues.

The present review deals with syntheses, properties and applications of the main leucine containing polymers and materials made of them that up to now have been studied.

INTRODUCTION

Leucine (abbreviated as Leu or L) is a branched-chain (contains isobutyl side chain) hydrophobic α-amino acid with the chemical formula depicted below:

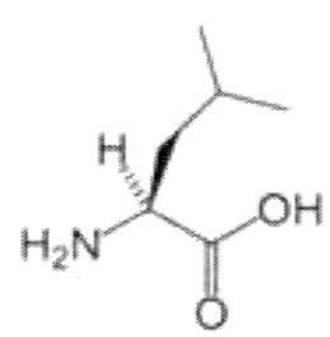

In the nature, this amino acid is a major component of various proteins such as the subunits in ferritin, astacin, other buffer proteins. Leucine is an essential amino acid meaning that the human body cannot synthesize it, and it therefore must be ingested.

Synthetic polymers composed of naturally occurring α-amino acids have become increasingly popular for both numerous biomedical and non-biomedical applications [1-11], and leucine, one can say without exaggeration, represents one of the most important building blocks for constructing such kind of polymeric materials that will be demonstrated below. Synthetic polymers composed of leucine could be classified as (I) polymers with the amino acid's residues in the polymeric backbones and (II) polymers with the amino acid's residues in the lateral chains.

I. Polymers with Leucine in the Backbones

Polymers with amino acid leucine in the backbones, in turn, can be subdivided into (A) polymers with conventional "head-to-tail" orientation of the amino acid (so called AB-type polymers) and (B) polymers with non-conventional "head-to-head" and "tail-to-tail" orientation of the amino acid (so called AA-BB-type polymers). To the first, AB-type belong numerous polypeptides (poly(amino acids)) designed mostly for biomedical applications, to the second, AA-BB-type – various hetero-chain polymers designed for both biomedical and technical applications. AB-type polymers are basically obtained *via* chain-growth polymerization, whereas the most applicable method to AA-BB-type polymers is step-growth polymerization (polycondensation).

I.A. AB Type Polymers Composed of Leucine

I.A.1. Synthesis and Applications of Leucine Based Polypeptides

The copolypeptides made of α-amino acids have a potential for biomedical applications such as polypeptide adhesives for use as surgical sealant, antimicrobial polypeptides for the treatment of autoimmune disorders, polypeptide membranes and smart systems for therapeutic delivery, polypeptide hydrogels for tissue engineering applications, etc. Polypeptides show also interesting physical-chemical and catalytic properties that expand scopes of their practical applications.

The most commonly used method for synthesizing amino acid based polypeptide (polyamino acid) is the ring-opening polymerization (ROP) of corresponding *N*-carboxyanhydride (NCA). This cyclic compound can be prepared by reaction of the amino acid with phosgene [12] or an equivalent compound (e.g., diphosgene or triphosgene) (Figure 1) [13]. Cyclization of the *N*-alkoxycarbonyl-protected amino acid using thionyl chloride has also been reported as an alternative method [14]. Polymerization of NCAs is carried out by addition of a nucleophilic initiator (e.g., water or amine compounds) in a ratio that determines the average polymer length [15,16].

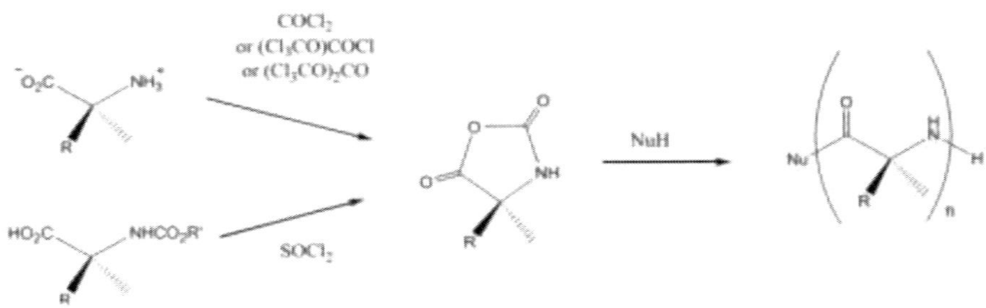

Figure 1. Synthesis of polypeptides from *N*-carboxyanhydrides.

ROP of L-leucine N-carboxyanhydride (NCA) from rehydrated hydrotalcite (HT) have recently been proposed and compared with those performed with typical initiators [17]. In general, alpha-helical chains containing 5–40 monomers with NCA end-groups *via* a monomer-activated mechanism were obtained. The synthetic HT clay was revealed an efficient initiator but also was useful as a support. Thus, polymers could be immobilized in HT through an anion-exchange procedure and specifically polyleucine chains were stabilized in the interlayer spacing of HT by hydrogen bonding with hydroxyl groups of the clay.

The ROP of leucine NCAs, depicted above, leads to the formation of either isotactic (when homo-polymerize L- or D-NCAs) or atactic (when polymerize DL-NCA) polyleucine [18]. It is interesting to note here that syndiotactic polyleucine has been synthesized as well. However, for the preparation of syndiotactic polymer an alternative synthetic strategy - step-growth polymerisation (AP-type polycondensation) was applied. According to this alternative scheme, N-(*tert*-Butoxycarbonyl)-L-Leu-D-Leu-L-Leu-D-Leu tetrapeptide was synthesized from L- and D-leucines using conventional in peptide chemistry coupling methods. This N-protected tetrapeptide was transformed into an activated tetrapeptide by conversion to the pentachlorophenyl (PCP) ester. Subsequent removal of the *tert*-butoxycarbonyl group by the treatment with HCl/Dioxane solution gave tetrapeptide $HCl.H_2N$-L-Leu-D-Leu-L-Leu-D-Leu-COO-PCP. This tetrapeptide was subjected to polycondensation in organic solvents in the presence of triethylamine and N,N-diethyl-1-amino-2-propanol as HCl acceptors, resulting in poly(tertrapeptide)s – in fact syndiotactic polyleucine [19] (Figure 2):

$$\text{n HCl.H}_2\text{N-L-}\mathbf{Leu}\text{-D-}\mathbf{Leu}\text{-L-}\mathbf{Leu}\text{-D-}\mathbf{Leu}\text{-COO-PCP} \xrightarrow{\textbf{Base}}$$

$$\text{-[-HN-L-}\mathbf{Leu}\text{-D-}\mathbf{Leu}\text{-L-}\mathbf{Leu}\text{-D-}\mathbf{Leu}\text{-CO-]}_n\text{-}$$

Figure 2. Synthesis of syndiotactic polyleucine.

Molecular weights of the obtained polymers were $\geq$ 34000 Da. Optical rotation measurements in trifluoroacetic acid showed that the polymer was highly syndiotactic, though a small degree of isomerization (0,9–2,1%) occurred and the polymer was somewhat enriched with L-leucine residues. The degree of isomerization decreased with decreasing base strength and concentration and with decreasing polarity of the solvent; these effects could be connected with a reversible C-H proton abstraction.

It is worthwhile to note here that Honda et al. [20] developed a microfluidic system (microreactor) for polymerization and copolymerization of amino acid NCAs and compared the properties of the products with those obtained by batchwise system under various experimental conditions. It was found that the polymers, including polyleucine and its copolymers with other amino acids, produced by the microreactor were characterized by narrower molecular weight distribution compared with polymers obtained by the batchwise system. Also, the molecular mass of the polymer produced using the microreactor was simply governed by the flow (pumping) rate. These results indicated that the microreactor could be a model for synthesis of amino acid polymer with highly controllable average molecular weight and molecular weight distribution.

Polyleucine and polypeptides containing polyleucine blocks have found interest in different applications, being the most interesting those related to catalysts for enantiomeric reactions, polymersomes and drug delivery carriers, hydrogels and wound dressing materials.

Polyleucine has application in organic syntheses since appears as an efficient catalysts for promoting highly enantioselective reactions (e.g., Julia-Colonna epoxidation of electron-deficient alkenes [21,22]). These peptides showed a high chiral amplification that bases the opinions concerning their role in the prebiotic origin of life [23-25].

Polymeric vesicles (polymersomes) have high potential applications in medicine, pharmacy, and biotechnology due to their small dimensions and particular aggregation form. These nanoparticles have fluid-filled cores with walls that consist of entangled chains [26]. Polymersomes are able to encapsulate hydrophilic molecules within the aqueous interior and also integrate hydrophobic molecules within the membrane.

Polymersomes are generally prepared from amphiphilic molecules of high molecular weight that allows getting membranes thicker, stronger, tougher and with higher stability than conventional liposomes. Responsive polymersome systems have been designed to obey certain stimuli (e.g., temperature, pH, light or redox potential, Figure 3) in order to facilitate their formation by changing the solubility of one block (A), to be dissolved/disintegrated (B) to release the cargo or to create pores in the membrane starting the drug delivery (C) [27].

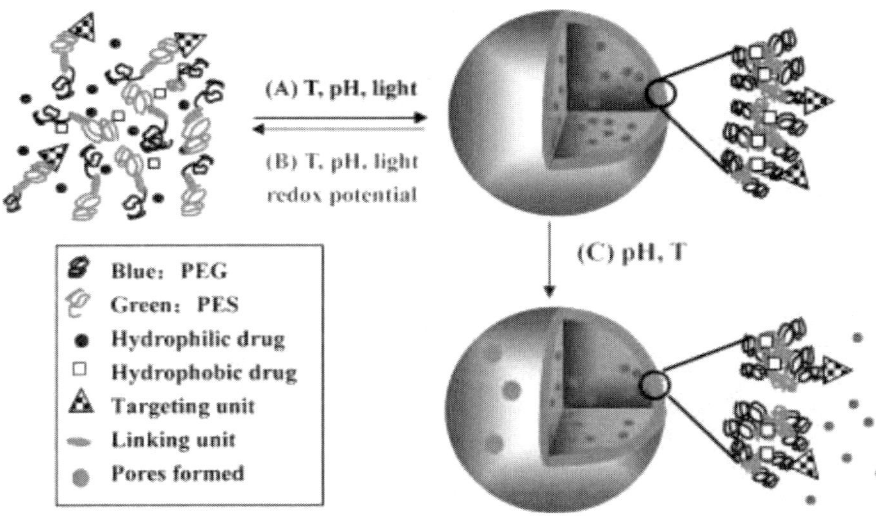

Figure 3. Formation and dissociation of stimuli-sensitive polymersomes. Reproduced with permission from Ref. [27].

Diblock peptides have been considered to prepare polymersomes and specifically leucine blocks have been used for the hydrophobic part. Thus, polyleucine-*b*-polylysine (PLeu-PLys) showed a pH-dependent release of encapsulated drugs. At low pHs polylysine was protonated, leading to the solubilization of the membrane and an instantaneous drug release [28,29]. However, these vesicles were not significantly internalized by cells, constituting a problem for delivery of drugs with intracellular targets. Polyarginine-*b*-polyleucine (PArg-PLeu) was also reported as pH-sensitive polymersome, being the vesicular assemble and the ability to transport cargos across cell membranes driven by the polyarginine block [30]. Specifically, the capability to enter cells is a consequence of the presence of multiple guanidinium groups in the arginine residues. Vesicles from poly(L-arginine)$_{60}$-*b*-poly(L-leucine)$_{20}$ were obtained with controlable size, showed good encapsulating efficience for hydrophilic cargos and high capability to cross the cell membrane and transport encapsulated cargos into different cell lines [30]. Endocytosis and intracellular trafficking behavior of vesicles was also evaluated. Results indicated that vesicles remained intact even after release from the endosomes, a feature that may difficult the reléase of encapsulated drugs and points out the importance of incorporating pH responsive residues into the hydrophobic domains to assist in vesicle disruption [31].

Nowadays the development of versatile drug delivery carriers (i.e., with capacity to encapsulate varied types of guest molecules) is of great interest for compound medicines. Star-block copolypeptides consisting of a hyperbranched polyethyleneimine (PEI) core, a hydrophobic polyleucine inner shell, and a negatively charged polyglutamate outer shell (Figure 4) were successfully synthesized by ROP of NCAs. These water soluble, amphiphilic copolypeptides were able to encapsulate versatile compounds ranging from hydrophobic to anionic and cationic hydrophilic guest molecules [32].

Double emulsions such as the water-in-oil-in-water can provide advantages over simple oil-in-water emulsions for encapsulation. For example, they can carry both polar and non-polar molecules and provide also a better control of the release [33]. Preparation of double emulsions is complicated and usually requires mixtures of surfactants for stability [34]. Nevertheless, the use of block copolypeptide surfactants such as poly(L-lysine.HBr)x-b-poly(racemic-leucine)y (x and y ranging from 20 to 100 and from 5 to 30 residues, respectively) was demonstrated to be highly efficient to prepare highly stable double emulsions in a simple process and using a single component [35]. Poly(*rac*-leucine) was fundamental because its disordered chain conformation improved solubility and helped to promote surface activity, while its peptidic nature stabilized the droplet interfaces through interchain hydrogen-bonding in the oil phase.

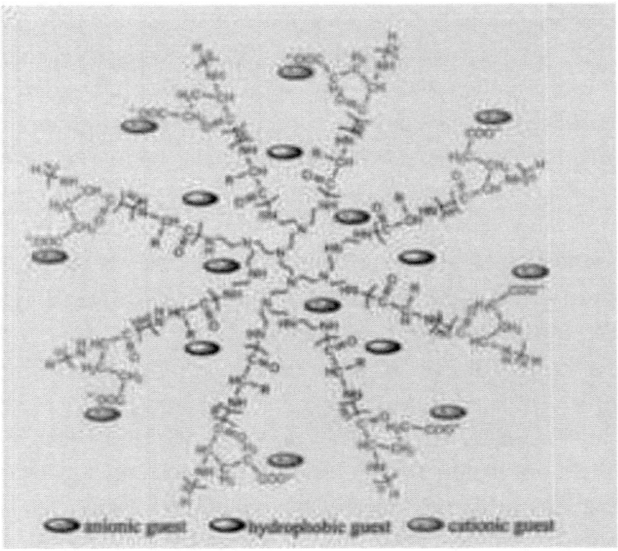

Figure 4. Idealized structure of star-block copolypeptides (R: hydrophobics ide chain of L-leucine) having only eight arms and ionic guest molecules. Reproduced with permission from Ref. [32].

Low molecular weight diblock copolypeptides with poly(L-lysine-HBr) or poly(L- glutamate-Na) as the hydrophilic block and poly(L-leucine), poly(L-valine) or poly(D/L-leucine) as the hydrophobic block were able to associate into rigid hydrogels at very low polymer concentrations (> 0.25 wt-%) in deionized water [36], which is an interesting feature for design of drug-delivery systems [37]. The rodlike chain conformations of the hydrophobic segments were found to be crucial to hydrogel formation since they direct assembly of these polymeric surfactants into highly anisotropic network structures rather than spherical micelles [38]. These hydrogels showed highly interesting properties for biomedical applications: High stability to heat (e.g., no loss in strength was observed at temperatures of 80-90 ºC) that allows autoclaving the material; rapid recovery of gel strength after shear thinning that allows their application *via* injection trough a needle; functionality of gel surface (e.g., lysine residues may provide cell adhesion points or sites for attachment of drugs or growth factors) and finally biodegradability. Furthermore, it was possible to adjust the copolymer composition and molecular weight to keep the hydrogel strength in high salt concentrations (i.e., 0.5 M NaCl) as well as in cell growth media and aqueous buffers [39]. This point is significant since polyelectrolites usually collapsed when salts were added into the media. Salts screen the charges along the polymeric chains, weakening the forces that support the gel network (i.e., those responsible for chain stretching and interchain repulsion).

Polylysine-*b*-polyleucine-*b*-polylysine triblock copolypeptides were found to allow preparation of samples with greater gel strength and better salt tolerance than diblock samples at similar concentrations [40]. Such triblock structures were characterized by a packing of hydrophobic α-helical oligoleucine segments perpendicular to extended fibrils in lamellar structure. Results indicated that hydrogel properties could be fine tuned by means of the block architecture beyond variation of polypeptide composition, concentration and chain length.

Hydrogels based on hyaluronan (HA) have broad applications in the biomedical field [41-43] (e.g., ophthalmologic surgery, tissue engineering and drug delivery) due to its unique structure and properties (e.g., biocompatibility and degradability). Polyleucine grafted HA was an interesting derivative considering the above explained positive results obtained with synthetic block copolypeptides that incorporated the hydrophobic polyleucine segment. The grafted samples were successfully synthesized *via* a Michael addition reaction between primary amine terminated poly(L-leucine) and acrylate-functionalized HA [44].

Poly(L-leucine) has been proposed as a wound dressing material since offer several advantages derived from its high biocompatibility, its decomposition into non-toxic compounds, the porous structure that can be attained by liophilization and the capability to incorporate antimicrobial drugs such as silver sulfadiazine [45,46]. In order to improve biocompatibility several effords have been carried out to get ROP avoiding the use of chemical initiators. Therefore, poly(L-leucine) was also prepared in the solid state by irradiation with γ-ray for about 1.3 kGy dose. The polymerization yield was close to 100% and non-toxic residue was detected [47].

I.A.2. Synthesis and Applications of Polydepsipeptides Containing Leucine

Polydepsipeptides (PDP) are alternating copolymers of α-hydroxy acids and α-amino acids that have interesting properties such as degradability and non-toxicity. These polymers degrade giving rise to acid and basic metabolites that could be recognized by the human body. Furthermore, degradation cause a low pH decrease which constitute an advantage respect to traditional polyesters that give only rise to acidic compounds.

The main synthetic route is the ROP of morpholine-2,5-dione derivatives, which can be used as a single monomer to render PEAs with a regular chemical sequence. Alternatively, random and even block copolymers can be prepared according to the selected synthetic procedure (e.g., one step or two step synthesis) and using the appropriate combination of morpholine-2,5-diones. In addition, morpholine-2,5-diones can also be polymerized with other cyclic monomers (e.g., lactones like glycolide, lactide or ε-caprolactone) to get new random copolymers having a higher ester content [48-50].

The 6-membered ring of morpholine-2,5-dione can be synthesized by three main procedures (Figure 5): [10,51] (a) Cyclization of N-(α-haloacyl)-α-amino acid salts, (b) Intramolecular transesterification of N-(α-hydroxyacyl)-α-amino acid esters and (c) Cyclization of O-(α-aminoacyl)-α-hydroxycarboxylic acids. Main problems of synthesis are the moderate yield and in general the occurrence of some racemization when optically active monomers were used.

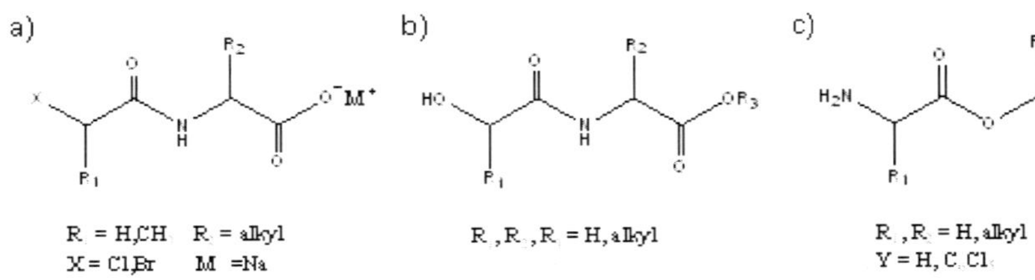

Figure 5. Monomers used in the preparation of morpholine-2,5- dione derivatives.

Polymerization of morpholine-2,5-diones are usually carried out in bulk at reaction temperatures close to the melting temperature [52]. Stannous octoate [$Sn(Oct)_2$] and water or alcohol molecules, being generally employed as catalyst and initiator, respectively. Enzymes such as lipases were another type of efficient catalyst [53] for polymerizations that, in this case, proceeded through ring-opening at the ester bond and consequently were strongly influenced by the configuration of the hydroxy acid moiety.

ROP of morpholine-2,5-diones has some limitations that are mainly associated with a low monomer yields and severe reaction conditions that may lead to racemization and the occurrence of secondary reactions. Therefore, other synthetic methods have been proposed to get PEAs having α-hydroxy acid and α-amino acid units. These alternative approaches result in the AA-BB type polymers with "head-to-head" and "tail-to-tail" orientations of leucine residues in the backbones and will be discussed in the Section I.B below.

The hydrophilicity and degradability of polydepsipeptides can be adjusted by the introduction of a hydrophilic block of polyethylene glycol (PEG) by means of employing an amino-terminated PEG as initiator of the ROP [54].

Regular polydepsipeptides offer several advantages as compared to randomly polymerized polypeptides since they are soluble in a wide range of organic solvents, may undergo conformational transitions in single solvents, and have a well-defined sequence. Polydepsipeptides melt at lower temperatures and have a lower trend to adopt helical conformations than polypeptides due to the lower number of hydrogen bonding interactions that could be established (i.e., the ester groups cannot form the hydrogen bonds through the carbonyl oxygen, because of lack of the hydrogen atom). Nevertheless, helix-coil transitions were observed as consequence of thermal denaturation, being evaluated the relative stability imparted to ordered helical structures of the different alkyl groups of α–amino acid units. Interestingly, it was determined from circular dichroism measurements that leucine (Leu) imparted in organic solvents a greater helix stability to polydepsipeptides than alanine (Ala), which is a typical α-helix forming amino acid [54].

Conformational analysis at the atomic level have been performed in order to reveal the effect of α-hydroxy acids in the stability of helical structures and to verify the helical structure found for leucine derivatives in non-polar solvents [55,56]. The crystal structure of the Boc–(Leu–Leu–Lac)$_3$–Leu–Leu–OEt was solved by X-ray diffraction and applying direct methods, being revealed an α-helix conformation despite three ester bonds were present in the main chain [57]. A slightly distorted α-helix conformation was also determined for the Boc-(Leu-Leu-Ala)$_2$-(Leu-Leu-Lac)$_3$-OEt oligomer, being suggested that the loss of hydrogen bonds were not sufficient to break the helical structures probably as a consequence of a significant flexibility of the main chain around the lactide residues [58].

Currently great efforts are focused to get in situ gelling polymers having a temperature-responsive sol-gel transition near the temperature of the human body and a biodegradable nature. These can be used as injectable polymers for biomedical applications (e.g., drug delivery systems and tissue engineering). Systems based on polypeptides have been developed but their use is scarce due to their low degradability under non-enzymatic conditions. Therefore, other systems like block copolymers consisting of hydrophilic polyethers (e.g., PEG) and aliphatic polyesters (e.g., polylactide) [59] or PEAs have been considered. The first showed problems concerning the weak mechanical properties of the gel formed above the sol-gel transition temperature and therefore triblock copolymers containing PDP blocks were evaluated due to their capability to establish hydrogen bonding intermolecular interactions. The temperature sensitivity of these triblock copolymers was strongly correlated with their micelle formation behavior in water, which is mainly regulated through the hydrophobic–hydrophilic balance. Microstructure (e.g., PDP-PEG-PDP or PEG-PDP-PEG distribution), PDP length and amino acid composition were factors that could be easily varied in order to control final properties. Both ABA- and BAB-type triblock copolymers containing glycolic acid (Glc) and hydrophobic L-leucine amino acid units showed temperature-responsive sol-gel transition [60].

Biodegradable anti-adhesive membranes were developed using an amphiphilic ABA-type triblock copolymer of poly[Lac-*co*-(Glc-Leu)]-PEG-poly[Lac-*co*-(Glc-Leu)]. Films were degraded and depleted gradually *in vivo* without inflammation, being the degradation rate highly dependent on the molecular architecture. Specifically, the molecular weight of the hydrophilic PEG segment and the depsipeptide unit content in the hydrophobic segment were the most significant factors [61].

Shape-memory capabilities and good elastic properties between 25 and 75 ºC have been found for thermoplastic phase-segregated multiblock copolymers with poly(ε-caprolactone) (PCL) and poly(isobutylmorpholine-2,5-dione) segments. A temporary shape with an almost fixation of the mechanical deformation and a quantitative recovery of the permanent shape with a switching temperature around body temperature were observed. Polymers were hydrolytically degradable at 37 ºC in a pH 7.4 phosphate buffer solution, mainly through the PDP sequence. Advantages of depsipeptide-based multiblock copolymers compared to polyester based shape-memory materials may result from their higher biocompatibility. Cytocompatibility of extractable substrates were therefore evaluated using a continuous fibroblast like cell line (L929). Cell viability, functional integrity of cell membranes, cell activity and assembly of the actin- and vinculin cytoskeleton indicated no incompatibilities between the extracts and L929 cells. Thus, the studied depsipeptide-based multiblock copolymers were proposed as promising candidates for soft multifunctional implant materials [62].

I.A.3. Self-Assembled Peptidic Structures

Peptides are nowadays interesting molecular entities to be used in the design of new self-assembled materials since they had noticeable advantages. These are mainly linked to their capability to adopt different secondary structures, the possibility of incorporating functional groups, the control of hydrophobic/hydrophilic characteristics through the selection of constitutive amino acids, their chemical versatility and their biodegradable and biocompatible character. In fact amino acids can provide interesting structural features such as: chirality, ionizable groups, electrostatic dipolar interactions, hydrogen-bond donors and acceptors, hydrophobic and hydrophilic domains and metal binding ligands [63].

The α-helix and the β-sheet are the main secondary structures that are arranged to give rise to the tertiary structure of proteins. For example, amyloid fibers are composed of well-ordered β-sheet structures, which are implicated in Alzheimer disease. These structures are characterized by a strong and stable aggregation that appears also interesting for the design of new nanoscale self-assembling materials. In the same way, α-helices can be organized giving rise for example to coiled-coil structures [63], in which helices with a seven-residue repeat sequence are coiled intertwiningly. These structures are characteristic of α-keratin [64] and also of leucine zippers that are involved in the dimerization of several transcription factors [65].

Coiled-coil helical assemblies tend to be more reversible than β-sheet structures and specifically are widely employed to get superhelical assemblies where 2-5 helices are wrapped around each other. A heptad periodicity of the amino acid sequence (i.e., $-(a\text{-}b\text{-}c\text{-}d\text{-}e\text{-}f\text{-}g)_n-$), is required to determine the aggregation number of helices and their relative orientation through the selected amino acids placed in the *a*/*d* and *e*/*g* positions, respectively (Figure 6) [66,67]. Hydrophobic non-polar residues (e.g., valine, isoleucine and leucine) are involved in the interactions between constituent helices whereas reversibility of the final assembly is usually achieved through electrostatic interactions (interchain salt bridges between residues placed in *e* (e.g., lysine) and *g* (e.g., glutamic acid) positions). The side chains of hydrophobic residues influence the packing geometry of the core and consequently they determine the association number. Coiled-coils can be formed by homo- and hetero-oligomers, a feature that mainly depends on the electrostatic interactions. Thus, homodimers are favoured when oppositely charged amino acids are placed in positions *e* and *g*, whereas amino acids with the same charge in these positions would repulse each other [68]. A heterodimer could be formed if another peptide with oppositely charged e' and g' residues becomes available [69].

The stability of coiled-coil structures increases with hydrophobic interactions and therefore minimum length of the peptide sequence in necessary (i.e., a peptide adopts the two-stranded α-helical coiled coil conformation in aqueous medium when it has more than three heptads (six helical turns)). Nevertheless, short coiled-coils (e.g., 35 residues) can be more stable than the much longer tropomyosin (284 residues) [70].

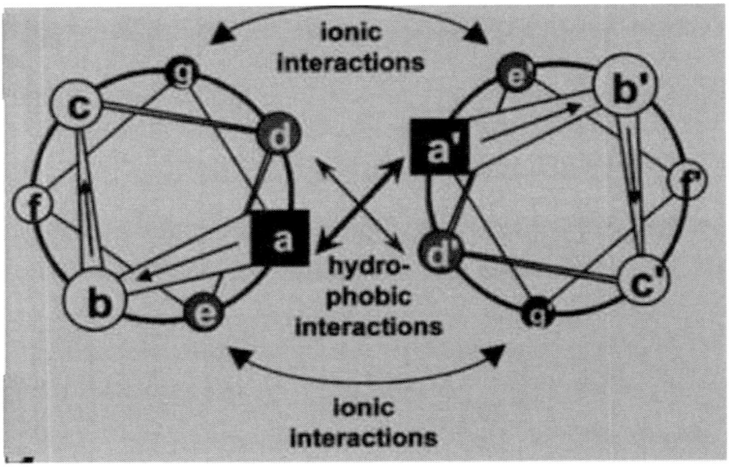

Figure 6. Helical-wheel diagram of a dimeric coiled-coil, showing the helical cross-section.

Peptide and protein structures are generally more stable when they are associated in an oligomeric state [71]. Thus, the process of self-assembly provides an excelent opportunity to design new materials. α-Helical coiled-coil peptides can be used as building blocks for biomaterials as also observed in nature [72]. Fibers can for example be derived from two interacting coiled-coil helices that are set ''out of phase'' within a single peptide (Figure 7a) [73]. Thus, the two halves that cannot interact intramolecularly must be arranged intermolecularly and form fibers. Self-assembling fibers (SAF) between heterodimeric coiled-coil building blocks were also obtained by means of using two complementary leucine-zipper-based sequences to assemble offset rather than blunt ended (Figure 7b) [74,75]. Appropriate selection of amino acid residues allows also establishing interactions between the outer surfaces of coiled-coils giving rise to highly ordered structures (Figure 8). Coiled coils acted in this case as stiff rods that packed in a hexagonal array, whereas peptides became arranged somehow to give a banding pattern perpendicular to the long axis of the fiber and a periodicity that was in agreement with the expected length of the involved peptide [76].

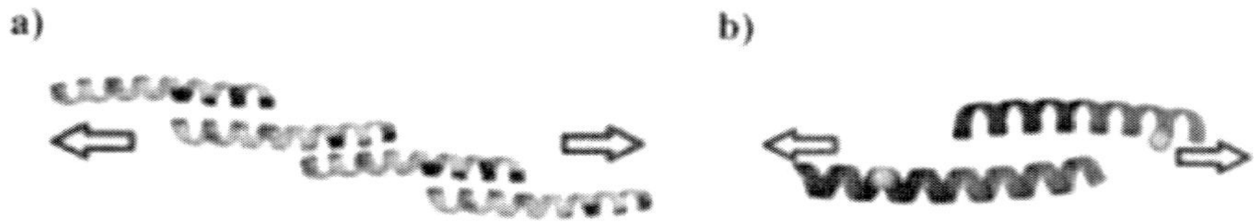

Figure 7. a) Assembly of coiled coils with out-of-phase hydrophobic faces (black and grey shading). b) The SAF design in which a sticky-ended heterodimer is prescribed using complementary g:e pairs and offset buried Asn side chains (sphere).

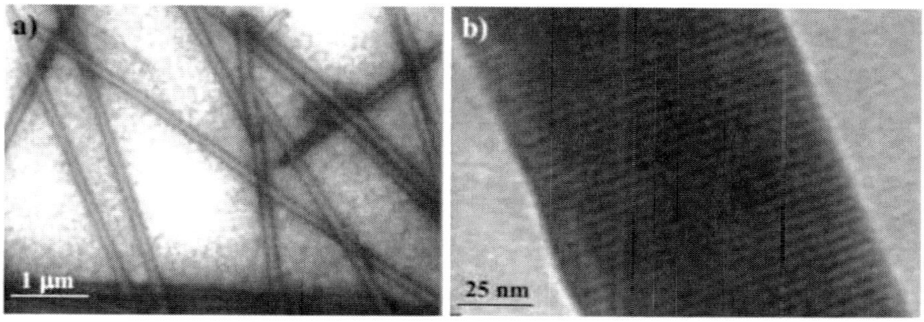

Figure 8. a) Electron micrograph showing fibers constituted by coiled-coils and made using the SAF system. b) High-resolution image showing a 4.2 nm banding pattern associated with crystallization of the α-helical building blocks within the fibers. Reproduced with permission from Ref. [75].

Potentiality of helix assemblies are now just been exploited for different applications that mainly involved the biomedical field. For example, a hybrid comprising an autophagy-inducing peptide (AIP) and a cell-penetrating peptide (CPP) connected *via* heterodimeric leucine zippers was generated and delivered into cells. This hybrid was found to successfully induce autophagy without significant cell death, while the same AIP directly connected to a CPP caused autophagy but also a significant cell death [77].

Drug delivery is one of the potential applications of coiled-coils since a controlled release of chemicals and medical agents can be derived from conformational changes. The stability of coiled-coils depend on factors like temperature, pH and ionic strength between others Therefore, stimuli-sensitive polymers can be easily obtained from the corresponding peptidic sequence [78]. Specifically, this can be employed to modulate the assembly, swelling and contraction of polymer networks.

In vivo dimerization of the strands of a coiled-coil can be exploited to get a targeted delivery of agents in a way that toxicity to normal tissues can be reduced and the therapeutic effect enhanced [79,80]. To this end, a therapeutic moiety (e.g., radionuclide) and a targeting moiety for site recognition and localization (e.g., antibody) can be respectively attached to the two strands of the coiled-coil structure to be *in vivo* formed.

Liposomes are phospholipid-based delivery systems that have been approved for treatment of cancer. The clinical use of phosholipid-based delivery systems is based on reduced toxicity to normal tissues with a simultaneous increase in tumor accumulation, being possible to take advantage of the enhanced permeability and retention effect of malformed blood vessels. However, drug release from conventional liposomes is a slow process that depends on passive leakage. In order to increase the therapeutic activity, liposomes able to release the loaded drug in response to external stimuli are extensively investigated [81]. Thermal energy in the form of local hyperthermia is a promising external trigger for cancer therapy. Poly- and oligopeptides can be synthesized with a defined sequence and chain length in such a way that can offer a good control of transition temperatures and afford for directional and reversible characteristics. Dissociative/unfolding properties of leucine zipper sequence may allow a good control, modulation, and timing of drug release under mild hyperthermia while improving *in vivo* drug retention. Therefore, leucine zipper peptides have successfully been incorporated into lipid bilayers during liposome preparation [82]. These peptide hybrids delivered high levels of anticancerigen drugs (e.g., doxorubicin) following localized mild hyperthermia (Figure 9). Specifically, the leucine zipper peptide was in an unfolded state at temperatures higher than 40 ºC.

Membrane proteins participate in numerous physiological processes (e.g., transport of solutes and macromolecules, intercellular signaling and cell-cell recognition). Many membrane proteins are composed of bundles of hydrophobic α-helical membrane-spanning segments. These could provide a low energy pathway for ion translocation through extruded liposomes and planar lipid membranes. Liposomes with incorporated long-chain poly-L-leucine were found to have proton permeability coefficients 5 to 7 times greater than control liposomes, whereas potassium permeability was not increased markedly. It was concluded that hydrophobic polyamino acids produce proton-conducting defects in lipid bilayers that may be used to model functional proton channels in biological membranes [83].

Leucine zipper coiled-coil domains are also highly interesting for the formation of advanced functional hydrogels. Thus, BAB triblock copolymers constituted by two terminal domains (B blocks) made of the heptad repeats characteristic of the coiled-coil motif and a central polyelectrolite domain (PEG) were able to self-assemble into stimuli-sensitive hydrogels. Gelation was reversible under suitable pH and temperature conditions as a consequence of the balance between the oligomerization of the helical ends and the swelling of the central polyelectrolyte domain [84].

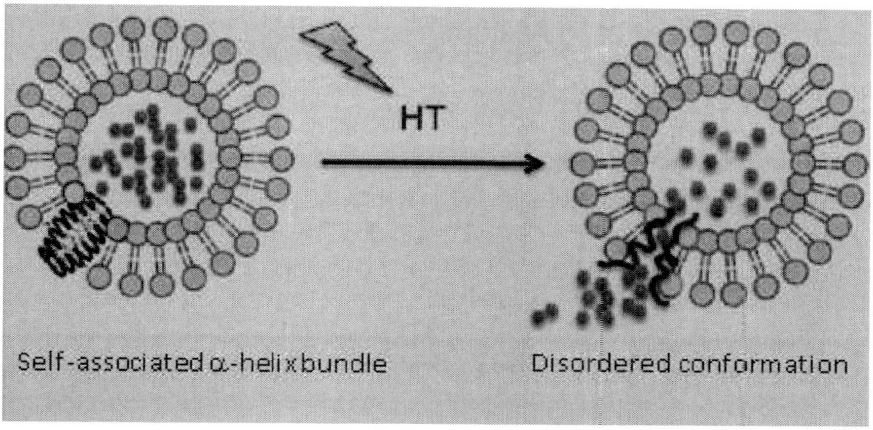

Figure 9. Thermoresponsive liposomes based on the incorporation of a leucine zipper sequence for the delivery of anticancer drugs. Reproduced with permission from Ref. [82].

I.A.4. Polypeptide-Polymer Hybrids Containing Leucine

Hybrid architectures constituted by peptide and synthetic or natural polymer blocks are receiving great interest since they open new challenges in areas like nanotechnology and biotechnology (e.g., biosensors, drug delivery systems, tissue engineering) [4,85-87]. In this way, hybrids can take profit of the indicated self-assembling capacity of the peptide building blocks, as well as the characteristics of the selected synthetic block which may afford solubility, versatility, rubber elasticity or even crystalline order. With a macroinitiator (e.g., a primary amine), polyleucine-based hybrid block copolymers can easily be prepared by ROP of NCA. Amine hydrochloride end-functionalized poly(ethylene glycol) (PEG) is one of the most used macroinitiators to endow polypeptide-based hybrid block copolymers with water-soluble and biocompatible properties [88,89].

The most part of the works focused on the supramolecular organization of peptide-synthetic polymer hybrids have selected poly(γ-benzyl-L-glutamate) (PBLG) or poly(Z-lysine) (PZlys) as the peptide block. These blocks have a capability to easily adopt α-helices or β-sheets by just adjusting the distribution of D/L units in their sequences. Moreover, both peptides have the possibility to be hydrolyzed resulting in poly(L-glutamic acid) (PLGA) and poly(L-lysine) (PLLys), which show pH-responsive behavior displaying coil to helix transitions depending on the pH of the medium. Specifically, PLGA adopts α-helical conformations at low pHs (i.e., pH ≤ 6.3), while PLL becomes α-helical at high pHs (i.e., pH ≥ 8.4).

Stimuli-responsive polypeptide block copolymers, as a type of block copolymers, are very appealing systems due to their capability to adopt different morphologies in selective solvents (e.g., micelles, star micelles, vesicles, tubules, fibrils or other kind of supramolecular aggregates). The most frequently used stimuli in these kinds of systems are pH, light, temperature, redox potential, magnetic field and ultrasounds.

Cationic micelles based on hybrid polypeptide copolymers have been demonstrated as effective non-viral gene vectors. This kind of micelles tends to generate compact nucleic acid complexes, which should be favourable to in vivo gene transfection. Specifically, tuneable and stable cationic micelles were prepared from poly(ethylene glycol)-*b*-poly(L-lysine)-*b*-poly(L-leucine) [90]. The three block copolymer formed micelles with well-controlled sizes from 40 to 90 nm that promoted gene expression both *in vitro* and *in vivo* with minimal adverse effects.

Transfection efficiency and biocompatibility were controlled by varying the lengths of PLLys and Poly(L-leucine) (PLLeu) blocks. High efficiency was a consequence of the capability of cationic micelles to promote DNA condensation and cell internalization. The copolymer adopts a core–shell structure in aqueous medium with hydrophobic segments as the core and cationic hydrophilic segments as the shell (Figure 10).

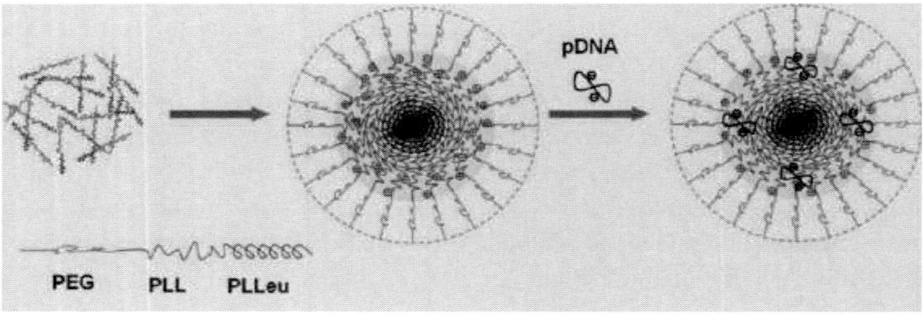

Figure 10. Formation of cationic micelles for DNA delivery from PEG-PLL-PLLeu copolymers. Reproduced with permission from Ref. [90].

Nanoparticles of the triblock copolymer poly(ethylene glycol)-*b*-poly(L-lysine)-*b*-poly(L-leucine) were used to deliver docetaxel (DTX) and small interfering RNA (siRNA-Bcl-2). This effective drug/gene vector provides a combination of therapeutic strategies with different mechanisms that can cooperatively impede tumor growth. The hydrophobic polyleucine core entrapped the anticancer drug, while the polylisine cationic backbone allowed for electrostatic interaction with the negatively charged siRNA [91].

Diblock and triblock copolymers of poly(ethylene glycol) and poly(DL-valine-*co*-DL-leucine) were also synthesized by ring-opening polymerization of the mixture of N-carboxyanhydride (NCA)s of DL-valine and DL-leucine initiated by amine- or diamine-terminated PEGs [92]. It was found that self-assembling of amphiphilic diblock structures led to micellar nanoparticles, while the triblock copolymers formed multi-lamellar structures.

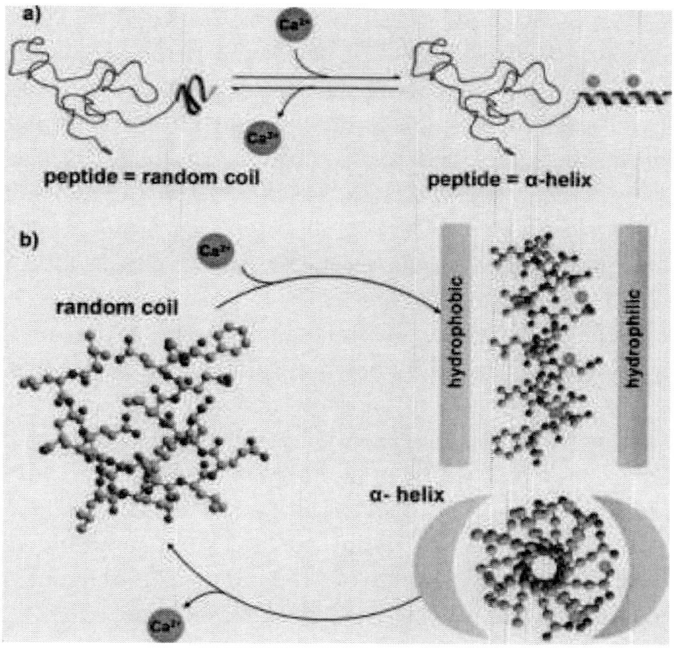

Figure 11. (a) Schematic representation of the calcium ion induced coil-to-helix transition in peptide-polymer conjugates. (b) Idealized mechanism for the calcium-ion regulated coil-to-helix transition of the peptide segment conformation. Reproduced with permission from Ref. [93].

Another recent smart strategy was based on the use of calcium ions as bioinspired trigger to reversibly control the coil to helix transition in peptide-

PEG conjugates having different block lengths (Figure 11) [93]. Established rules for the design of amphipathic α-helices were applied to get a peptide segment capable to perform a reversible coil-to-helix transition. Therefore, a specific pattern of polar (i.e., glutamic acid, *p*) and hydrophobic (i.e., leucine, *h*) amino acids (i.e., $[hpphppp]_x$) was selected. The peptide, as well as, PEG conjugates, adopted in the solution a statistical chain segment conformation prior to the addition of calcium ions. However, the addition of different concentrations of calcium chloride stimulated (as detected by circular dichroism) the coil to α-helix transition. This transition was reversible since a coil conformation was again detected after the subsequent addition of strong competitive Ca^{2+} cations. In terms of the effect of the synthetic polymer length, it was observed that on increasing the molecular weight of PEG block the random coil structure was more favored.

Racemic hybrid poly(ethylene glycol)-*b*-poly(racemic-leucine) (PRL) copolymers with different leucine residues were synthesized and characterized. Spherical micelles with tunable size from 170 to 250 nm and loaded with DTX were prepared by dialysis. The size of micelles was well controlled by changing the length of the hydrophobic polyleucine block. The racemic hybrids showed a higher drug-loading ability and an easier drug-loading process compared with the corresponding levo forms. All drug-loaded micelles exhibit a sustained release, a good biocompatibility and were more toxic than the free drug against human breast cancer cells, being consequently of high interest for anticancer therapy [94].

Polysaccharide–polypeptide hybrid microspheres (Figure 12) with a hollow structure were prepared by one-pot ring-opening polymerization of L-leucine NCA initiated by partially deacetylated chitin. Synthesis was carried out *via* an interfacial strategy under mild reaction conditions [95]. The water-soluble chitin (degree of deacetylation 0.5) was first dissolved in an aqueous phase and then emulsified with ethyl acetate to form water-in-oil (W/O) emulsion. The L-leucine *NCA* in oil phase was then introduced. Chitin initiated the graft copolymerization of NCA through its amino groups in a process that started at the interface and rendered the copolymer grafted with polyleucine side chains. This copolymer was insoluble in both phases and consequently precipitated, resembling gel formation and giving rise to hollow microspheres.

Amphiphilic hybrid diblock copolymer constituted by polyleucine and PEG and having terminal mannose groups were synthesized and their aggregation behavior studied [96]. Aggregates with mannose groups on their surface are interesting since they appear suitable for recognition behavior toward certain proteins and have been proved to be potentially useful for the development of vaccination vectors [97]. To this end the hydrochloride of an α-azido, ω-amino PEG was employed as initiator of the ROP of L-leucine NCA. Mannose was subsequently conjugated at the end of the PEG chain by click chemistry. The amphiphilic copolymers self-assembled giving rise to morphologies that were depended on the chain length of polyleucine since it influenced both the hydrophobic interactions and the secondary structure. The formed aggregates were able to bind reversibly the lectin Concanavalin A.

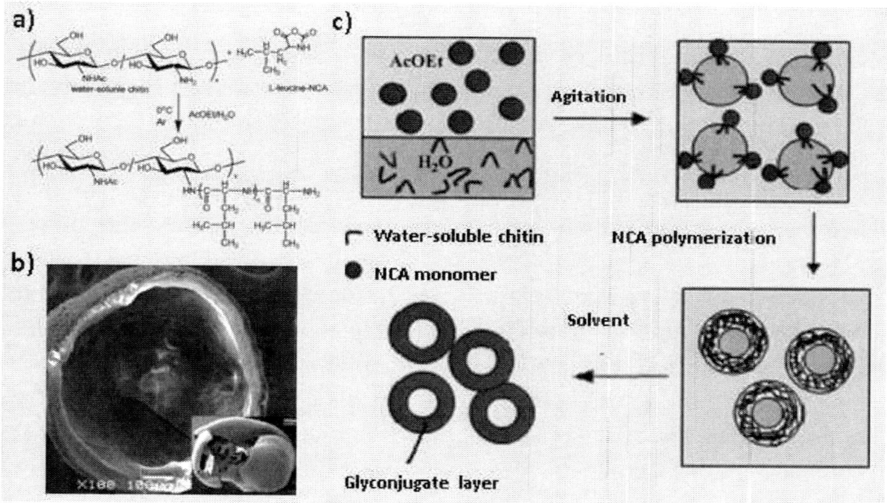

Figure 12. a) Reaction between water-soluble chitin and L-leucine *N*-carboxyanhydride. b) Cross-sectional SEM image of of chitin-*g*–polyleucine microsphere. The inset shows a close-up of a broken sphere, c) Scheme for the *in situ* synthesis of hollow microspheres via interfacial polymerization approach. Reproduced with permission from Ref. [95].

Strong pH dependence was observed for polymers which peptidic chains had not an intrinsic pH sensitivity. Thus, the amphiphilic copolymer, poly(L-leucine) grafted polyallylamine showed pH reversible conformational and morphological changes [98]. Under acidic conditions the polymer formed amyloid-like fibrils that subsequently grew into a planar plate composed of stacked β-sheets. Amine groups were protonated at low pHs, being a rapid aggregation of the poly(L-leucine) grafted chains hindered by the electrostatic repulsions. An antiparallel β-sheet structure with intermolecular hydrogen bonds among the grafted chains was in this case favoured. On the other hand, under basic conditions, the polyallylamine main chain was deprotonated and took a compact conformation. Globular aggregates were formed and specifically an α-helical conformation was favoured for polyleucine chains.

The use of polyleucine as a wound dressing [45,46] has several limitations that concern low solubility in organic solvents (soluble in toxic benzene), slow release of antibacterial agents, low bioadhesion and slow biodegradation. Hybrids containing PEG segments (e.g., polyleucine-*b*-polyethylene glycol-*b*-polyleucine) were considered as an alternative since they can provide a better control on the release of the incorporated drugs. Furthermore, the triblock copolymer may have additional advantages like increase of solubility, bioadhesion and flexibility, control of biodegradation, absorption of body fluid and non-immunogenic effect. Samples impregnated with silver sulfadiazine (AgSD) showed that suppression of bacterial proliferation in the wound dressing was dependent upon the PEG content. In fact the release of AgSD was increased in proportion to the PEG content in the copolymer. In summary, properties of these materials were found adequate for their use as artificial skin [99].

Polyleucine-*b*-polyethylene glycol-*b*-polyleucine was probed to be useful for the extraction of Au^{3+} ions from dilute aqueos solutions. The procedure consisted on extraction with a chloroform solution of the copolymer followed by the separation of the chloroform gel phase induced by the thermo-sensitive sol-gel transition. The Au^{3+} ion was captured by the amino group of the chain end of the peptide copolymer with a high specificity, being the equilibrium constant of the binding larger than that of other ions like Pd^{2+} and Pt^{4+} [100].

I.A.5. Leucine-Based Co-Polyamides and Polyesters

Biodegradable AB-type copolyamides (co-poly(amino acids), Figure 13) were prepared by melt polycondensation of 6-aminocaproic acid and L-leucine by Zhang et al. [101].

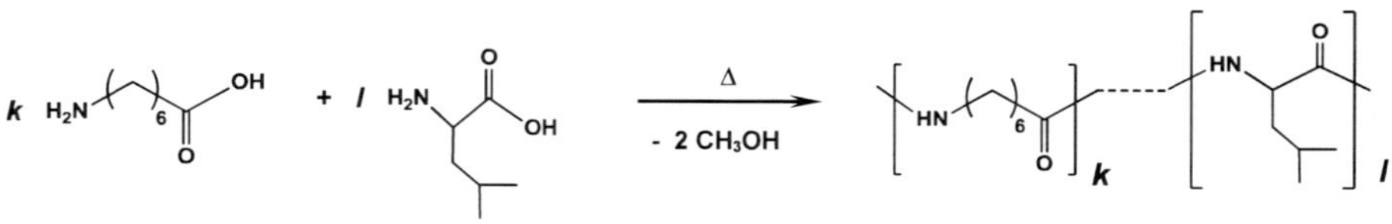

Figure 13. Synthesis of copolyamides on the basis of 6-aminocaproic acid and L-leucine.

It was shown that by increasing leucine content of the comonomers, the intrinsic viscosity, melting point, and melting enthalpy of copoly(amino acid)s decrease while the density increases. The enzymatic degradation of the polymers films was tested using papain; results showed that the copoly(amino acid)s are degradable and the enzymatic degradation rate increases with increasing leucine content in the comonomers. Considering good biodegradability, copoly(amino acid)s are perfectly suited for use as biomedical materials and biodegradable packaging materials, as well as in many other applications. In an alternative approach copolyamides for the same applied purposes were obtained by melt copolycondensation of fatty diacids and diamines with α-amino acids [102]; it should be noted here that in such case the amino acids' orientation within the polymeric backbone would be of both type "head-to-tail" and "head-to-head". In other words, the polyamides' backbones represent a mixture of AB and AA-BB type structures with respect to α-amino acid.

Interesting method of the synthesis of α-amino acid-based poly(ester urethane) was developed by Jayakannan *et al.* [103] using amino acids' based hetero-bifunctional monomers. For this, amino acids were converted into corresponding methyl ester hydrochloride salts which then were reacted with methyl chloroformate in the presence of HCl acceptor resulting in the goal ester-urethane monomers as depicted in Figure 14. Hetero-bifunctional monomers on the basis various amino acids including L-leucine were synthesized.

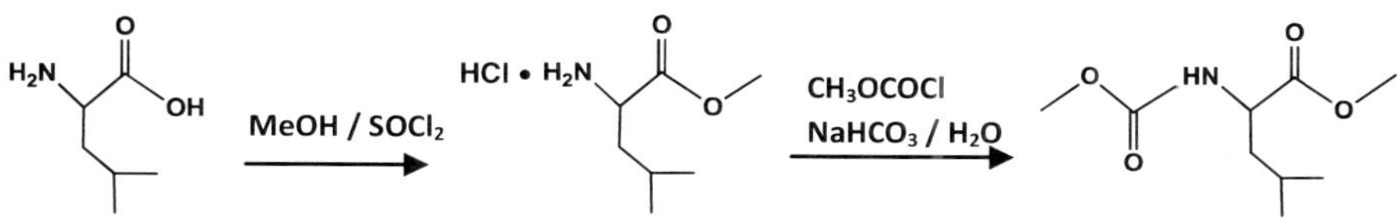

Figure 14. Synthesis of hetero-bifunctional ester-urethane monomers from leucine.

The goal poly(ester urethane)s were obtained by melt polycondensation (150°C) of hetero-bifunctional ester-urethane monomers with fatty diols (Figure 15) using titanium tetrabutoxide as a catalyst.

The poly(ester urethane) obtained from leucine-based monomer and 1,12-dodecanediol under these conditions had M_w 11,000 Da and PDI 1.31. It has to be noted that in this case, like the copolyamides above [102], the polymer's backbone represents a mixture of AB and AA-BB type structures with respect to α-amino acid.

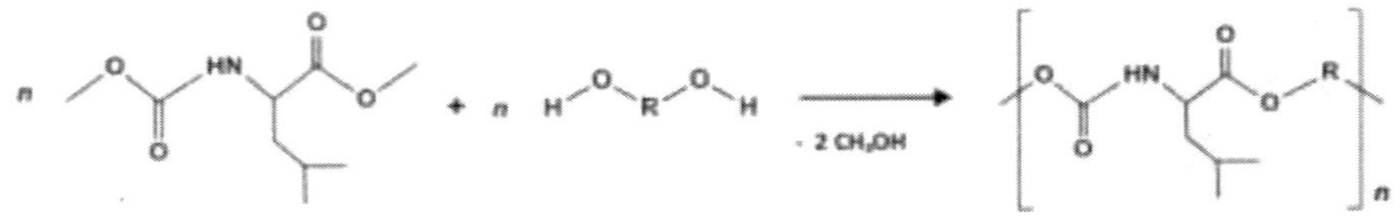

Figure 15. Synthesis of poly(ester urethane)s *via* melt polycondensation of hetero-bifunctional ester-urethane monomers (shown for leucine-based monomer).

Cohen-Arazi et al. [104] transformed natural α-amino acids into α-hydroxy acids using a straightforward reliable and inexpensive method of diazotization of α-amino acids (Figure 16) [105,106]:

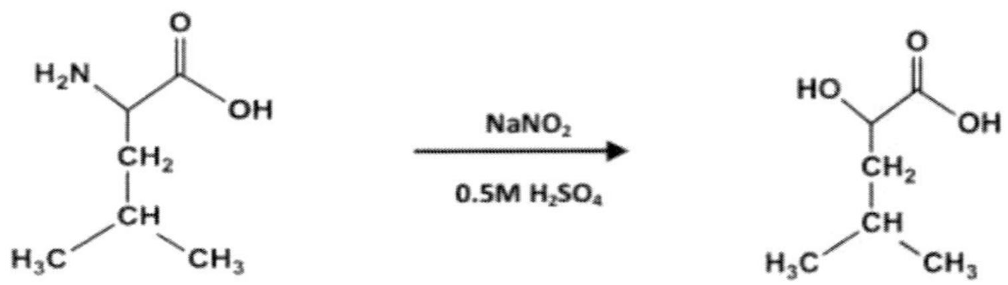

Figure 16. Synthesis of the α-hydroxy leucine from Leucine.

The obtained α-hydroxy leucine was transformed into corresponding homopolyester by thermal polycondensation in bulk using p-toluenesulfonic acid as a catalyst (Figure 17):

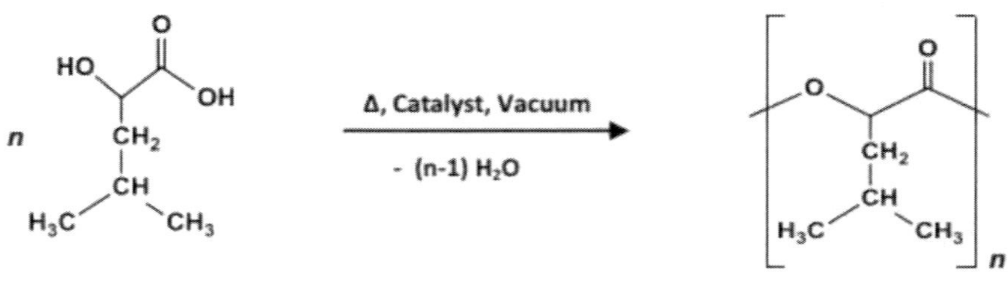

Figure 17. Synthesis of homopolyester by direct condensation of α-hydroxy leucine.

Under the used conditions low-molecular-weight homopolymers (M_w up to 2000) were obtained. The copolycondensation with lactic or glycolic acid slightly increased the molecular weights of the polymers. The highest molecular weight (M_w = 5400 Da) was achieved after copolycondensation of α-hydroxy leucine with 6-hydroxyhexanoic acid, presumably due to its long aliphatic chain.

The obtained polyesters were screened for acute toxicity using a lactate dehydrogenase (LDH) assay and were found to be nontoxic. The biocompatibility of the newly synthesized polymers was assessed on a neuronal (PC12 pheochromocytoma) cell line following differentiation and an endothelial (bEnd.3) cell line following adhesion and proliferation. Several polyesters induced neuronal differentiation, and none of the tested polymers reduced the adhesion of endothelial cells.

I.A.6. Inorganic/Organic Nanocomposites from Polymers Containing Polyleucine Blocks

Hydroxyapatite (HAp: $Ca_{10}(PO_4)_6(OH)_2$) is a calcium salt of orthophosphoric acid with compositional and biological similarities to the mineral phase of human bone. Biominerilazation during bone formation leads to a nanocomposite with layers of collagen molecules alternating with layers of nanocrystals of HAp (mostly in the carbonated hydroxide-deficient from) [107]. This combination of inorganic and organic phases provides increased toughness compared to HAp alone [108]. Research on the formation of biocomposites of HAp in conjunction of either synthetic or natural polymers is nowadays a topic of great interest. Different procedures for preparing such biocomposites have been developed being one of them the self-assembling using block copolypeptide templates. For example, self-assembling thermo-reversibly gelling anionic and zwitterionic block copolymers conjugated to hydroxyapatite nucleating peptides were used as templates for precipitation of calcium phosphate nanostructures [109].

Efforts are now mainly focused to reduce the polymer concentration (usually close to 25 wt%) needed for gel formation since the weight fraction of the incorporated inorganic phase will be logically limited. The block copolypeptide constituted by lysine and leucine units ($Lys_{170}Leu_{30}$) is able to form gels at very low weight fractions (0.5–4 wt%) and consequently was investigated as a template for biomineralization. Results were highly promising since for the first time the inorganic weight fraction could be increased to over 50%, which is very close to the content determined in natural bone [110].

Biomimetic synthesis of highly ordered biosilica from synthetic precursors such as tetraethylorthosilicate (TEOS) and tetramethylorthosilicate (TMOS) is currently receiving great attention. The employment of biomolecules as templates can offer a precise control over the nanostructure of the obtained inorganic materials as well as on their final properties. Synthetic polypeptides and block copolypeptides have been chosen to direct biosilicification and simultaneously tune the resulting nanostructures [111,112]. New hierarchical materials can be prepared by employing amphiphilic copolypeptides and their supramolecular assemblies as organic templates to direct the biomimetic synthesis of inorganic-organic hybrid materials [113]. Specifically, poly(L-lysine HBr)-*b*-poly(L-leucine) diblock copolypeptide (Lys-Leu) has been studied as template. The resulting biosilica morphology was controlled by molecular weight and the nature of counterions. Application of the sol–gel method could replicate typical Lys-Leu hydrogel networks to offer porous silica fibril networks. Furthermore, application of ultrasound to the silicification media could transform flexible silica fibrils to long and ordered silica rods by the so-called sono–gel process. The approach was highly efficient and required a minimum work on template synthesis to make hierarchical silica materials ordered at different levels [114].

A variety of organisms have a form of protein-mediated silica morphogenesis in which highly ordered structures made of silica (i.e., biosilica) are produced under mild, ambient conditions [115]. This feature contrasts with the extreme conditions that are usually required to control silica morphology by most industrial approaches. Therefore different efforts have been focused to mimic biosilica production capabilities by engineering peptides to resemble aspects of naturally occurring biomolecules. Weidner *et al.* [116] have recently demonstrated that different biosilica nanostructures could be formed in vitro from peptides composed solely of Lys and Leu amino acids. Morphologies were strongly dependent on the hydrophobic periodicity since it directs the formation of particular secondary structures at air-water interfaces and on solid, polar, and nonpolar surfaces [117,118]. Arrangements of Leu and Lys residues with a periodicities of 3.5, 3.0 and 2.0 favoured that peptides adopt the α-helical form, the 3_{10} helix and the β-strand, respectively. These secondary structures led to nanoparticle, elongated rod-like and wire-like silica morphologies (Figure 18).

Bionanocomposites (BNC)s containing amino acid groups were prepared by dispersion through ultrasonication technique of multi-walled carbon nanotubes (MWCNT)s into a poly(ester-imide) (PEI) matrix [119]. This was obtained from reaction of a L-leucine based diacid that provided biodegradable and biocompatible characteristics to complement those given by carbon nanotubes (e.g., excellent electronic properties, high tensile strength, ultralight weight, chemical and thermal stability [120]).

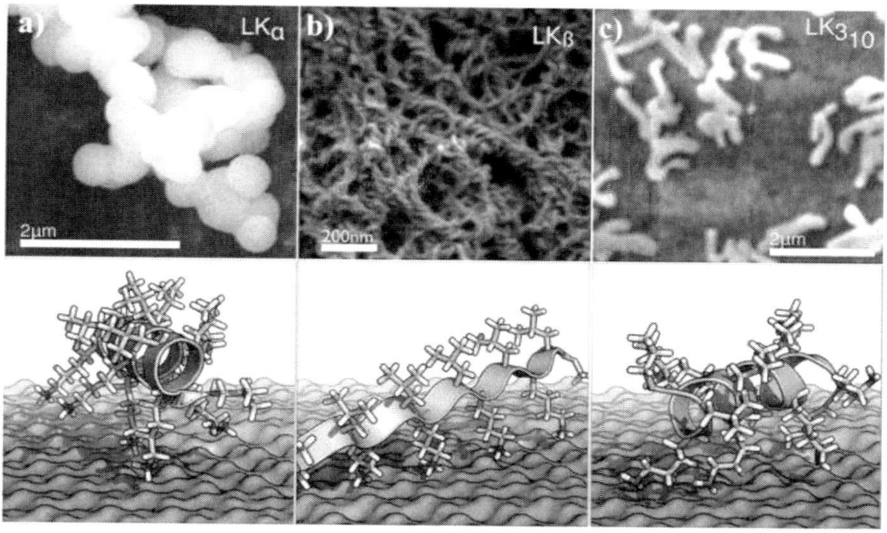

Figure 18. SEM micrograps (top) of the different morphologies of precipitates assembled by leucine-lysine peptides with different Leu-Lys periodicities and secondary structures (bottom). Reproduced with permission from Ref. [116].

Micro- and nanoparticles of chitosan are scarcely toxic for pulmonary epithelial cell lines and consequently appear safe for any formulation intended for inhalation and treatment of lung diseases (e.g., asthma and chronic obstructive pulmonary disease) [121]. Synthesis of a conjugate of chitosan with L-leucine and the preparation of nanoparticles from this conjugate for use in pulmonary drug delivery have recently been reported [122].

Conjugation of L-leucine to chitosan was interesting since created a hydrophobic environment around the chitosan backbone that reduced interparticle interactions and enhanced the dispersibility of particles from a dry powder inhaler.

I.A.7. Conducting and Magnetic Applications of Hybrids Containing Leucine Polymers

Conjugated organic materials have received great attention for the fabrication of efficient chemo- and bio-sensors [123-125]. These conjugated systems have an extended electronic delocalization spanning along the entire molecular backbone that should provide electrical conductivity, low ionization potential and low energy optical transitions [126]. Electron delocalization and consequently the indicated properties can be highly influenced by interactions with analyte species. Therefore, conjugated organic materials can efficiently convert the interaction with target analytes into an observable signal (e.g., optical, electrical or both).

Poly(aryleneethynylene)s (PAEs) are highly investigated for their potential applications as sensing materials. New derivatives carrying amino acidic (e.g., L-leucine) side arms as active sites for the capture of metal ions have been developed considering the coordinating ability of peptides and amino acids [127,128]. Specifically, new PAEs carrying leucine have been prepared and tested for their ability to bind Hg(II) ions. Good analytical sensitivity and excellent selectivity were demonstrated, appearing these materials very promising for the development of sensors for Hg(II) [129].

New polythiophenes bearing L-Leu as chiral center were designed to get chiral conducting surfaces that should be appropriate to recognize specific target molecules. The procedure involved the synthesis of L-leucine oligothiophenes and the subsequent electrochemical oxidation [130]. A chiral conducting surface is interesting since should display several advantages such as the ability to be regenerated by controlling the thickness of the deposited film layers and to recognize specific motifs. Synthesized polymers displayed excellent stability in both doped and undoped states and exhibited excellent adhesive properties on several electrodes (platinum, gold, and glassy carbon). Cyclic voltammetry experiments demonstrated that hydrogen bonds could be established between the chiral conducting surface and free amino acids such as L-Leu, D-Leu, D-Ala, L-Ala, and D/L-Ala. Unfortunately, responses (e.g., potential shift and current) were similar for the interactions between of chiral surface and both enantiomeric forms of amino acids.

Amino acids including L-Leu were employed as precursors in emeraldine base/DMF solutions prior to acid doping. Basic characterization of these materials indicated that the optical activity could be maintained for long times in both solid and solution phases due to hydrogen bonding interactions. These involved the amino-acid molecules and the amine (NH) and imine (-N=) sites on the emeraldine base form, while stereo-selectivity of the polymer chain could be preserved. Factors responsible for influencing the chiroptical properties in helical conducting polyaniline and the chiral discrimination of different amino acids have also discused [131].

Development of magnetic nanoparticles is current subject of interest due to their biomedical applications (e.g., in vitro diagnostics) and specifically beneficial factors such as the possibility of ubiquitous tissue accessibility. The utilization of magnetic iron oxide nanoparticles has the added advantage that the body is designed to process excess iron (i.e., it is firstly stored in the core of protein ferritin and subsequently metabolized into elemental iron and oxygen by hydrolytic enzymes) [132]. Novel magnetic core-shell iron oxide nanoparticles, covered with a polypyrrole functionalized by covalent attachment of glycyl-leucine or bovine serum albumin were synthesized using *N*-hydroxysuccinimide as activating reagent. The morphology of the functionalized magnetic nanoparticles was investigated by light scattering and high resolution electron microscopy, showing a core-shell structure [133].

Optically active particles are interesting due to their potential applications in chiral recognition/resolution, asymmetric catalysis, and enantioselective crystallization [134-136]. Composites consisting of Fe_3O_4 magnetic nanoparticles (shell) and porous optically active microspheres (core) have currently been prepared. The procedure consisted on preparing polystyrene (PS) template microspheres that were swelled by a chlorobenzene solution containing N-acryloyl-L-leucine methyl ester (M1) as a chiral monomer, divinylbenzene (DVD) as crosslinking agent and an initiator (BPO) for free radical polymerization. The PS template was removed after copolymerization took place inside the microspheres. Fe_3O_4 nanoparticles amino-modified by 3-aminopropyltriethoxysilane (APTES) could be succesfully anchored onto their surfaces after hydrolysis of the methyl ester groups (Figure 19) [137].

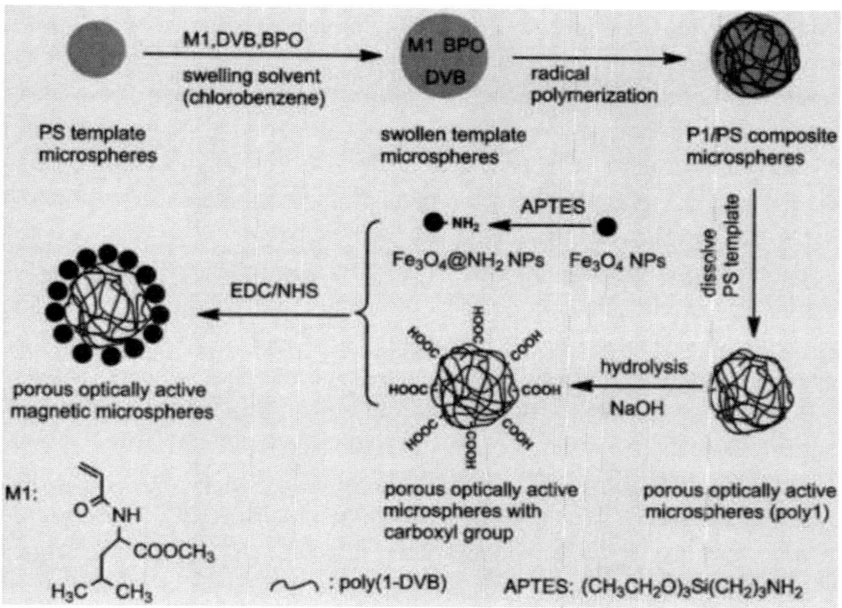

Figure 19. Synthesis of porous, optically active and magnetic composite microspheres. Reproduced with permission from Ref. [137].

Corrosion attack on iron in acidic medium can be reduced by means of organic polymer coatings that contain nitrogen [138]. The efficiency of these compounds as corrosion inhibitors is linked to the number of electrons available for donation and the distribution of the electron density around the donor atom [139]. Poly(vinyl alcohol-leucine) was synthesized and its influence on corrosion inhibition of mild steel in 1 M hydrochloric was demonstrated to be more than 95% [140].

A new poly(L-Leucine)-1,3-diamino propane–lithium iodide (LiI) sample was prepared to get a superionic lithium based solid polymer electrolyte. Polyleucine was chosen as the host material due to its molecular structure which can easily be complexed with lithium salts [141]. A powder mixture of the two component was ground in an agate mortar, pelletized and sintered at 90 °C for 3 days to get a material with 1.0×10^{-3} $S.cm^{-1}$ when 50 wt-% of LiI was added. The material was considered a promising candidate to be used in a solid-state battery or lithium-polymer batteries.

I.B. AA-BB Type Polymers Composed of Leucine

In the previous section AB-type leucine-based polymers with naturally occurring "head-to-tail" orientation in the polymeric backbones were discussed. The present section deals with the leucine-based AA-BB type polymers in which leucine's residues are incorporated in non-natural "head-to-head" and "tail-to-tail" manner. Different classes of AA-BB leucine-based polymers were designed for both biomedical and technical applications. The AA-BB type polymers are basically obtained *via* step-growth polymerization (polycondensation).

Series of heterochain polymers were synthesized on the basis of imide-dicarboxylic acids obtained by interaction of L-leucine with various cyclic anhydrides. Langer et al. [142], by condensation of equimolar amounts of leucine and trimellitic anhydride, obtained asymmetric imide-diacid containing both aromatic (attached directly to the benzene ring) and aliphatic (belongs to the leucine residue) carboxylic groups (Figure 20a, R=OH). This diacid was used for synthesizing aromatic-aliphatic polyanhydrides which have more reasonable hydrolysis (biodegradation) rate as compared with pure aromatic (slow hydrolysis) and aliphatic (fast hydrolysis) polyanhydrides.

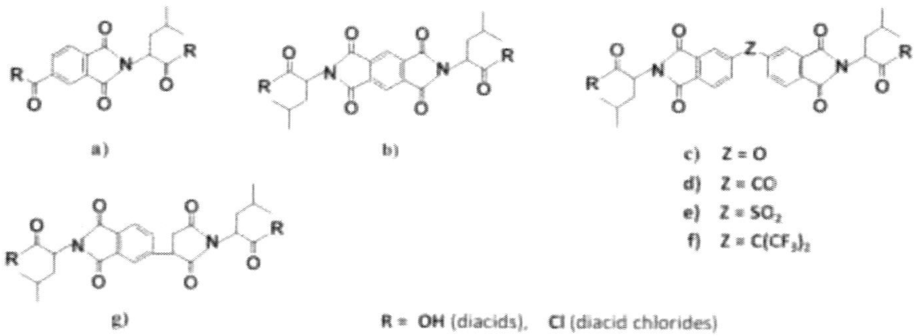

Figure 20. Imide containing diacid on the basis of leucine and cyclic anhydrides.

Leu-based imide-diacid, as well as trimellitimide-diacids on the basis of other amino acids [142,143] were transformed into polyanhydrides by melt polycondensation using mixed anhydrides diacids/acetic acid as monomers.

Mallakpour and co-workers [9,144-150] obtained a variety of optically active and biodegradable polymers on the basis of dicarboxylic acids made by condensation of L-leucine with various cyclic anhydrides. Along with imide-diacid given in Figure 20a, R=OH, this team used diimide-diacids made by condensation of leucine with different cyclic dianhydrides – pyromellitic dianhydride (Figure 20b, R=OH), *3,3',4,4'-diphenyloxidetetracarboxylic acid dianhydride* (Figure 20c, R=OH), 3,*3',4,4'*-***benzophenonetetracarboxylic*** *acid dianhydride* (Figure 20d, R=OH), 3,*3',4,4'-diphenylsulphonetetracarboxylic acid dianhydride* (Figure 20e, R=OH), *4,4'*-(hexafluoroisopropylidene)di*phthalic anhydride* (Figure 20f, R=OH), and Epiclon [5-(2,5-dioxotetrahydrofurfuryl)-3-methyl-3-cyclohexyl-1,2-dicarboxylic acid anhydride] (Figure 20g, R=OH).

The diacids depicted in Figure 20a-g, R=OH were applied for synthesizing optically active poly(amide imide)s (PAIs) without activation of carboxylic groups by interaction with aromatic diamines under the conditions of "Direct Polycondensation" [151] in the presence of condensing system triphenylphsipite/Pyridine/N-methylpyrrolidone/$CaCl_2$ [152] or triphenylphsipite/tetra-*n*-butylammonium bromide [145]. Optically active PAIs on the basis of free chiral imide-diacids, i.e., without activation of carboxylic groups, were also obtained by their interaction with different aliphatic and aromatic diisocyanates [144].

In another approach, for synthesizing optically active PAIs, the carboxyl groups of imide-diacids were activated beforehand by their transformation into corresponding chlorides [144,148,153] (Figure 20a-20g, R=Cl). The diacid chlorides were polycondensed with aromatic diamines under the conditions of either conventional solution polycondensation (in o-cresol or DMA) or microwave-assisted step-growth polymerization. The latter proceeded rapidly compared to solution polycondensation.

In an alternative approach the diacid chlorides were interacted with two moles of *p*-amino benzoic acid (Figure 21) that led to imide-diamide-diacids (bis-*p*-aminobenzoic acids) which represent aromatic diacids:

Figure 21. Synthesis of the imide-diamide-diacids on the basis of p-aminobenzoic acids.

The dimerized *p*-aminobenzoic acids were transformed into corresponding polyamides *via* eithet direct polycondensation [154,155] or interaction with diisocyanates [156].

Leu-based polymers obtained by Mallakpour's team are thermally stable, some of them are biodegradable in soil. The optically active polymers are promising as catalysts for asymmetric synthesis and as a stationary phase for separating optical isomers.

Another type of leucine-based cyclic monomer– diketopiperazine (cyclic dipeptide) was used by Rafiemanzelat *et al.* [156] as a bifunctional building block for synthesizing polymers containing leucine in the backbones, using a two-step polymerization method. At the first step, the cyclic dipeptide was reacted with a slight excess of 4,4′-methylene-bis-(4-phenylisocyanate) to produce isocyanate-terminated poly(imide-urea) oligomers (hard segment). At the second step the resulting telechelic diisocyanate was reacted with polyethyleneglycols of different molecular weights (PEG-400, PEG-600, PEG-1000 and PEG-2000) resulting in a series of new segmented poly(ether urethane urea)s, depicted in Figure 22.

The obtained copolymers have inherent viscosity up to 1.8 dL/g, were hydrolyzable, thermally stable and soluble in amide type solvents. Polymers containing different molecular weights of PEGs soft segments show different thermal stability, phase separation, hard segment cohesiveness and hydrolysis (biodegradation) rate. The multiblock polymers composed of diketopiperazines can be considered as potentially physiological active since the small cyclic peptides, which are ultimate products of their biodegradation, reveal remarkable physiological activity (see Ref. [157] and Refs. cited therein).

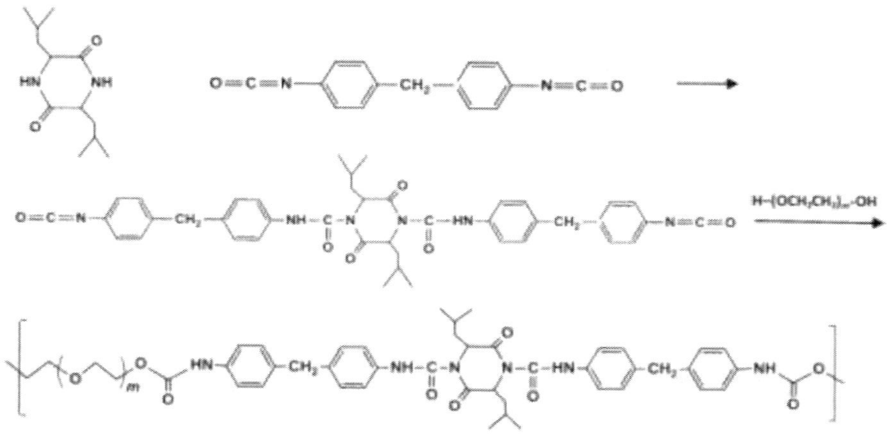

Figure 22. Segmented poly(ether urethane urea)s on the basis of leucine-diketopiperazine.

Biodegradable polymers, namely poly(ester amide)s (PEAs) containing Phe-Leu linear dipeptides in the backbones, were obtained by Kize et al. [158] by direct condensation of N,N'-adipoyl-bis-L-phenylalanine with di-*p*-toluenesulfonic acid salt of bis-(L-leucine)-1,4-butylene diester in DMF using 1-ethyl-3-(3-dimethylaminopropyl)carbodiimide hydrochloride as a coupling agent, and triethylamine as *p*-toluensulfonic acid acceptor. A low-molecular-weight polymer (M_n = 5,600 Da) with a moderate yield (77%) was obtained under this reaction conditions. In an alternative approach such kind of PEAs with dipeptide links in the backbones were synthesied *via* so called Solution Active Polycondensation (SAP) by the interaction of activated di-*p*-nitrophenyl ester of N,N'-adipoyl-bis-L-phenylalanine with di-*p*-toluenesulfonic acid salt of bis-(α-amino acid)-1,4-butylene diesters [159]. The interesting feature of the suggested synthetic methods [158,159] consists in the formation of dipeptide fragments during the chain growth reaction, i.e., there is no need in the synthesis of dipeptide-containing monomers for which many reactions for protection and activation of amino or carboxyl groups would be required.

The method of SAP [5,160] was successfully applied to the synthesis of biodegradable poly(ester amide)s (PEAs) having high molecular weights (M_w $\leq$ 167,000) and valuable material properties [161,162]. These polymers were obtained by polycondensation of activated dinitrophenyl esters of dicarboxylic acids with key amino acid based monomers - di-*p*-toluenesulfonic acid salts of bis-(α-amino acid)-1,4-butylene diesters. It should be emphasized that, in spite of rather tedious procedure of purification of the PEAs from low-molecular-weight by-product - *p*-nitrophenol, the SAP is more universal compared to other polycondensation methods since can be applied to many dicarboxylic acids, regardless of their chain length and constitution, e.g., to succinic and other 1,2-dicarboxylic acids. The SAP can be applied also to the synthesis of amino acid based biodegradable polymers (AABBPs) of related classes [163] - poly(ester urethane)s using activated diols bis-carbonates, and poly(ester urea)s using activated diphenylcarbonates. It has to be noted here that some limitations of SAP connected with the polymers' purification can be sorted out using diacids' activating compounds which are much more easily removable from the polymers by simple washing with water, e.g., N-hydroxysuccinimide.

For synthesizing PEAs on the basis of hydrophobic diacids such as sebacic acid and higher diacids preference should be given to diacids' dichlorides and more fast and facile Interfacial Polycondensation (IP) [164,165]. The latter gives better results for synthesizing high-molecular-weight poly(ester urea)s using phosgene (or triphosgene) as key bis-electrophilic monomers [166]. For synthesizing amino acid based PEAs thermal polycondensation of dimethyl esters of N,N'-diacyl-bis-amino acids with diols was successfully applied as well [164].

α-Amino acid L-leucine turned out one of the most valuable building blocks for constructing AABBPs destined for biomedical applications. Leucine-based regular PEAs [162], in contrast to semicrystalline L-phenylalanine-based analogues, are amorphous that is important for even biodegradation of both surgical and drug-eluting implants. It was reported that the high initial crystallinity form large amounts of tiny and rough highly resistant particles when the amorphous phase was degraded; such particles are now known as very inflammatory and thus well explain the late dramatic inflammatory response (See Ref. [167] and Refs. cited therein). Leu-based PEAs are soluble in medically acceptable organic solvent such as ethanol and form highly flexible films. Ethanol solution of co-PEA composed of sebacic acid, bis-(L-leucine)-1,6-hexylene diester and bis-(L-phenylalanine)-1,6-hexylene diester is promising as a liquid (spray) wound dressing [168]. The preparation sprayed onto the wound forms a thin, elastic, and transparent film that accelerates healing of superficial wounds, ulcers, and burns. This liquid dressing can be loaded with various drugs including antimicrobials such as chlorohexidine, silver sulfadiazine, nanosilver, etc.

The functional biodegradable co-PEAs composed of sebacic acid, bis-(L-leucine)-1,6-hexylene diester and L-lysine [169,170] revealed excellent elasticity (elongation at break $\geq$ 800%) and adhesion to stainless steel, is biocompatible and do not elicit an inflammatory reaction [171,172]. The polymer is almost ideal for the application as drug eluting vascular stent coating [173], as well as for other biomedical applications such as wound care polymer compositions, vaccine and polymer particles delivery compositions, delivery of ophthalmologic agents to the exterior or interior of the eye, therapeutic polymers, etc. (See Ref. [5] and Refs. cited therein). Some leucine-based PEAs turned out very effective as medicated wound dressings in the management of infected superficial wound and ulcers [174,175].

Leucine-based PEAs composed of fatty diacids and diols for some applications are too ductile due to wholly amorphous structure and low glass transition temperature (T_g). To impart some rigidity to these polymers, they could be combined (*via* either blending or copolymerization) with leucine-based PEAs made of "rigid" bicyclic diols – dianhydrohexitoles [176,177], which show T_g within 89-91 °C.

Increasing the macromolecular rigidity of leucine-based PEAs can also be achieved by incorporating in the polymeric backbones unsaturated fumaric acid's residues [178]. These PEAs are not as ductile as saturated analogous, however, are flexible enough to form elastic films, in contrast to L-phenylalanine-based analogous [179] that are too rigid and form brittle films. The unsaturated PEAs were further chemically modified (with SH- and NH_2-reagents) into functional derivatives and subjected to thermal and photochemical transformations (curing).

Another kind of high-molecular-weight (M_w up to 67.000) functional leucine-based PEAs with desirable material properties were obtained *via* SAP of di-*p*-nitrophenyl-*trans*-epoxy succinate with di-*p*-toluenesulfonic acid salt of bis-(L-leucine)-alkylene diesters. The epoxy-PEAs were chemically modified under mild conditions: oxirane groups along the backbone were reacted with both nucleophilic and electrophilic reagents, and also subjected to thermal and chemical curing [180]. The capability of these polymers to be chemically or thermally modified substantially expands material properties and, hence, the scopes of potential applications of PEAs as absorbable surgical devices and drug carriers.

Polydepsipeptides (PDPs), as it was discussed above, represent one of the most promising polymers for constructing biodegradable biomaterials. These polymers degrade with releasing α-hydroxy and α-amino acids that could be assimilated by the organism *via* metabolic means. ROP of morpholine-2,5-

diones, that represent a basic method of the synthesis of PDPs, has some limitations that are mainly associated with the use of toxic tin-organic catalyst, low yields of starting monomers – morpholine-2,5-diones (9-33 %), formation in some cases of either low-molecular-weight oligomers or polymers with unfavourable mechanical properties. Therefore, other synthetic methods such as step growth polymerization (polycondensation) have been proposed to get a class of PEAs having α-hydroxy acid and α-amino acid units. High-molecular-weight (M_w up to 52,500 Da) AA-BB type PDPs were synthesized by SAP of di-*p*-toluenesulfonic acid salts of bis-(L-leucine)-alkylene diesters with activated di-*p*-nitrophenyl esters of O,O'-diacyl-bis-glycolic acid (R = O-*p*-C_6H_4-NO_2) [181], as depicted in Figure 23.

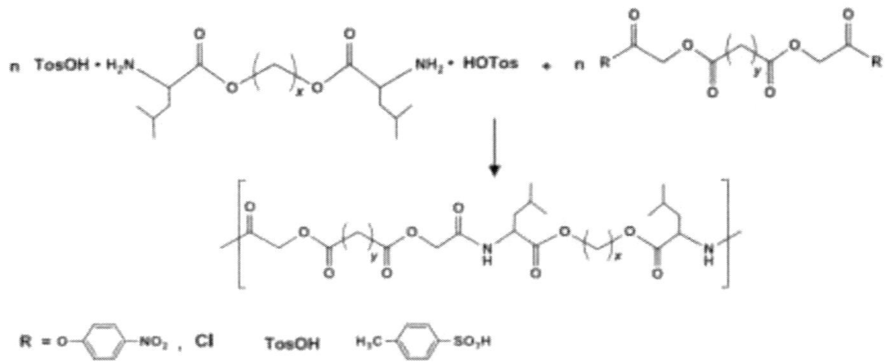

Figure 23. Synthesis of AA-BB type polydepsipeptsides.

The AA-BB type PDPs with valuable material properties ($M_w \leq 30{,}000$ Da) were also obtained *via* IP using corresponding diacid chlorides (R = Cl in Figure 23) [182]. It has to be noted here that telechelic diacid chlorides (oligomeric polylactide terminated by the residues of sebacoyl chloride) can also be used for synthesizing polymers composed of α-hydroxy and α-amino acids [183] though in such case α-hydroxy and α-amino acids are not in close vicinity to each other, i.e., do not form depsipeptide units. Such kind of units are crucial in determining specific properties of PDPs since ester bonds in these units are highly polarized owing to the influence of neighbor amide groups (like ester groups in poly(lactide/glycolide polymers) that causes high rates of hydrolysis as compared with regular ester bonds in many polyesters such as poly(alkylene dicarboxylate)s [1], polycaprolactone, etc.

A new and less studied, but highly promising class of biodegradable biomaterials is amino acid based ~~biodegradable~~ poly(ester urea)s (PEUs) first synthesized in 1977 *via* SAP [184]. Later it was shown that PEUs with better material properties could be obtained *via* IP of di-*p*-toluenesulfonic acid salts of bis-(amino acid)-alkylene diesters with phosgene or triphosgene [166]. The PEU obtained in such way on the basis of di-*p*-toluenesulfonic acid salts of bis-(L-leucine)-1,6-hexylene diester (labeled as 1L6), to our best knowledge, represents the only AABBP which recrystallizes after melting. Therefore, the 1L6 is promising candidate for obtaining resorbable surgical sutures which, in contrast to sutures made of lactide/glycolide and related polymers, could be more compatible with tissues and more stable on storage. The 1L6 PEU was successfully electrospun to microfibrous scaffolds [185]. A high solubility of the polymer allowed using appropriate solvents to load different antibacterial drugs (biguanides) and even enzymes without being denatured. Adhesion and proliferation of both epithelial-like (Vero) and fibroblast-like (Cos-7) cells onto films and scaffolds demonstrated the biocompatibility of the new PEU.

Properties of PEUs such as solubility, crystallinity, elasticity, etc. can be modified by synthesizing their copolymers with PEAs. For example, copoly(ester urea amide) composed of 93 mol% of 1L6 and 7% of 8L6[1] was successfully used as a carrier to get electrospun scaffolds constituted by homogeneous and continuous microfibers with a high content of poly(3-thiophene methyl acetate) as a conducting polymer [186]. This should contribute with specific properties like biocompatibility and an enhancement of cell growth by electrostimulation, leading to a material with clear advantages for the biomedical field. Physicochemical characterization of the hybrid scaffolds revealed a low crystallinity, a high thermal stability and a remarkable electrochemical response.

II. Polymers with Leucine in the Lateral Chains

As noted above many biological processes are governed by macromolecular interactions. Multivalent binding has an important role in regulating the specificity and avidity of other biological processes as well. In the previous paragraphs bioactivities of polymers containing amino acid leucine's residues in the backbones have been discussed.

However, biologically active polymers can also be designed by incorporating the amino acid in the lateral chains. Such kind of polymers can be obtained *via* three different ways: (i) chain growth polymerization including

[1] PEA derived from sebacic acid (8), leucine (L) and 1,6-hexanediol (6).

radical polymerization (RP), atom transfer radical polymerization (ATRP) and ring opening metathesis polymerization (ROMP), (ii) step growth polymerization (polycondensation) and (iii) polymer-analogous transformations.

The polymers are of interest for biomedical applications, however, optically active representatives could be promising as catalysts of asymmetric synthesis, as well as a stationary phase to separate optical isomers.

II.A. Polymers *via* Chain-Growth Polymerization

Design and synthesis of pH responsive polymeric materials has become an important subject due to their applications in diverse fields including controlled drug delivery, biomedical applications, membrane science, sensors and actuators, oil recovery, colloid stabilization, etc. Efforts have been made to incorporate stimuli-responsive biomolecules in synthetic polymers to develop pH responsive "smart" non-biological hybrid macromolecules with high water solubility, enhanced biocompatibility, bio-mimetic structure and properties. One of the most promising approaches is the design of side-chain amino acid-based pH responsive polymers *via* polymerization of corresponding vinyl monomers.

A series of polymers carrying COOH groups along with the temperature-sensitive groups (amido and isopropyl) were synthesized by RP of acryloyl and methacryloyl L-leucine with isopropylacrylamide initiated with azobis(isobutyronitrile) at 65°C in two different solvents - DMSO and water, and thermodynamic behavior of the obtained polyelectrolytes was studied [187-189].

A number of optically active polyanions - poly(N-acryloyl amino acids) bearing various functional groups along with lipophilic ones were synthesized from the radical polymerization of *N*-acryloyl amino acid monomers including N-acryloyl L-leucine [190]. Acrylic monomers were prepared by the interaction of acryloyl chloride with amino acids. The synthesized linear homopolyanions were tested for heparin-like activities: (i) inhibition of heparanase enzyme, (ii) release of basic fibroblast growth factor from the extracellular matrix, and (iii) inhibition of smooth muscle cell proliferation. Polymers based on leucine and tyrosine were highly active in all three tests (microgram level).

Optically active polymers were also obtained by ATRP of N-methacryloyl L-leucine methyl ester (PMALM) [191]. The influence of varying initiator, catalyst, ligand, solvent, and temperature on ATRP was studied. Optimized polymerization conditions involved tris[2-(dimethylamino)ethyl]amine (Me_6TREN) as ligand, CuBr as catalyst, methyl 2-bromopropionate as initiator, and toluene or benzene as solvent. Under these conditions the PMALM samples produced possess tunable molecular weight and low polydispersity. The obtained polymers were characterized by differential scanning calorimetry, thermogravimetry, and polarimetry.

A review on the recent advances in side-chain amino acid-based pH responsive polymers, their synthesis and potential applications was published very recently [192].

To design artificial analogs to biologically active peptides, amino acid-functionalized (including leucine) norbornene derivatives were used as monomers in ROMP reaction that kinetically belongs to chain-growth polymerization. For example, leucine-functionalized 5-norbornene-2,3-dicarboximide monomers have been polymerized with Mo–carbene complexes to obtain polymers depicted in Figure 24a.

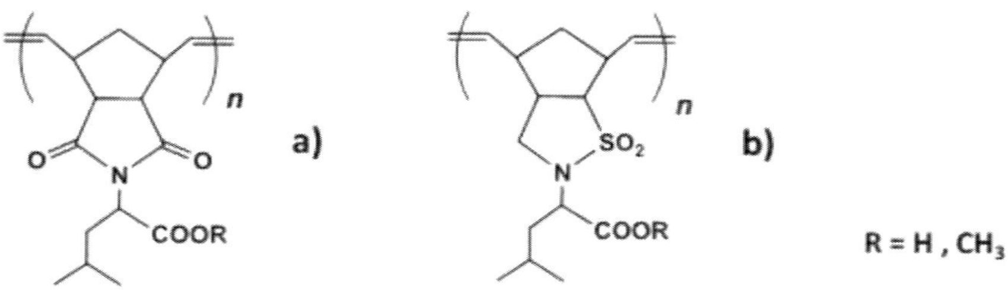

Figure 24. Leucine containing polymers synthesized by ROMP.

The analogous polymers possessing amino acid-derived carboxyl groups can also be synthesized by ROMP using Grubbs first-generation catalyst without the need to protect the carboxyl groups (R=H). Biologically active polymers depicted in Figure 24b containing sulfonamide groups have been synthesized by ROMP with the both Grubbs first- and second-generation catalysts (See Ref. [193] and Refs. cited therein).

II.B. Polymers *via* step-Growth Polymerization

Polymers with leucine residues in the lateral chains were obtained also *via* step-growth polymerization techniques. A basic principles of the synthesis of such kind of polymers are discussed in the reviews [9,144]. As an example below is given the synthesis of optically active polyamides with leucine residues in the lateral chains. For this, Mallakpour *et al.* [194] using multistage synthetic strategy, obtained N-substituted 5-aminoisophthalic acid containing N-phthalyl-L-leucine attached *via* *p*-aminobenzoic acid spacer (Figure 25).

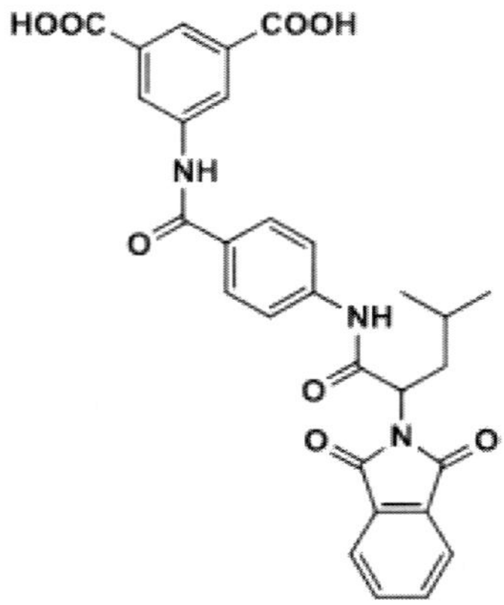

Figure 25. A derivative of 5-aminoisophthalic acid containing N-phthalyl-L-leucine in the lateral chain.

The obtained diacid monomer was transformed into corresponding polyamides by interaction with different aliphatic and aromatic diisocyanates as it was discussed for the related diacids in the Section I.B.

II.C. Polymers *via* Polymer-Analogous Transformations

Several samples of the polymers with leucine in the lateral chains, obtained *via* polymer-analogous transformations, have been discussed in the Paragraph I.A.4 above: polyleucine grafted hyaluronic acid (HA) samples were synthesized *via* a Michael addition reaction between primary amine terminated poly(L-leucine) and acrylate-functionalized HA [44]; another type of polysaccharide/polypeptide hybrid such as chitin grafted with polyleucine side chains - was prepared by one-pot ring-opening polymerization of L-leucine NCA initiated by partially deacetylated chitin (Figure 12) [95]. The allylamine grafted by poly(L-leucine) was reported as well [98].

Brush-like polymers/copolymers bearing polypeptide side chains (including poly(L-leucine)) have been synthesized [193] *via* grafting of the polypeptides using NCAs to the functional prepolymers obtained *via* ROMP and bearing the trimethylsilylamino groups at the side chain, as depicted in Figure 26:

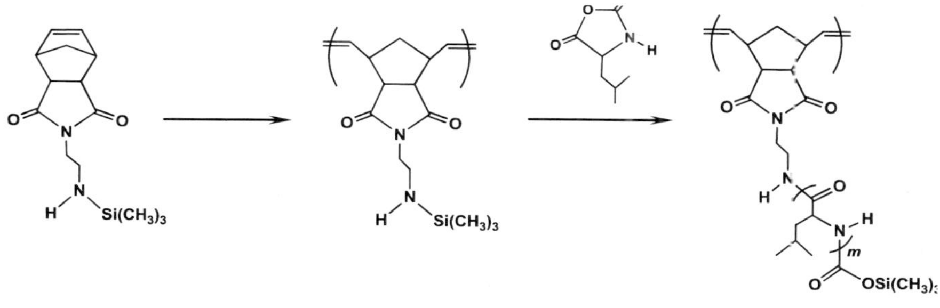

Figure 26. Synthesis of polymers bearing brush-like polypeptides by integration of NCA polymerization with ROMP.

Chitosan-leucine conjugate (Figure 27) was synthesized by multistage acylation of chitosan's amino groups with leucine [195]. The nanoparticles (10-30 nm) obtained from both chitosan and the conjugate were evaluated for toxicity and inflammatory effects using bronchial epithelial cell line, BEAS-2B. The nanoparticles were tested also for their effects on cell viability by MTT assay, on trans-epithelial permeability by using sodium fluorescein as a fluid phase marker. The conjugate nanoparticles had a low overall toxicity (half maximal inhibitory concentration IC_{50} = 2 mg/mL following 48 h exposure; no induction of IL-8 release at 0.5 mg/mL concentration), suggesting that they may be safe for pulmonary drug delivery applications.

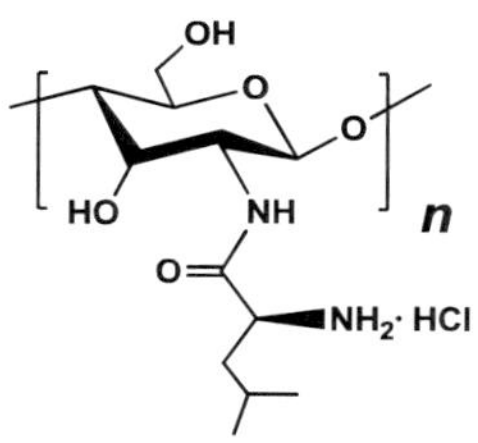

Figure 27. Chitosan-leucine conjugate.

pH-Sensitive unimolecular reversed micelles consisting of a hyperbranched D-glucan core and L-leucine ethyl ether shell were obtained by interaction of D-glucan with N-carbonyl L-leucine ethyl ester (L-leucine based isocyanate) [196]. The size-selective encapsulations and pH-sensitive release properties of leucine-modified *D*-glucan toward water-soluble dyes were discussed from the viewpoint of the unimolecular reversed micelle properties of the modified polysaccharide.

A library of novel well-defined bis-amide based copolymers was prepared through postpolymerization modification of poly(pentafluorophenyl acrylate) (obtained by RAFT process) with amines including ammonia and amide derivatives of amino acids like L-leucine amide [197]. Some of bis-amide copolymers displayed rich phase behavior in aqueous solution such as thermoreversible gelation at low temperature and high concentration while other samples displayed inverse temperature dependent solubility (lower critical solution temperature) behavior. A hydrophobically modified polyacrylamide copolymer displayed upper critical solution temperature (UCST) behavior in aqueous solution. Significantly, driven by polymer-polymer hydrogen bonding, copolymers self-associated into highly ordered, regular structures of several tens to hundreds of micrometers in size. Morphologies included sheet-like, rod-like and honeycomb-like structures and depended strongly on the chemical composition of copolymers.

REFERENCES

[1] A. Díaz, R. Katsarava, J. Puiggalí. "Synthesis, properties and applications of biodegradable polymers derived from diols and dicarboxylic acids: from polyesters to poly(ester amide)s". *Int. J. Mol. Sci.* 2014, 15, 7064-7123.

[2] A.C. Fonseca, M.H. Gil, P.N. Simões. "Biodegradable poly(ester amide)s – A remarkable opportunity for the biomedical area: Review on the synthesis, characterization and applications". *Progr. Polym. Sci.*, 2014, 39, 1291–1311.

[3] S.K. Murase, J. Puiggalí. "Poly(Ester Amide)s: Recent Developments on Synthesis and Applications". In: *Natural and Synthetic Biomedical Polymers*, Ch. 8, Kumbar, S., Laurencin, C., Deng, M., Eds., Elsevier Inc., 2014.

[4] M. Morell, J. Puiggalí. "Hybrid block copolymers constituted by peptides and synthetic polymers: An overview of synthetic approaches, supramolecular behavior and potential applications". *Polymers*, 2013, 5, 188-224.

[5] R. Katsarava, Z. Gomurashvili. "Biodegradable Polymers Composed of Naturally Occurring α-Amino Acids". In: *Handbook of Biodegradable Polymers - Isolation, Synthesis, Characterization and Applications*, Lendlein, A. and Sisson, A., Eds., Wiley-VCH, Verlag GmbH & Co. KGaA. 2011, 107-131.

[6] A. Rodriguez-Galan, L. Franco, J. Puiggali. "Degradable Polyurethanes and Poly(ester amide)s". In: *Handbook of Biodegradable Polymers - Isolation, Synthesis, Characterization and Applications*. Lendlein, A. and Sisson, A., Eds., Wiley-VCH, Verlag GmbH & Co. KGaA. 2011, 133-154.

[7] H. Sun, F. Meng, A.A. Dias, M. Hendriks, J. Feijen, Z. Zhong. "α-Amino Acid Containing Degradable Polymers as Functional Biomaterials: Rational Design, Synthetic Pathway, and Biomedical Applications". *Biomacromolecules*, 2011, 12, 1937–1955.

[8] W. Khan, S. Muthupandian, S. Farah, N. Kumar, A.J. Domb, "Biodegradable Polymers Derived From Amino Acids". *Macromol. Biosci.* 2011, 11, 1625–1636.

[9] S. Mallakpour, M. Dinari, "Progress in Synthetic Polymers Based on Natural Amino Acids". J. *Macromol. Sci., Part A: Pure and Appl. Chem.*, 2011, 48, 644–679.

[10] A. Rodriguez-Galan, L. Franco, J. Puiggali. "Degradable Poly(ester amide)s for Biomedical Applications". *Polymers*, 2011, 3, 65-99.

[11] R. Katsarava, D. Tugushi. "Non-conventional polymers composed of naturally occurring α-amino acids". In: *Unique Properties of Polymers and Composites: Pure and Applied Science Today and Tomorrow*. V.1. Bubnov, Yu. N., Vasnev, V.A., Askadskii A.A., Zaikov, G.E., Eds., NOVA Sci. Publ., 2011, 113-131.

[12] F. Fuchs. "Über *N*-Carbonsäure-anhydride". *Ber.*, 1922, 2943.

[13] R. Katakai, Y. Iizuka. "An improved rapid method for the synthesis of N-carboxy alpha-amino acid anhydrides using trichloromethyl chloroformate". *J. Org. Chem.*, 1985, 50, 715–716.

[14] H. Leuchs, W. Geiger. "Über die Anhydride von α-Amino-*N*-carbonsäuren und die von α-Aminosäuren". *Ber.*, 1908, 41, 1721–1726

[15] H. Sekiguchi. "Mechanism of N-carboxy-α-amino acid anhydride (NCA) polymerization". *Pure Appl. Chem.*, 1981, 53, 1689–1714.

[16] H. R. Kricheldorf, "Anionic Ring-opening Polymerization: N-Carboxyanhydrides". In: *Comprehensive Polymer Science*. G. Allen, J. C. Bevington, Eds. Pergamon. Oxford, 1989, 3, 531–551.

[17] R.A. Miranda, E. Finocchio, J. Llorca, F. Medina, G. Ramis, J.E. Sueirasab, A.M. Segarra. "Synthesis and characterization of poly-L-leucine initialized and immobilized by rehydrated hydrotalcite: Understanding stability and the nature of interaction". *Phys. Chem. Chem. Phys.*, 2013, 15, 15645-15659.

[18] J. Stulz, D. Müller, W.E. Hull, H.R. Kricheldorf. "Secondary structure of peptides, 10. FT-IR and 13C NMR CP/MAS study on tacticity and secondary structure of poly(D,L-leucine)s". *Makromol. Chem.*, 1983, 184, 1311–1322.

[19] F.G. Fick, J. Semen, H.G. Elias. "Syndiotactic poly(leucine)". *Makromol. Chem.*, 1978, 179, 579–590.

[20] T. Honda, M. Miyazaki, H. Nakamura, H. Maeda. "Controllable polymerization of N-carboxy anhydrides in a microreaction system". *Lab. Chip.*, 2005, 5, 812–818.

[21] S. Juliá, J. Guixer, J. Masana, J. Rocas, S. Colonna, R. Annuziata, H. Molinari. "Synthetic enzymes. Part 2. Catalytic asymmetric epoxidation by means of polyamino-acids in a triphase system". *J. Chem. Soc., Perkin Trans. 1.*, 1982, 1317-1324.

[22] M.J. Porter, J. Skidmore. "Asymmetric epoxidation of electron-deficient olefins". *Chem. Commun.*, 2000, 1215–1225.

[23] M.J. Porter, S.M. Roberts, S.M., J. Skidmore "Polyamino acids as catalysts in asymmetric synthesis". *Bioorganic & Medicinal Chem.*, 1999,7, 2145–2156.

[24] G. Carrea, S. Colonna, D.R. Kelly, A. Lazcano, G. Ottolina, S.M. Roberts, "Polyamino acids as synthetic enzymes: mechanism, applications and relevance to prebiotic catalysis". *TRENDS in Biotechnology* 2005, 23, 507-513.

[25] S. Colonna, D. Perdicchia, E. Di Mauro. 'Enantioselective reactions catalyzed by synthetic enzymes. A model for chemical evolution". *Tetrahedron: Asymmetry,* 2009, 20, 1709-1714.

[26] G. Battaglia, A.J. Ryan. "Bilayers and Interdigitation in Block Copolymer Vesicles". *J. Am. Chem. Soc.* 2005, 127, 8757–8764.

[27] F. Meng, Z. Zhong, J. Feijen. "Stimuli-Responsive Polymersomes for Programmed Drug Delivery". *Biomacromolecules*, 2009, 10, 197-209.

[28] E.G. Bellomo, M.D. Wyrsta, L. Pakstis, D.J. Pochan, T.J. Deming, "Stimuli-responsive polypeptide vesicles by conformation-specific assembly". *Nat. Mater.* 2004, 3, 244–248.

[29] E.P. Holowka, D.J. Pochan, T.J. Deming. "Charged Polypeptide Vesicles with Controllable Diameter". *J. Am. Chem. Soc.* 2005, 127, 12423–12428.

[30] E.P. Holowka, V.Z. Sun, D.T. Kamei, T.J. Deming. "Polyarginine segments in block copolypeptides drive both vesicular assembly and intracellular delivery". *Nat. Mater.* 2007, *6*, 52–57.

[31] V.Z. Sun, Z. Zhibo Li, T.J. Deming, D.T. Kamei. "Intracellular fates of cell- penetrating block copolypeptide vesicles". *Biomacromolecules*, 2011,12, 10-13.

[32] W. Zhuang, L. Liao, H. Chen, J. Wang, Y. Pan, L. Zhang, D.L iu. "Water soluble star-block copolypeptides: Towards biodegradable nanocarriers for versatile and simultaneous encapsulation". *Macromol. Rapid Commun.*, 2009, 30, 920–924.

[33] K. Pays, J. Giermanska-Kahn, B. Pouligny, J. Bibette, F. Leal-Calderon. "Double emulsions: how does release occur?". *J. Control. Release*, 2002, 79, 193–205.

[34] N. Garti. "Double emulsions: Scope, limitations and new achievements". *Colloids Surf. A*, 1997, 123, 233–246.

[35] J.A. Hanson, C.B. Chang, S.M. Graves, Z.B. Li, T.G. Mason, T.J. Deming. "Nanoscale double emulsions stabilized by single-component block copolypeptides". *Nature*, 2008, 455, 85–89.

[36] A.P. Nowak, V. Breedveld, L. Pakstis, B. Ozbas, D.J. Pine, D. Pochan, T.J. Deming. "Rapidly recovering hydrogel scaffolds from self-assembling diblock copolypeptide amphiphiles". *Nature*, 2002, 417, 424–428.

[37] J. Kopecek, L. Sprincl, H. Bazilova, J. Vacik, "Biological tolerance of poly(N-substituted acrylamides)". *J. Biomed. Mater. Res.*, 1973, 7, 111–121.

[38] D.J. Pochan, L. Pakstis, B. Ozbas, A.P. Nowak, T.J. Deming. "SANS and Cryo-TEM Study of Self-Assembled Diblock Copolypeptide Hydrogels with Rich Nano- through Microscale Morphology". *Macromolecules* 2002, 35, 5358-5360.

[39] A.P. Nowak, V. Breedveld, D.J. Pine, T.J. Deming. "Unusual salt stability in highly charged diblock co-polypeptide hydrogels". *J. Am. Chem. Soc.* 2003, 125, 15666-15670.

[40] A.P. Nowak, J. Sato, V.Breedveld, T.J. Deming. "Hydrogel formation in amphiphilic triblock copolypeptides". *Supramolecular Chemistry*,18, 2006, 423-427.

[41] Y.H. Yun, D.J. Goetz, P. Yellen, W.Chen. "Hyaluronan microspheres for sustained gene delivery and site-specific targeting". *Biomaterials*, 2004, 25, 147-157.

[42] P. Aragona, G. Di Stefano, F. Ferreri, R. Spinella, A. Stilo. *Br. J. Ophthalmol.* 2002, 86, 879–884.

[43] S. Ohya, Y. Nakayama, T. Matsuda. "Thermoresponsive Artificial Extracellular Matrix for Tissue Engineering: Hyaluronic Acid Bioconjugated with Poly(*N*-isopropylacrylamide) Grafts". *Biomacromolecules*, 2001, 2, 856–863.

[44] X. Wang, J. Messman, J.W. Mays, D. Baskaran. "Polypeptide grafted hyaluronan: Synthesis and characterization". *Biomacromolecules*, 2010, 11, 2313–2320.

[45] Y. Kuroyanagi, E. Kim, N. Shioya. "Evaluation of a synthetic wound dressing capable of releasing silver sulfadiazine". *J. Burn Care Rehabil.*, 1991, 12, 105-115.

[46] Y. Kuroyanagi, E. Kim, M. Kenmochi, H. Kakeyama, M. Nakamure, A. Takeda, N. Shioya. "A silver-sulfadiazine-impregnated wound dressing composed of poly-L-leucine sponge matrix: an evaluation of clinical cases". *J. Appl. Biomater.*, 1992, 3, 153-161.

[47] C.H. Len. "A study on the solid-state polymerization of poly(L-leucine) initiated by γ-ray". *Radiat. Phys. Chem.* 1995, 46, 851-854.

[48] P.J. Dijkstra, J. Feijen. "Synthetic pathways to polydepsipeptides". *Macromol. Symp.* 2000, 153, 67-76.

[49] J. Helder, J. Feijen, S.J. Lee, S.W. Kim. "Copolymers of D,L-lactic acid and glycine". *Makromol. Chem. Rapid Commun,* 1986, 7, 193–198.

[50] T. Ouchi, T. Nozaki, A. Ishikawa, I. Fujimoto, Y. Ohya. "Synthesis and enzymatic hydrolysis of lactic acid depsipeptide copolymers with functionalized pendant groups". *J. Polym. Sci. A: Polym. Chem.*, 1997, 35, 377–83.

[51] Y. Feng, J. Guo. "Biodegradable Polydepsipeptides". *Int. J. Mol. Sci.*, 2009, 10, 589-615.

[52] D. Nissen, C. Gilon, M. Goodmann. "Polydepsipeptides. 4. Synthesis of the Alternating Polydepsipeptides Poly(Ala-Lac) and Poly(Val-Lac)". *Macromol. Chem. Supplement*, 1975, 1, 23-53.

[53] Y. Feng, J. Knüfermann, D. Klee, H. Höcker. "Enzyme-catalyzed Ring-opening Polymerization of 3(S)-isopropyl-morpholine-2,5-dione". *Macromol. Rapid Comm.* 1999, 20, 88-90.

[54] T. Tian, Q.Chen, C. Yu, J. Shen. "Amino-terminated poly(ethylene glycol) as the initiator for the ring-opening polymerization of 3-methylmorpholine-2,5-dione". *Eur. Polym. J.*, 2003, 39, 1935-1938.

[55] W.J. Becktel, G. Wouters, D.M. Simmons, M. Goodman. "Polydepsipeptides. 11. Conformational Analysis of Polydepsipeptides Containing Methyl, Isopropyl, and Isobutyl Side Chains". *Macromolecules*, 1985, 18, 630-634.

[56] R. Katakai, K. Kobayashi, N. Yonezawa, M. Yoshida. "Hexane-soluble octadecadepsipeptide forming hydrophobic α-helical rod end-capped by the C-terminal γ-turn". *Biopolymers*, 1996, 38, 285–290.

[57] T. Ohyama, H. Oku, M. Yoshida, R. Katakai. "Crystal structure of a depsipeptide, Boc–(Leu–Leu–Lac)3–Leu–Leu–OEt". *Biopolymers*, 2001, 58, 636–642.

[58] T. Ohyama, H. Oku, A. Hiroki, Y. Maekawa, M. Yoshida, R. Katakai. "The crystal structure for a depsipeptide Boc–(Leu–Leu–Ala)$_2$–(Leu–Leu–Lac)$_3$–OEt with a 3_{10}-helical segment", *Biopolymers*, 2000, 54, 375–378.

[59] B. Jeong, S.W. Kim, Y.H. Bae, "Thermosensitive sol–gel reversible hydrogels". *Adv. Drug Deliv. Rev.*, 2002, 54, 37–51.

[60] Y. Ohya, H. Yamamoto, K. Nagahama, T. Ouchi. "Effects of polydepsipeptide side-chain groups on the temperature sensitivity of triblock copolymers composed of polydepsipeptides and poly(ethylene glycol)". *J. Polym. Sci: Part A: Polym. Chem.*, 2009, 47, 3892–3903.

[61] Y. Ohya, T. Nakai, K. Nagahama, T. Ouchi, S. Tanaka, K. Kato. "The synthesis and biodegradability of poly(lactide-random-depsipeptide)-PEG-poly(lactide-random-depsipeptide) ABA-type triblock copolymers". *J. Bioact. Compat. Polym.*, 2006, 21, 557-577.

[62] A. Battig, B. Hiebl, Y. Feng, A. Lendlein, M. Behl. "Biological evaluation of degradable, stimuli-sensitive multiblock copolymers having polydepsipeptide and poly(ε-caprolactone) segments in vitro". *Clinical Hemorheology and Microcirculation.*, 2011, 48, 161-172.

[63] Lupas. "Coiled coils: new structures and new functions". *Trends Biochem. Sci.*, 21 (1996) 375-382.

[64] F.H.C. Crick, "Is α-Keratin a Coiled Coil?". Nature. 1952, 170, 882-883.

[65] Z. Ruszczak, "Effect of collagen matrices on dermal wound healing". *Adv. Drug Deliv. Rev.*, 2003, 55, 1595-1611.

[66] J.Y. Su, R.S. Hodges, C.M. Kay, "Effect of Chain Length on the Formation and Stability of Synthetic alpha-Helical Coiled Coils". *Biochemistry*, 1994, 33, 15501-15510.

[67] Y.B. Yu. "Coiled-coils: stability, specificity, and drug delivery potential". *Adv. Drug Deliv. Rev.*, 2002, 54, 1113–1129.

[68] M. Pechar, P. Kopeckova, L. Joss, J. Kopecek. "Associative diblock copolymers of poly(ethylene glycol) and coiled-coil peptides". *Macromol. Biosci.*, 2002, 2, 199–206.

[69] D.A. Lindhout, J.R. Litowski, P. Mercier, R.S. Hodges, B.D Sykes. "NMR solution structure of a highly stable de novo heterodimeric coiled-coil". *Biopolymers*, 2004, 75, 367–375.

[70] S.Y.M. Lau, A.K. Taneja, R.S. Hodges. "Synthesis of a model protein of defined secondary and quaternary structure". *J. Biol. Chem.* 1984, 259, 13253–13261.

[71] K.H. Mayo. "Recent advances in the design and construction of synthetic peptides: for the love of basics or just for the technology of it". *TIBTECH*, 2000, 18, 212–217.

[72] D.E.Wagner, C.L. Phillips, W.M. Ali, G.E. Nybakken, E.D. Crawford, A.D. Schwab, W.F. Smith, R. Fairman. "Toward the development of peptide nanofilaments and nanoropes as smart materials". *Proc. Natl. Acad. Sci. USA*, 2005, 102, 12656–12661.

[73] A.N. Lupas, M. Gruber. "The structure of α-helical coiled coils". *Adv. Prot. Chem.*, 2005, 70, 37–78.

[74] M.J. Pandya, G.M. Spooner, M. Sunde, J.R. Thorpe, A. Rodger, D.N. Woolfson. "Sticky-End Assembly of a Designed Peptide Fiber Provides Insight into Protein Fibrillogenesis". *Biochemistry*, 2000, 39, 8728–8734.

[75] D.N. Woolfson. "Building fibrous biomaterials from alpha-helical and collagen-like coiled-coil peptides". *Biopolymers*, 2010, 94, 118-127.

[76] D. Papapostolou, A.M. Smith, E.D.T. Atkins, S.J. Oliver, M.G. Ryadnov, L.C. Serpell, D.N. Woolfson. "Engineering nanoscale order into a designed protein fiber". *Proc. Natl. Acad. Sci. USA*, 2007, 104, 10853–10858.

[77] Y. Hakata, S. Tsochiya, H. Michiue, T. Ohtsuki, H. Matsui, M. Miyazawa, M. Kitamatsu. "A novel zipper motif-based hybrid peptide delivers a functional cargo inside cells". *Chem.Commun.*, 2015, 51, 413-416.

[78] W.A. Petka, J.L. Harden, K.P. McGrath, D. Wirtz, D.A. Tirrell. "Reversible hydrogels from self-assembling artificial proteins". *Science*, 1998, 281, 389–392.

[79] J.R. Moll, S.B. Ruvinov, I. Pastan, C. Vinson. "Designed heterodimerizing leucine zippers with a range of PIs and stabilities up to 10–15 M". *Protein Sci.*, 2001, 10, 649–655.

[80] D.M. Goldenberg. "Targeted therapy of cancer with radiolabeled antibodies". *J. Nucl. Med.*, 2002, 43, 693–713.

[81] G.A. Koning, A.M.M. Eggermont, L.H. Lindner, T.L.M. Hagen. "Hyperthermia and Thermosensitive Liposomes for Improved Delivery of Chemotherapeutic Drugs to Solid Tumors". *Pharm. Res.* 2010, 27, 1750–1754.

[82] Z.S. Al-Ahmady, W.T. Al-Jamal, J.V. Bossche, T.T. Bui, A.F. Drake, A.J. Mason, K. Kostarelos. "Lipid–Peptide Vesicle Nanoscale Hybrids for Triggered Drug Release by Mild Hyperthermia *in Vitro* and *in Vivo*". *ACS Nano*, 2012, 6, 9335–9346.

[83] A.E. Oliver, D.W. Deamer. "Alpha-helical hydrophobic polypeptides form proton-selective channels in lipid bilayers". *Biophys. J.*, 1994, 66, 1364–1379.

[84] S. Banta, I.R. Wheeldon, M. Blenner. "Protein engineering in the development of functional hydrogels". *Ann. Rev. Biomed. Engin.*, 2010, 12, 167-186.

[85] H. Schlaad, M. Antonietti. "Block copolymers with amino acid sequences: molecular chimeras of polypeptides and synthetic polymers". *Eur. Phys. J. E*, 2003, 10, 17–23.

[86] H.G. Börner. "Strategies exploiting functions and self-assembly properties of bioconjugates for polymer and materials sciences". *Prog. Polym. Sci.*, 2009, 34, 811–851.

[87] J.F. Lutz, H.G. Börner. "Modern trends in polymer bioconjugates design". *Prog. Polym. Sci.*, 2008, 33, 1–39.

[88] I.V. Dimitrov, I.V. Berlinova, P.V. Iliev, N.G. Vladimirov. "Controlled synthesis of peptide-based amphiphilic copolymers". *Macromolecules*, 2008, 41, 1045–1049.

[89] J.F. Lutz, D. Schutt, S. Kubowicz. "Preparation of well-defined diblock copolymers with short polypeptide segments by polymerization of N-carboxyanhydrides". *Macromol. Rapid Commun.*, 26 (2005), 23–28.

[90] J. Deng, N. Gao, Y. Wang, H. Yi, S. Fang, Y. Ma, L. Cai, "Self-Assembled Cationic Micelles Based on PEG-PLL-PLLeu Hybrid Polypeptides as Highly Effective Gene Vectors". *Biomacromolecules*, 2012, 13, 3795–3804.

[91] C. Zheng, Z M. Heng, P. Gong, J. Deng, H. Yi, P. Zhang, Y. Zhang, P. Liu, Y. Yifan Ma, L. Cai. "Polypeptide cationic micelles mediated co-

delivery of docetaxel and siRNA for synergistic tumor therapy". *Biomaterials*, 2013, 34, 3431-3438.

[92] I. Cho, J.B. Kim, H.J. Jung. "Synthesis and characterization of di- and triblock copolymers of poly(ethylene oxide) and poly(DL-valine-co-DL-leucine)". *Polymer*, 2003, 44, 5497–5500.

[93] R.I. Kühnle, D. Gebauer, H.G. Borner, "Calcium ions as bioinspired triggers to reversibly control the coil-to-helix transition in peptide-polymer conjugates". *Soft Matter,* 2011, 7, 9616–9619.

[94] P.F. Gu, H. Xu, B.W. Sui, J.X. Gou, L.K. Meng, F. Sun, X.J. Wang, N.Qi, Y. Zhang, H.B. He, X. Tang. "Polymeric micelles based on poly(ethylene glycol) block poly(racemic amino acids) hybrid polypeptides: Conformation-facilitated drug-loading behavior and potential application as effective anticancer drug carriers". *Intern. J. Nanomed.*, 2012, 7, 109-122.

[95] J. Wang, C. Liu, P. Chi. "*In situ* preparation of glycoconjugate hollow microspheres mimics the extracellular matrix *via* interfacial polymerization". *Intern. J. Biol. Macromol.* 2008, 42, 450–454.

[96] R. Wang, G.T. Chen, F.S. Du, Z.C. Li. "Preparation and aggregation behavior of mannose-terminated poly(ethylene glycol)-*b*-poly(L-leucine) in wáter". *Colloids and Surfaces B: Biointerfaces*, 85, 2011, 56–62.

[97] J. Rieger, H. Freichels, A. Imberty, J.L. Putaux, T. Delair, C. Jerome, R. Auzely-Velty. "Polyester nanoparticles presenting mannose residues: toward the development of new vaccine delivery systems combining biodegradability and targeting properties", *Biomacromolecules*, 2009, 10, 651–657.

[98] M. Higuchi, T. Inoue, H. Miyoshi, M. Kawaguchi. "pH-induced reversible conformational and morphological regulation of polyleucine grafted polyallylamine assembly in solution". *Langmiur*, 2005, 21, 11462-11467.

[99] H.J. Kim, E.Y. Choi, J.S. Oh, H.C. Lee, S.S. Park, C.S. Cho. "Possibility of wound dressing using poly(L-leucine)/poly(ethylene glycol)/poly(L-leucine) triblock copolymer". *Biomaterials*, 2000, 21, 131-141.

[100] S. Tanimoto, T. Okunishi, H. Yamaoka, Y. Hirokawa. "Specific gold ion scavenger by means of sol-gel transition of peptide copolymer". *Polymer Journal*, 2011, 43, 313-316.

[101] W. Zhang, Y. Huang, "Biodegradable Copoly(Amino Acid)s Based on 6-Aminocaproic Acid and L-Leucine". *J. Polym. Environ.*, 2011, 19, 177–181.

[102] I. Arvanitoyannis, E. Psomiadou, N. Yamamoto, E. Nikolaou, J.M.V. Blanshard. "Novel copolyamides based on adipic acid, 1,6-hexanediamine and α-amino acids: 2. Study of properties and their biodegradability for food-packaging applications". *Polymer*, 1995, 36, 2957-2961.

[103] S. Anantharaj, M. Jayakannan. "Polymers from Amino acids: Development of Dual Ester-Urethane Melt Condensation Approach and Mechanistic Aspects". *Biomacromolecules*, 2012, 13, 2446–2455.

[104] N. Cohen-Arazi, A.J. Domb, J. Katzhendler. "New Biocompatible Polyesters Derived from α-Amino Acids: Hydrolytic Degradation Behavior". *Polymers,* 2010, *2*, 418-439.

[105] S. Deechongkit, S.L. You, J.W. Kelly. "Synthesis of all nineteen appropriately protected chiral alpha-hydroxy acid equivalents of the alpha-amino acids for Boc solid-phase depsi-peptide synthesis". *Org. Lett.*, 2004, *6*, 497–500.

[106] T. Bauer, J. Gajewiak. alpha-Hydroxy carboxylic acids as ligands for enantioselective diethylzinc additions to aromatic and aliphatic aldehydes. *Tetrahedron* 2004, 60, 9163–9170.

[107] J.Y. Rho, L. Kuhn-Spearing, P. Zioupos. "Mechanical properties and the hierarchical structure of bone". Medical engineering & physics, 1998, 20, 92–102.

[108] Z.Li, L. Yubao, Y. Aiping, P. Xuelin, W. Xuejiang, Z. Xiang, "Preparation and *in vitro* investigation of chitosan/nano-hydroxyapatite composite used as bone substitute materials". *J. Mater. Sci. Mater. Med.,* 2005, 16, 213–219.

[109] M. Kanapathipillai, Y. Yusufoglu, A. Rawal, Y.Y. Hu, C.T. Lo, P. Thiyagarajan, Y. E. Kalay, M Akinc, S. Mallapragada, K. Schmidt-Rohr. "Synthesis and Characterization of Ionic Block Copolymer Templated Calcium Phosphate Nanocomposites". *Chem. Mater.*, 2008, 20, 5922–5932.

[110] Y.Y. Hu, Y. Yusufoglu, M. Kanapathipillai, C.Y. Yang, Y. Wu, P. Thiyagarajan, T. Deming, M. Akinc, K. Schmidt-Rohr, S. Mallapragada. "Self-assembled calcium phosphate nanocomposites using block copolypeptide templates". *Soft Matter*, 2009, 5, 4311–4320.

[111] F. Rodriguez, D.D.Glawe, R.R. Naik, K.P. Hallinan, M.O. Stone. "Study of the Chemical and Physical Influences upon in Vitro Peptide-Mediated Silica Formation". *Biomacromolecules*, 2004, 5, 261-265.

[112] M.M. Tomczak, D.D. Glawe, L.F. Drummy, C.G. Lawrence, M.O. Stone, C.C. Perry, D.J. Pochan, T.J. Deming, R.R. Naik, "Polypeptide-

Templated Synthesis of Hexagonal Silica Platelets". *J. Am. Chem. Soc.*, 2005, 127, 12577-12582.

[113] S. Hayashi, K. Ohkawa, H. Yamamoto, "Random and Sequential Copolypeptides Containing *O*-Phospho-L-Threonine and L-Aspartic Acid: Roles in $CaCO_3$ Biomineralization". *Macromol. Biosci.*, 2006, 6, 228-240.

[114] L. Xia, Y. Liu, Z. Li. "Biosilicification templated by amphiphilic block copolypeptide assemblies". *Macromol. Biosci.*, 2010, 10, 1566–1575.

[115] C.L. Chen, N.L. Rosi. "Peptide-based methods for the preparation of nanostructured inorganic materials". *Angew. Chem. Int. Ed.*, 2010, 49, 1924-1942.

[116] J.E. Baio, A. Zane, V. Jaeger, A.M. Roehrich, H. Lutz, J. Pfaendtner, G.P. Drobny, T. Weidner. "Diatom mimics: Directing the formation of biosilica nanoparticles by controlled folding of lysine-leucine peptides". *J. Am. Chem. Soc.* 2014, 136, 15134−15137.

[117] W.F. DeGrado, J.D. Lear. "Induction of peptide conformation at apolar water interfaces. 1. A study with model peptides of defined hydrophobic periodicity". *J. Am. Chem. Soc.* 1985, 107, 7684-7689.

[118] J.R. Long, N. Oyler, G.P. Drobny, P.S. Stayton. "Assembly of alpha-helical peptide coating on hydrophobic syrfaces". *J. Am. Chem. Soc.* 2002, 124, 6297-6303.

[119] S. Mallakpour, S. Soltanian. "Functionalized multi-wall carbon nanotube reinforced poly(ester-imide) bionanocomposites containing L-leucine amino acid units". *J. Polym. Res.*, 2014, 21, 335.

[120] B. Scheibe, E. Borowiak-Palen, R.J. Kalenczuk."Oxidation and reduction of multiwalled carbon nanotubes-preparation and characterization". *Mater. Charact.*, 2010, 61, 185–191.

[121] A. Grenha, C.I. Grainger, L.A. Dailey, B. Seijo, G.P. Martin, C. Remunan-Lopez, B. Forbes. "Chitosan nanoparticles are compatible with respiratory epithelial cells *in vitro*". *Eur. J. Pharm. Sci.*, 2007, 31, 73−84.

[122] M.D.A. Muhsin, G. George, K. Beagley, V. Ferro, C.Armitage N., Islam. "Synthesis and toxicological evaluation of a chitosan-L-leucine conjugate for pulmonary drug delivery applications". *Biomacromolecules*, 2014, 15, 3596−3607.

[123] H. Peng, L. Zhang, J. Spires, C. Soeller, J. Travas-Sejdic. "Synthesis of a functionalized polythiophene as an active substrate for a label-free electrochemical genosensor". *Polymer*, 2007, 48, 3413–3419.

[124] R. Phillips, I.B. Kim, B.E. Carson, B. Tidbeck, Y. Bai, T.L. Lowary, L.M. Tolbert, U.H.F. Bunz. "Sugar-substituted poly(p-phenyleneethynylene)s: sensitivity enhancement toward lectins and bacteria". *Macromolecules*, 2008, 41,7316–7320.

[125] H. Jiang, P. Taranekar, J.R. Reynolds, K.S. Schanze. "Conjugated Polyelectrolytes: Synthesis, Photophysics and Applications". *Angew. Chem. Int. Ed.*, 2009, 48, 4300–4316.

[126] T.A. Skotheim, E.R.L. lsenbaumer, J.R. Reynolds, Eds., *Handbook of conducting polymers*. Marcel Dekker, New York, 1998.

[127] M.D. Shults, D.A. Pearce, B. Imperiali. "Modular and tunable chemosensing scaffold for divalent zinc". *J. Am. Chem. Soc.*, 2003, 125, 10591–10597.

[128] A.D. Aguilar, E.S. Forzani, X. Li, N. Tao, L.A. Nagahara, I. Amlani, R. Tsui. "Chemical sensors using peptide-functionalized conducting polymer nanojunction arrays". *Appl. Phys. Lett.*, 2005, 87, 193108/1–193108/3.

[129] D. Compagnone, A. Ricci, M.Del Carlo, M. Chiarini, A. Pepe, C. Lo Sterzo. "New poly(aryleneethynylene)s as optical active platforms in biosensing. Selective fluorescent detection of Hg(II) obtained by the use of aminoacidic groups anchored on conjugated backbones". *Microchimica Acta*, 2010, 170, 313-319.

[130] C.D. McTiernan, K. Omri, M. Chahma. "Chiral conducting surfaces via electrochemical oxidation of L-leucine-oligothiophenes". *J. Org. Chem.*, 2010, 75, 6096–6103.

[131] Sudha, D. Kumar, M. Iwamoto. "Investigation of the chiroptical behavior of optically active polyaniline synthesized from naturally occurring amino acids". *Polymer Journal.*, 45, 2013, 160-165.

[132] D. Brigger, C.Dubernet, P. Couvreur, "Nanoparticles in cancer therapy and diagnosis". *Adv. Drug Delivery Rev.*, 2002, 54, 631-651.

[133] A. Nan, R. Turcu, L. Bratu, C. Leostean, O. Chauvet, E. Gautron, J. Liebscherc. "Novel magnetic core-shell Fe3O4 polypyrrole nanoparticles functionalized by peptides or albumin". *Arkivoc*, 2010, 185-198.

[134] C. Song, C.H. Zhang, F.J. Wang, W.T. Yang, J. P. Deng. "Chiral polymeric microspheres grafted with optically active helical polymer chains: a new class of materials for chiral recognition and chirally controlled reléase". *Polym. Chem.*, 2013, 4, 645-652.

[135] D.Y. Zhang, H.Y. Zhang, C. Song, W.T. Yang, J.P. Deng. "Chiral microspheres constructed by helical substituted polyacetylene: A new

class of organocatalyst toward asymmetric catalysis". *Synth. Met.* 2012, 162, 1858-1863.

[136] B. Chen, J.P. Deng, W.T. Yang. "Hollow Two-Layered Chiral Nanoparticles Consisting of Optically Active Helical Polymer/Silica: Preparation and Application for Enantioselective Crystallization". *Adv. Funct. Mater.*, 2011, 21, 2345-2350.

[137] D. Liu, J. Deng, W. Yang. "A facile method for preparing porous, optically active, magnetic Fe3O4@poly(N-acryloyl-leucine) inverse core/shell composite microspheres". *Macromol. Rapid Commun.* 2014, 35, 91−96.

[138] C. Jeyaprabha, S. Sathiyanarayanan, K.L.N. Phani, G. Venkatachari. "Influence of poly (aminoquinone) on corrosion inhibition of iron in acid media". *Appl. Surf. Sci.*, 2005, 252, 966-975.

[139] W. Li, X. Zhao, F. Liu, B. Hou, "Investigation on inhibition behavior of S-triazole–triazole derivatives in acidic solution". *Corros. Sci.*, 2008, 50, 3261-3266.

[140] A. Ali Fathima Sabirneeza, S. Subhashini, R. Rajalakshmi. "Water soluble conducting polymer composite of polyvinylalcohol and leucine: An effective acid corrosion inhibitor for mild steel". *Materials and Corrosion*, 2013, 64, 74-82.

[141] N.H. Kaus, N. Lahazan, A.H. Ahmad. "Ionic conductivity in poly(L-leucine)1,3-diamino propane–lithium iodide solid polymer electrolyte". *Polym. Adv. Technol.*, 2009, 20, 156–160.

[142] A. Staubli, E. Mathiowitz, M. Lucarelli, R. Langer. "Characterization of Hydrolytically Degradable Amino Acid Containing Poly(anhydride-co-imides)". *Macromolecules*, 1991,24, 2283-2290.

[143] K.E. Uhrich, A. Gupta, T.T. Thomas, C.T. Laurencinf, R. Langer. "Synthesis and Characterization of Degradable poly(anhydride-co-imides)". *Macromolecules*, 1995, 28, 2184-2193.

[144] S.E. Mallakpour, A. Zadehnazari. "Advances in synthetic optically active condensation polymers – A review". *eXPRESS Polymer Letters*, 2011, 5, 142–181.

[145] S.E. Mallakpour, K. Banihassan, M.R. Sabzalian. "Novel Bioactive Chiral Poly(amide–imide)s Containing Different Amino Acids Linkages: Studies on Synthesis, Characterization and Biodegradability". *J. Polym. Environ.* 2013, 21, 568–574.

[146] S.E. Mallakpour, F. Rafiemanzelat. "Preparation and Properties of New Copoly(amide-imide-ether-urethane)s based on Bis(*p*-amido benzoic

acid)-N-trimellitylimido-L-leucine by Two Different Polymerization Methods". *Polym. Bull.*, 2007, 58, 339–350.

[147] S.E. Mallakpour, E. Kowsari. "Soluble Novel Optically Active Poly(amide–imide)s Derived from N,N-(4,4'-Oxydiphthaloyl)-bis-L-leucine Diacid Chloride and Various Aromatic Diamines: Synthesis and Characterization. *J. Appl. Polym. Sci.*, 2005, 96, 435–442.

[148] S.E. Mallakpour, A.R. Hajipour, M.R. Zamanlou. "Synthesis of optically active poly(amide-imide)s derived from N,N'-(4,4'-carbonyldiphthaloyl) -bis-L-leucine diacid chloride and aromatic diamines by microwave radiation". *Journal of Polymer Science Part A: Polymer Chemistry*, 2001, 39, 177–186.

[149] S.E. Mallakpour, E. Kowsari. "Microwave-Assisted and Conventional Polycondensation Reaction of Optically Active N,N-(4,4-Sulphonediphthaloyl)-bis-L-Leucine Diacid Chloride with Aromatic Diamines". *J. Appl. Polym. Sci.*, 2004, 91, 2992–3000.

[150] S.E. Mallakpour, A.R. Hajipour, M.R. Zamanlou. "Microwave-Assisted Synthesis of Optically ActivePoly(amide imide)s Derived from Diacid Chloride Containing Epiclon and L-Leucine with Aromatic Diamines". *J. Polym. Sci. Polym. Chem.*, 2003, 41, 1077–1090.

[151] R. Katsarava. "Condensing agents in polycondensation". *Russian Chem. Rev.* (Translated by British Library), 1989, 58, 891-903.

[152] S.E. Mallakpour, A.R. Hajipour, R. Roohipour-fard. "Direct polycondensation of N-trimellitylimido-L-leucine with aromatic diamines". *Europ. Poly. J.*, 2000, 36, 2455–2462.

[153] S.E. Mallakpour, A.R. Hajipour, S. Habibi. "Facile synthesis of new optically active poly(amide–imide)s derived from N,N'-(pyromellitoyl)-bis-L-leucine diacid chloride and aromatic diamines under microwave irradiation". *Europ. Polym. J.*, 2001, 37, 2435–2442.

[154] S.E. Mallakpour, A.R. Hajipour, R. Vahabi. "Synthesis and characterization of novel poly(amide imide)s based on bis(*p*-amidobenzoic acid)-*n*-trimellitylimido- L-leucine". *J. Appl. Polym. Sci.*, 2002, 84, 35–43.

[155] S.E. Mallakpour, A.R. Hajipour, S. Khoee. "Rapid synthesis of optically active poly(amide–imide)s by direct polycondensation of aromatic dicarboxylic acid with aromatic diamines". *European Polymer Journal*, 2002, 38, 2011–2016.

[156] S.E. Mallakpour, F. Rafiemanzelat. "Microwave-assisted and classical heating polycondensation reaction of bis(p-amido benzoic acid)-N-trimellitylimido-L-leucine with diisocyanates as a new method for

preparation of optically active poly(amide imide)s. *J. Appl. Polym. Sci.*, 2004, 93, 1647–1659.

[157] F. Rafiemanzelat, E. Abdollahi, "Synthesis and characterization of hydrolysable poly(ether-urethane-urea)s derived from L–leucine anhydride cyclopeptide; a green synthetic method for monomer". *Polym. Degrad. Stab.*, 2010, 95, 901-911.

[158] Y. Fan, M. Kobayashi, H. Kise. "Synthesis and biodegradability of new polyesteramides containing peptide linkages". *Polym. J.*, 2000, 32, 817-822.

[159] D. Kharadze, T. Omiadze, G. Tsitlanadze, Ts. Goguadze, N. Arabuli, Z. Gomurashvili, R. Katsarava. "New biodegradable polymers derived from N,N'-diacyl-bisphenylalanine". *Polym. Sci. Ser A (Russia)*, 1994, 36, 1214-1218.

[160] R. Katsarava, Active Polycondensation – from peptide chemistry to amino acid based biodegradable polymers, In: *H.R. Kricheldorf (Ed.), "Polycondensation 2002", Macromolecular Symposia.* Wiley-VCH. 2003, 199, 419-429.

[161] N. Arabuli, G. Tsitlanadze, L. Edilashvili., D. Kharadze, Ts. Goguadze, V. Beridze, Z. Gomurashvili, R. Katsarava."Heterochain polymers based on natural α-amino acids. Synthesis and enzymatic hydrolysis of regular poly(ester amide)s based on bis(L-phenylalanine)-α,ω–alkylene diesters and adipic acid". *Macromol. Chem. Phys.*, 1994, 195, 2279-2289.

[162] R. Katsarava, V. Beridze, N. Arabuli, D. Kharadze, C.C. Chu, C.Y. Won. "Amino acid based bioanalogous polymers. Synthesis and study of regular poly(ester amide)s based on bis(α-amino acid)-α,ω–alkylene diesters and aliphatic dicarboxylic acids". *J. Polym. Sci. Polym. Chem.*, 1999, 37, 391-407.

[163] T. Kartvelishvili, G. Tsitlanadze, L. Edilashvili, N. Japaridze, R. Katsarava. "Amino acid based bioanalogous polymers. Regular poly(ester urethane)s and poly(ester urea)s based on bis(phenylalanine)-α,ω–alkylene diesters". *Macromol. Chem. Phys.*, 1997, 198, 1921-1932.

[164] L. Asín, E. Armelin, J. Montane, A. Rodríguez-Galán, J. Puiggalí. "Sequential poly(ester amide)s based on glycine, diols, and dicarboxylic acids: Thermal polyesterification *vs.* interfacial polyamidation". *J. Polym. Sci. A*, 2001, 39, 4283–4293.

[165] D.K. Knight, K. Mequanint, E.R. Gillies. Strategies in functional poly(ester amide) syntheses to study human coronary artery smooth muscle cell interactions. *Biomacromolecules*, 2011, 12, 2475–2487.

[166] Z.D. Gomurashvili, R. Katsarava, D. Tugushi. "Poly(ester urea) polymers and methods of use". U.S. Patent # 8,765,164 (2014).

[167] M. Vert. "Degradable and bioresorbable polymers in surgery and in pharmacology: beliefs and facts". *J. Mater. Sci. Mater. Med.,* 2009, 20, 437–446.

[168] www.polymerpharm.ge

[169] G. Jokhadze, M. Machaidze, H. Panosyan, C.C. Chu, R. Katsarava. "Synthesis and characterization of functional elastomeric poly(ester amide) copolymers". *J. Biomater. Sci. Polym. Ed.*, 2007, 18, 411-438.

[170] C.C. Chu, R. Katsarava. "Elastomeric functional biodegradable copolyester amides and copolyester urethanes". U.S. Patents ##: 6,503,538 (2003), 7,304,122 (2007), 7,408,018 (2008).

[171] K.M. DeFife, K. Grako, G. Cruz - Aranda, S. Price, R. Chantung, K. Macpherson, R. Khoshabeh, S. Gopalan, W.G. Turnell. "Poly(ester amide) co-polymers promote blood and tissue compatibility. *J. Biomater. Sci.*, 2009, 20, 1495-1511.

[172] S.H. Lee, I. Szinai, K. Carpenter, R. Katsarava, G. Jokhadze, C.C. Chu, Y. Huang, E. Verbeken, O. Bramwell, I. De Scheerder, M.K. Hong. "In - vivo biocompatibility evaluation of stents coated with a new biodegradable elastomeric and functional polymer". *Coron. Artery Dis.*, 2002, 13, 237-241.

[173] Z. Gomurashvili, H. Zhang, J. Da, T.D. Jenkins, J. Hughes, M. Wu, L. Lambert, K. A. Grako, K.M. DeFife, K. Macpherson, V. Vassilev, R. Katsarava, W.G. Turnell. From drug-eluting stents to biopharmaceuticals: poly(ester amide) a versatile new bioabsorbable biopolymer, *ACS Symposium Series 977: Polymers for Biomedical Applications,* A. Mahapatro, A.S. Kulshrestha, Eds. Oxford University Press. 2008, 10-26.

[174] K. Markoishvili, G. Tsitlanadze, R. Katsarava, J.G. Morris, A. Sulakvelidze. "A Novel Sustained-Release Matrix Based on Biodegradable Poly(Ester Amide)s and Impregnated with Bacteriophages and an Antibiotic Shows Promise in Healing Wounds Infected with Various Pathogenic Bacteria". *Intern. J. Dermatology*, 2002, 41, 453-458.

[175] D. Jikia, N. Chkhaidze, E. Imedashvili, I. Mgaloblishvili, G. Tsitlanadze, R. Katsarava, J.G. Morris,Jr., A. Sulakvelidze. "The use of PhagoBioDerm, a novel biodegradable preparation capable of the sustained release of bacteriophages and ciprofloxacin, in the complex treatment of *Staphylococcus aureus*-infected local radiation injuries

caused by the exposure to Sr^{90}". *Clinical and Experimental Dermatology*, 2005, 30, 23-26.

[176] Z. Gomurashvili, H.R. Kricheldorf, R. Katsarava. "Amino acid based bioanalogous polymers. Synthesis and study of new regular poly(ester amides)s composed of hydrophobic α-amino acids and dianhydrohexitoles". *J. Macromol. Sci. Pure and Appl. Chem.* 2000, 37, 215-227.

[177] M. Okada, M. Yamada, M. Yokoe, K. Aoi. "Biodegradable polymers based on renewable resources. V. Synthesis and biodegradation behavior of poly(ester amide)s composed of 1,4:3,6-dianhydro-D-glucitol, α-amino acid, and aliphatic dicarboxylic acid units". *Journal of Applied Polymer Science*, 2001, 81, 2721-2734.

[178] E. Chkhaidze, D. Tugushi, D. Kharadze, Z. Gomurashvili, C.-C. Chu, R. Katsarava, New unsaturated biodegradable poly(ester amide)s composed of fumaric acid, L-leucine and α,ω-alkylene diols, *J. Macromol. Sci., Part A, Pure & Appl. Chem.* 2011, 48, 544-555.

[179] K. Guo, C.C. Chu, E. Chkhaidze, R. Katsarava. "Synthesis and Characterization of Novel Biodegradable Unsaturated Poly(Ester-Amide)s". *J. Polym. Sci. Part A: Polym. Chem.*, 2005, 43, 1463-1477.

[180] N. Zavradashvili, G. Jokhadze, M. Gverdtsiteli, G. Otinashvili, N. Kupatadze, Z. Gomurashvili, D. Tugushi, R. Katsarava. "Amino Acid Based Epoxy-Poly(Ester Amide)s - a New Class of Functional Biodegradable Polymers: Synthesis and Chemical Transformations". *J. Macromol. Sci., Part A, Pure & Appl. Chem.*, 2013, 50, 449-465.

[181] N. Ochkhikidze, E. Razmadze, D. Tugushi, N. Kupatadze, Z. Gomurashvili, R. Katsarava. "AABB-poly(depsipeptide)s – a new class of amino acid based biodegradable polymers". *International symposium "Polycondensation-2008"*, 8-11 September, 2008, 72, Tokyo, Japan.

[182] N. Ochkhikidze, R. Katsarava. Unpublished data.

[183] A.C. Fonsecaa, J.F.J. Coelhoa, J.F.A. Valenteb, T.R. Correiab, I.J. Correiab, M.H. Gila, P.N. Simõesa. "Poly(ester amide)s based on (L)-lactic acid oligomers and α-amino acids: influence of the α-amino acid side chain in the poly(ester amide)s properties". *J. Biomat. Sci.,Polym. Ed.*, 2013, 24, 1391–1409.

[184] T. Kartvelishvili, G. Tsitlanadze, L. Edilashvili, N.Japaridze, R. Katsarava. "Amino acid based bioanalogous polymers. Regular poly(ester urethane)s and poly(ester urea)s based on bis(phenylalanine)-α,ω–alkylene diesters. *Macromol. Chem. Phys.*, 1997, 198, 1921-1932.

[185] A. Díaz, L.J. del Valle, D. Tugushi, R. Katsarava, J. Puiggalí. "New poly(ester urea) derived from L-leucine: electrospun scaffolds loaded with antibacterial drugs and enzymes". *Mater. Sci. Engineering C,* 2015, 46, 450–462.

[186] M. Planellas, M.M. Pérez-Madrigal, L.J. delValle, S. Kobauri, R. Katsarava, C. Alemán, J. Puiggalí. "Microfibres of conducting polythiophene and biodegradable poly(ester urea) for scaffolds". *Polym. Chem.*, 2015, 6, 925–937.

[187] M. Casolaro."Vinyl Polymers Containing L-Valine and L-Leucine Residues: Thermodynamic Behavior of Homopolymers and Copolymers with N-Isopropylacrylamide". *Macromolecules,* 1995, 28, 2351-2358.

[188] M. Casolaro, R. Barbucci. "Thermodynamic Behavior of Polyelectrolytes with the Lower Critical Solution Temperature (LCST) Phenomenon". *Polymers for Advanced Technologies.* John Wiley & Sons, Ltd., 1996, 7, 831-838.

[189] M. Casolaro. "Solution behaviour of poly(N-acryloyl+leucine) and its copolymers with N-isopropylacrylamide". *Polymer,* 1997, 38, 4215-4222.

[190] A. Bentolila, I. Vlodavskii, R. Ishai-Michaeli, O. Kovalchul, C. Haloun, A.J. Doimb. "Poly(N-acryl amino acids): A New Class of Biologically Active Polyanions". *J. Med. Chem.*, 2000, 43, 2591–2600.

[191] L. Feng, J. Hu, Z. Liu, F. Zhao, G. Liu. "Preparation and properties of optically active poly(N-methacryloyl L-leucine methyl ester)". *Polymer,* 2007, 48, 3616-3623.

[192] S.G. Roy, P. De. "pH Responsive Polymers with Amino Acids in the Side Chains and Their Potential Applications". *J. Appl. Polym. Sci.*, 2014, 131, 41084.

[193] S. Sutthasupa, M. Shiotsuki, F. Sanda. "Recent advances in ring-opening metathesis polymerization, and application to synthesis of functional materials". *Polymer Journal*, 2010, 42, 905–915.

[194] S. Mallakpour. A. Zadehnazari. "Use of Ionic Green Solvent for the Synthesis of Optically Active Aromatic Polyamides Containing a L-Leucine Moiety under Microwave Irradiation". *Designed Monomers and Polymers* 2009, 12, 589–604.

[195] M.D.A. Muhsin, G. George, K. Beagley, V. Ferro, C. Armitage, N. Islam. "Synthesis and Toxicological Evaluation of a Chitosan - L - Leucine Conjugate for Pulmonary Drug Delivery Applications". *Biomacromolecules*, 2014, 15, 3596-3607

[196] T. Satoh. "Unimolecular micelles based on hyperbranched polycarbohydrate cores". *Soft Matter*, 2009, 5, 1972–1982.
[197] Y. Zhu, A.B. Lowe, P.J. Roth. "Postpolymerization synthesis of (bis)amide (co)polymers: Thermoresponsive behavior and self-association". *Polymer*, 2014, 55, 4425-4431.

In: Leucine
Editor: Sophia R. Newman
ISBN: 978-1-63482-593-1

Chapter 3

An Overview on Leucine Metabolites Alpha-Ketoisocaproate (KIC) and Beta-Hydroxi-Beta-Methyl Butyrate (HMB) in Skeletal Muscle Function and Sports Performance

***Lucas Guimarães-Ferreira*[1]*, Jason Cholewa*[2]*, Isadora Clivatti Furigo*[3]*, João Alfredo Bolivar Pedroso*[3] *and Nelo Eidy Zanchi*[3]**

[1]Center of Physical Education and Sports, Federal University of Espirito Santo, Vitoria, Brazil
[2]Department of Kinesiology Recreation and Sport Studies, Coastal Carolina University, Conway, SC, US
[3]Department of Physiology and Biophysics, University of Sao Paulo, Sao Paulo, Brazil

Abstract

Alpha-ketoisocaproate (KIC) and Beta hydroxy- beta methylbutyrate (HMB) are metabolites of the amino acid leucine. The first step in HMB metabolism is the reversible transamination of leucine to KIC, which occurs mainly extrahepatically. Following this enzymatic reaction, KIC is converted to HMB by two routes: through the cytosolic enzyme KIC dioxygenase; or through formation of isovaleryl-CoA by branched-chain ketoacid dehydrogenase (BCKD) in liver, which after some steps results in HMB formation. Under normal conditions the majority of KIC is converted into isovaleryl-CoA, with approximately 5% of leucine consumed metabolized into HMB. HMB and KIC have been proposed to enhance physical performance and muscle hypertrophy via their effects on muscle proteolysis and protein synthesis. Studies investigating the mechanisms by which HMB promotes increases in strength and muscle hypertrophy are still in preliminary stages. Based on the results of these studies, it is postulated that enhancements from leucine metabolite supplementation involve the following mechanisms: improvements in sarcolemal integrity; upregulation of IGF-1 gene expression in skeletal muscle and IGF-1 production by liver; stimulation of protein synthesis by activating the mTOR signaling pathway; suppression of skeletal muscle proteolysis by the inhibition of the ubiquitin–proteasome system; and modulation of lipopolysaccharides, tumor necrosis factor alpha (TNF-alpha) and angiotensin 2, attenuating protein synthesis depression in catabolic conditions. In turn, KIC has been used in conjunction with glycine and arginine (marketed as GAKIC). Acute GAKIC ingestion has been demonstrated to significantly enhance human performance during intense muscle contractions and repeated cycling sprints. The proposed mechanism of action of GAKIC includes: removal of deaminated nitrogen waste produced during exercise; regeneration of L-leucine via leucine dehydrogenase enzyme; promoting HMB synthesis via KIC dioxygenase enzyme; reducing ammonia levels during intense exercise; and increasing ATP levels through Isovaleryl-CoA and Acetyl-CoA formation. Interestingly, KIC ingestion does not result in the same effects of GAKIC, suggesting that glycine and arginine are both necessary for the enhanced performance. In this chapter the effects of HMB and KIC on muscle function and sports performance will be discussed, based on recent research findings.

INTRODUCTION

Nutritional sport supplements are widely consumed by athletes and non-athletes to enhance performance and body composition (Bernstein et al., 2003). In the United States alone, more than 3 million people use or have used nutritional supplements (Molinero and Marquez, 2009). Among these supplements amino acids and their metabolites are most commonly purchased. Data suggests that the amino acid leucine and its metabolites alpha-ketoisocaproate (KIC) and beta-hydroxi-beta-methyl butyrate (HMB) inhibits skeletal muscle proteolysis and promotes muscle protein synthesis (Hider et al., 1969; Tischler et al., 1982; Nissen and Abumrad, 1997). Recent studies present some evidence that KIC and HMB supplementation may increase muscle performance during high-intensity exercise, and HMB in particular may have anti-catabolic properties during periods of energy restriction or in the treatment of skeletal muscle wasting conditions.

An Overview on KIC and HMB Metabolism

KIC and HMB are metabolites of the essential branched chain amino acid (BCAA) leucine. This amino acid is converted in KIC via the enzyme branched chain amino acid transferase mainly extrahepatically in a transamination reaction (Block and Buse, 1990). KIC can follow two different pathways. In the first, HMB is produced by the cytosolic KIC dioxygenase reaction in liver. In the second, Isovaleryl-CoA is synthetized by the branched-chain ketoacid dehydrogenase (BCKD) enzyme in liver mitochondria and, after several further steps, beta-hydroxy-beta-methylglutaryl-CoA (HMG-CoA) is produced via the enzyme HMG-CoA synthase (Zanchi et al., 2011). The main fate of HMB is the conversion to HMG-CoA in the liver, which is converted to cholesterol (Nissen and Abumrad, 1997). An overview on KIC and HMB metabolism is illustrated in figure 1.

The endogenous synthesis of HMB is approximately 0.2 to 0.4 g of HMB per day in a 70 kg individual. The major site of HMB production is the liver, presenting high dioxygenase activity, but studies suggest that muscle and other tissues can produce HMB in low amounts (Nissen and Abumrad, 1997). It was estimated that 2 to 10% of leucine is converted to HMB within the body (Van Koevering and Nissen, 1992).

The time to reach peak plasma HMB concentration is 1 and 2 h following the ingestion of 1 and 3 g of HMB, respectively, with a half-life of ~2.5 h (Vukovich et al., 2001). Only 14 to 29% of HMB is excreted in urine, which suggests that 70 to 86% is retained in the body for further metabolism. Plasma HMB returns to baseline levels approximately 9 h following ingestion (Vukovich et al., 2001). The common form of HMB available on the commercial market is a calcium salt and the majority of studies have administered HMB in this form. The ingestion of HMB as free acid gel results in higher plasma concentrations (+185%) with 25% greater tissue uptake and utilization due to improved clearance; however, urinary losses are not different between the two HMB forms (Fuller et al., 2011).

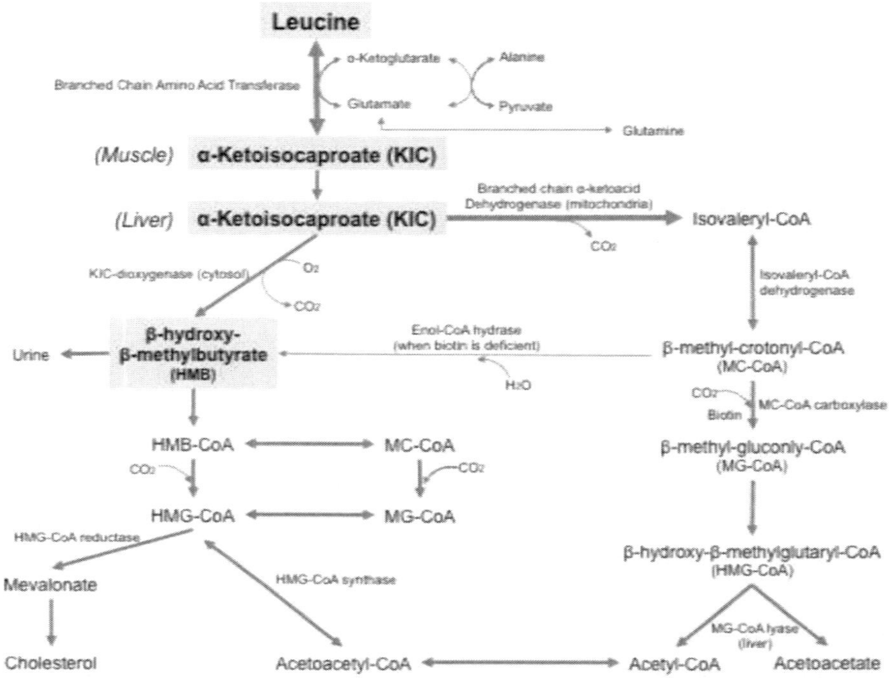

Figure 1. HMB and KIC metabolism (Adapted from Nissen and Abumrad, 1997).

EFFECTS OF KIC SUPPLEMENTATION ON PERFORMANCE

KIC may improve exercise capacity reducing intramuscular urea as leucine dehydrogenase consumes an NH_3 to aminate KIC back to leucine in skeletal muscle (Matthews et al., 1999). Yarrow et al. (2007) investigated the effects of KIC (1.5 g and 9.0 g) ingestion 45 and 25 min prior to the performance of leg and chest press repetitions to failure (at 50% 1RM) and 30 s repeated bouts of vertical jumping. Compared to an isoenergetic placebo no significant differences were observed between conditions or doses in repetitions to failure or peak and mean vertical jump performance. This study suggests that acute KIC ingestion prior to a single bout of high intensity exercise does not increase performance. Further studies are required to evaluate the efficacy of acute and chronic KIC supplementation on higher volume traiing sessions and across a training cycle, respectively.

The addition of glycine and arginine (marketed as GAKIC) may increase the potential of KIC as an ergogenic aid. Two studies have investigated the effects of GAKIC ingestion on repeated bouts of anaerobic cycling performance. Buford and Koch (2004) demonstrated that GAKIC ingestion (11.2 g) 45 min prior to exercise attenuates the decline in mean power output during five supramaximal 10-s cycle ergometer sprints separated by 1-min rest intervals in untrained individuals. On the contrary, Beis et al. (2011) founded no effect of GAKIC ingestion on repeated anaerobic cycling performance in highly trained athletes. Peak and mean power declined from the 1st sprint to the 10th with no difference between conditions. These results suggest that GAKIC has positive effect on repeated bouts of high-intensity cycling exercise performance in untrained individuals, but not in highly trained athletes. Endurance training attenuates leucine oxidation and branched-chain 2-oxoacid dehydrogenase activation (McKenzie et al., 2000), therefore future studies with longer duration training protocols are required to investigate the ergogenic efficacy of GAKIC and endurance trained athletes.

Stevens et al. (2000) investigated the effects of GAKIC ingestion (11.2 g) on fatigue resistance index (as described by Kaminski et al., 2000) and total work on during concentric and eccentric knee extension work on an isokinetic dynamometer. GAKIC ingestion increased fatigue resistance by up to 28% and total muscle work by 10.5% when compared to isoenergetic control condition. Wax et al. (2013a) recently investigated the effects of acute GAKIC ingestion (10.2 g) on resistance training performance in men and women. Seven resistance-trained men performed 5 sets at 75% of 1 repetition maximum (1RM) to failure in leg press exercise after GAKIC or placebo ingestion. GAKIC ingestion increased total load volume 22.5% compared to placebo condition (Wax et al., 2013a). In the second study, nine resistance-trained college-aged females ingested 10.2 g of GAKIC or placebo and performed six sets to failure of leg extensions at 50% 1RM. GAKIC supplementation significantly increased leg extension total load volume by 16.40%; however, no differences between conditions were reported for post exercise glucose, lactate, or maximal heart rate (Wax et al., 2013b). The results of these studies suggest that GAKIC supplementation may improve work capacity during acute resistance training, but further research is required to investigate the effects of chronic GAKIC supplementation on resistance training adaptations.

In conclusion, the current body of research suggests that KIC ingestion in conjunction with arginine and glycine (10.2 or 11.2 g) may be necessary to exert positive effects on performance. GAKIC ingestion also appears more effective in untrained or moderate-trained individuals than in highly trained subjects. Future studies using KIC and GAKIC supplementation are required to determine its efficacy as an ergogenic aid and to investigate its mechanism of action.

Mechanisms of Action of GAKIC

The mechanism of action by which GAKIC ingestion enhances acute performance is not fully understood. Several possible mechanisms have been previously suggested (Yarrow et al., 2007), however none of these suggestions have been rigorously investigated and all require further study. These mechanism include: removal of ammonia during high-intensity exercise bouts, reducing fatigue; regeneration of leucine via the enzyme branched chain aminotransferase; HMB formation via KIC dioxygenase enzyme; ATP formation from isovaleryl-CoA; increased creatine synthesis from glycine and arginine; and, nitric oxide formation from arginine.

Effects of HMB Supplementation on Performance

Individuals engaged in high intensity and high volume training, especially activities with high force eccentric components such as resistance training and team sports, often incur a high degree of muscle damage and delayed onset muscle soreness (DOMS). In untrained populations high degrees of DOMS can lead people to avoid engaging in regular exercise, and in highly trained populations DOMS can lead to interruptions in training, which may reduce adaptation by interfering with the principal of continuous loads demands of training.

Several studies have demonstrated the potential for HMB to reduce muscle damage and DOMS following intense training. Nissen et al. (1996) demonstrated that three weeks of chronic HMB at 1.5 g/day or 3.0 g/day resulted in improved strength with the greatest improvements in strength observed at 3.0 g/day. Urinary 3 methyl-histadine (3-MH) was reduced compared to placebo and to greater extent than following a single week of supplementation (Kreider et al., 1999). Additionally, HMB supplementation did not able to significantly reduce CK concentrations (creatine kinase), an intra-cellular enzyme that links from muscle to the blood when there is muscle damage, until the third week of training. These effects appeared to be greater when ingesting 3 g/day of HMB compared to 1.5 g/day of HMB. Other investigations that have supplemented with HMB for two (Vukovich and Dreifort, 2001) or more weeks (Nissen and Abumrad 1997) have generally reported that HMB supplementation lowers indices of skeletal muscle damage and soreness, while those supplementing for shorter periods of time have not (Table 1) (Kreider et al., 1999). Therefore, HMB supplementation may be most useful when ingestion begins two weeks prior to the onset of a new training period or change in training workload.

The acute effects of HMB supplementation have not been well investigated. Wilson et al. (2009) examined the effects of an oral 3 g bolus of HMB on 16 untrained males 60 min prior to unilateral, isokinetic leg extension exercise. Acute HMB ingestion prevented a significant rise in lactate dehydrogenase, and tended to decrease soreness in the quadriceps. A second study by Wilson et al. (2013) demonstrated similar findings 20 resistance trained men performing a high-volume powerlifting training session. Townsend et al. (2013) demonstrated in 40 resistance trained men that HMB-free acid attenuates the inflammatory response to exercise following 4 sets of 10 repetition squats to failure, which may reduce the recovery time following intense exercise. In contrast, Paddon-Jones et al. (2001) reported 6 days of HMB supplementation had no effect on muscle damage following an acute bout of isokinetic eccentric elbow flexion.

The available data on the literature suggest that when supplemented chronically untrained individuals demonstrate a greater response to HMB supplementation than trained subjects (reviewed by Zanchi et al., 2011). A meta-analysis published by Rowlands and Thomson (2009), consisting of eleven studies, concluded that HMB supplementation induces small benefits to lower-body and average strength and negligible gains to upper-body strength in untrained individuals, while in resistance-trained individuals all strength outcomes were trivial. An exercise stimulus with higher intensity and volume may be necessary to induce skeletal muscle damage in trained individuals (Clarkson and Hubai, 2002). It was proposed that if these conditions were lacking, HMB supplementation would not be efficacious (Wilson et al., 2008). However, the influence of muscle damage to muscle hypertrophy and protein accretion is still questionable, and further investigations are needed to investigate this relationship. Athletes and well-trained individuals may also not respond as well to HMB supplementation because of training-induced suppressions of muscle protein breakdown. A summary of studies evaluating the effects of HMB supplementation on body composition, performance and muscle damage is presented on table 1.

Table 1. Summary of studies that evaluated the effects of HMB supplementation on muscle strength, muscle mass and muscle damage

Reference	Samples	Supplementation protocol		Results
Nissen et al. (1996)	Humans	1.5 or 3.0 g HMB day^{-1}	3 and 7 weeks with RT	Significant decrease in muscle proteolysis induced by exercise
Kreider et al. (1999)	Humans	3.0 or 6.0 g HMB day^{-1}	4 weeks with RT	No gain in strength or muscle mass
Kreider et al. (2000)	Humans (football players)	3.0 g HMB day^{-1}	28 days[a]	No differences in catabolism markers, body composition, or repetitive sprint performance
Panton et al. (2000)	Humans	3.0 g HMB day^{-1}	4 weeks with RT	Lower levels of muscle damage markers
Gallagher et al. (2000a)	Humans	3.0 g HMB day^{-1}	8 weeks with RT	Induced increases in muscle mass
Gallagher et al. (2000a)	Humans	6.0 g HMB day^{-1}	8 weeks with RT	No response to supplementation
Gallagher et al. (2000b)	Humans	6.0 g HMB day^{-1}	8 weeks with RT	No risk regarding renal and liver function or hematological parameters
Clark et al. (2000)	Humans	3 g of HMB + 14 g arginine + 14 g glutamine day^{-1}	8 weeks	In individuals with HIV, increased lean mass and improved immune system function
Slater et al. (2001)	Humans	3.0 g HMB day^{-1}	6 weeks with RT	No changes in body composition, muscular strength or biochemical markers
Jówko et al. (2001)	Humans	3.0 HMB day^{-1}	3 weeks with RT	No changes in muscle mass
Vukovich et al. (2001)	Humans	3.0 g HMB day^{-1}	8 weeks with RT	Increased body fat loss in elderly subjects

Reference	Samples	Supplementation protocol		Results
Vukovich and Dreifort (2001)	Humans (endurance-trained cyclists)	3.0 g HMB day^{-1}	2 weeks	Increased onset of blood lactate accumulation (OBLA) with no changes in lactate accumulation peak or VO_{2peak}.
Soares et al. (2001)	Rats	0.002 g HMB $kg^{-1}.day^{-1}$	7 days	Decreased muscle damage and increased muscle fiber diameter under hindlimb immobilization procedures
Ransone et al. (2003)	Humans (football players)	3.0 g HMB day^{-1}	4 weeks[a]	No gain in strength or muscle mass
Hoffman et al. (2004)	Humans (football players)	3.0 g HMB day^{-1}	10 days[a]	No gain in strength or muscle mass
Flakoll et al. (2004)	Humans	2.0 g HMB + 5.0 g arginine + 1.5 g lysine day^{-1}	12 weeks	Gain in maximum strength and muscle mass
Van Someren et al. (2005)	Humans	3.0 g HMB + 3.0 g KIC day^{-1}	2 weeks with RT	Gain in maximum strength and increased skeletal muscle repair
Thomson et al. (2009)	Humans	3.0 g HMB day^{-1}	9 weeks with RT	Increased lower-body strength with no effects on body composition
O'Connor & Crowe (2007)	Humans (rugby players)	3.0 g HMB day^{-1}	6 weeks[b]	No gain in strength, nor in anthropometrical parameters
Lamboley et al. (2007)	Humans	3.0 g HMB day^{-1}	5 weeks[c]	Aerobic capacity enhancement with no alterations in body composition
Portal et al. (2011)	Humans (adolescent volleyball players)	3.0 g HMB day^{-1}	7 weeks[d]	Greater increase in FFM; increase in knee extension isokinetic force and in lower-body anaerobic power. No alterations in knee extension or upper-body isokinetic force

Table 1. (Continued)

Reference	Samples	Supplementation protocol		Results
Gerlinger-Romero et al. (2011)	Rats	0.320 g HMB $kg^{-1}.day^{-1}$	28 days	No alterations in muscle weight; increases in GH and IGF-1
Pinheiro et al. (2012)	Rats	0.320 g HMB $kg^{-1}.day^{-1}$	30 days	Increase in maximal tetanic force and resistance to acute muscle fatigue
Townsend et al. (2013)	Humans	1 g HMB-FA	Acute	Attenuation of initial immune response to intense resistance exercise
Deutz et al. (2013)	Humans (older adults)	3 g HMB day^{-1}	10 days	Attenuation of lean body mass loss associated with bed rest
Alway et al. (2013)	Rats (aged)	0.340 mg $kg^{-1}.day^{-1}$	4 weeks	Enhanced muscle stem cell proliferation; prevention in the functional decline of maximal plantar flexion isometric force; increase in plantaris muscle mass and cross sectional area during reloading after hindlimb suspension
Wilson et al. (2013)	Humans	3 g HMB-FA	Acute	Attenuation of creatine kinase levels and improved the perceived recovery status after intense resistance exercise. No difference in 3-methylhistadine, testosterone and cortisol levels
Baptista et al. (2013)	Rats	0.600 mg $kg^{-1}.day^{-1}$	Up to 10 days	No attenuation effect on muscle cross sectional area during immobilization and dexamethasone treatment

Reference	Samples	Supplementation protocol		Results
Wilson et al. (2014a)	Humans	3 g HMB-FA day^{-1}	8 weeks	Increase in strength and vertical jump performance. Increased fat mass gain
Wilson et al. (2014b)	Humans	3 g HMB-FA + 400 mg ATP day^{-1}	12 weeks	Increase in strength, vertical jump performance and wingate power. Increased fat mass gain
Kraemer et al. (2014)	Humans	1.57 HMB day^{-1e}	12 weeks	No effect on citokyne response to training
Noh et al. (2014)	Rats	0.150 or 0.600 g HMB .kg^{-1}.day^{-1}	5 days	Attenuation of dexamethasone-induced reductions on grip strength, muscle damage and muscle wasting

FFM, fat free mass; HMB, beta-hydroxy-beta-methylbutyrate; HMB-FA, HMB free acid; KIC, α-keto isocaproate; RT, resistance training; a associated with specific exercise program for football players; b associated with specific exercise program for rugby players; c associated with an aerobic exercise training; d associated with specific exercise program for adolescent volleyball players ; eHMB as part of a supplement containing essential amino acids (l-glutamine, arginine) and taurine.

Table adapted from Zanchi et al. (2011).

In conclusion, HMB supplementation shows potential to decrease DOMS and muscle damage when HMB is supplemented before the onset of exercises containing a high volume of lower body training, especially training that has a high eccentric component. HMB doses of 3 g/day appears more efficacious than 1.5 g/day to decrease muscle damage, and 2-3 weeks of continuous supplementation may be required for results to manifest. Finally, although some research shows that HMB supplementation can positively affect trained individuals, untrained individuals appear to benefit most from HMB supplementation.

HMB Supplementation for Athletes Training in Energy Restricted State

Lean body compositions have been shown to positively correlate with athletic performance in a variety of activities such as water polo, soccer, and tactical fitness tests (Platanou & Varamenti, 2011; Silvestre et al., 2006; Sporis et al., 2011). Athletes involved in competitions that require a significant caloric restrition to make weight or achieve a very lean body composition, such as mixed martial arts or bodybuilding, whereby muscle mass and consequently perofrmance is reduced may also benefit from HMB supplementation. In a proof of concept study, Park et al. (2013) demonstrated HMB improved sensorimotor function during exercise and body composition, and attenuated muscle mass and strength loss during catabolic conditions in mice (Park et al. 2013). In humans, Hunga et al. (2010) reported that HMB supplementation resulted in greater reductions in body fat and a trend toward muscle sparing in female judo athletes on a severely energy restricted diet. Further research in healthy athletes engaged in caloric restriction is required to verify these hypotheses.

Mechanisms of Action of HMB

Several mechanisms were proposed for the action of HMB on muscle function and sports performance. Nissen et al. (1996) proposed a mechanism in which HMB acts as a precursor of structural components of the cell membrane, thus enhancing the sarcolemma repair after intense contractile activity. However, limited research data supports this theory. Our group has demonstrated that in rodents chronic HMB treatment led to an increase in muscle glycogen and ATP content (Pinheiro et al., 2012). HMB may accelerate the tricarboxylic acid cycle which leads to an increase in ATP content and malate-aspartate-shuttle activity, thereby providing substrate for glycogen synthesis post exercise. HMB may also increase fat oxidation during exercise and recovery, leading to improved aerobic performance. We recently demonstrated that HMB treatment in rats promotes an increase in hormone-sensitive lipase mRNA and protein expression and an increase in plasma non-esterified fatty acids levels (Yonamine et al., 2014). *In vitro* data also demonstrate that HMB supplementation increases fatty free acids oxidation in skeletal muscle (Cheng et al., 1997; 1998).

HMB may also directly increase muscular force production. Data from our research group demonstrate that treatment with HMB in rats for 4 weeks ($320mg.kg^{-1}.day^{-1}$) led to an increase in specific force generation and muscle quality (force normalized to muscle mass or muscle cross sectional area). We also observed an increased resistance to acute fatigue during repeated high-intensity muscle contractions without changes in contraction and relaxation times (Pinheiro et al., 2012). These results may not be generalizable however, and more research is needed in humans to investigate the effects of HMB supplementation on muscle quality.

HMB affects skeletal muscle protein metabolism by enhancing protein synthesis and inhibiting protein degradation. Kornasio et al. (2009) reported HMB treatment promoted an increase in DNA synthesis and stimulated myoblast proliferation and differentiation in chicken and human myoblasts. Always et al. (2013) demonstrated HMB treatment increases satellite cell proliferation in response to disuse injury in aging rats. These studies suggest that HMB may reduce recovery time by increasing the satellite cell response to exercise induced muscle damage.

HMB has also been shown to increase total mRNA expression *in vitro* (Kornasio et al., 2009) but not *in vivo* after 30 days of HMB treatment in soleus and EDL rat muscles (Gerlinger-Romero et al., 2011). In this later study, however, serum IGF-1 was increased in response to HMB treatment, as well as GH gene and protein expression on pituitary gland, suggesting a role of increased growth factors in mediating exercise recovery. In murine myoblasts, HMB stimulates the mTOR pathway of muscle protein synthesis via PI3K/Akt (Kornasio et al., 2009). Animal data also suggest that HMB treatment inhibits proteasome-dependent skeletal muscle proteolysis, acting as an anti-catabolic compound (Smith et al., 2005). Further studies, especially in humans, are necessary to better elucidate the effects of HMB on specific protein synthesis and degradation pathways *in vivo*. A summary of the proposed HMB mechanisms of action is illustrated in figure 2.

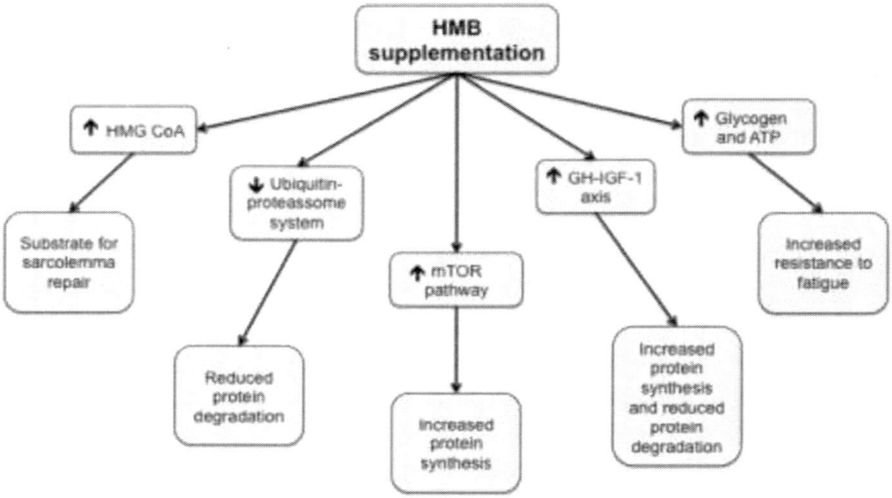

Figure 2. HMB mechanisms of action.

Safety of HMB Supplementation

Studies examining the safety of HMB supplementation did not demonstrate any side effects (Nissen and Abumrad, 1997; Gallagher et al., 2000; Nissen et al., 2000; Rathmacher et al., 2004; Baxter et al., 2005; Baier et al., 2009). For example, Baier et al. (2009) demonstrated that daily ingestion of 2-3 g of HMB in combination with amino acids for one year did not result in any difference in markers of hepatic and renal function, as well as on blood lipids levels. In fact, Nissen et al. (2000) demonstrated that the ingestion of 3.0 g HMB per day resulted in a decrease of total cholesterol (5.8%), low-density lipoprotein (LDL) cholesterol (7.3%), and systolic blood pressure (4.4 mmHg) when compared with placebo group. Despite human data indicating the safety of HMB supplementation, studies with rats indicate that a high dose chronic HMB treatment (320 $mg.kg^{-1}.day^{-1}$) resulted in increased fasting insulin levels and impaired insulin signaling, leading to insulin resistance (Gerlinger-Romero, 2011; Yonamine et al., 2014). Future studies are necessary to determine the effects of HMB on insulin sensitivity in humans.

Conclusion

The results of recent controlled studies suggest that KIC ingestion in combination with glycine and arginine may enhance performance during repeated bouts of high-intensity exercise. The effects of monotherapy and chronic KIC supplementation alone on muscle performance still need to be determined. Some studies demonstrate the potential for HMB supplementation to improve contractile function, muscle mass gain and fat loss, but these effects appear to be more pronounced in catabolic conditions and in untrained individuals subjected to intense exercise training. More research is required to determine the potential of HMB supplementation to improve strength and muscle gain in trained individuals during periods of caloric balance.

References

Alway, SE; Pereira, SL; Edens, NK; Hao, Y; Bennett, BT. β-Hydroxy-β-methylbutyrate (HMB) enhances the proliferation of satellite cells in fast muscles of aged rats during recovery from disuse atrophy. *Exp Gerontol,* 2013, 48(9), 973-84.

Baier, S; Johannsen, D; Abumrad, N; Rathmacher, JA; Nissen, S; Flakoll, P. Year-long changes in protein metabolism in elderly men and women supplemented with a nutrition cocktail of beta-hydroxy-beta-methylbutyrate (HMB); L-arginine; and L-lysine. *JPEN J Parenteral Enteral Nutr,* 2009, 33, 71-82.

Baxter, JH; Carlos, JL; Thurmond, J; Rehani, RN; Bultman, J; Frost, D. Dietary toxicity of calcium beta-hydroxy-beta-methyl butyrate (CaHMB). *Food Chem Toxicol,* 2005, 43, 1731-41.

Beis, L; Mohammad, Y; Easton, C; Pitsiladis, YP. Failure of glycine-arginine-alpha-ketoisocaproic acid to improve high intensity exercise performance in trained cyclists. *Int J Sport Nutr Exerc Metab*, 2011, 21(1), 33-39.

Bernstein, A; Safirstein, J; Rosen, JE. Athletic ergogenic aids. *Bull Hosp Jt Dis,* 2003, 61:164–171.

Block, KP; Buse, MG. Glucocorticoid regulation of muscle branched-chain amino acid metabolism. *Med Sci Sports Exerc,* 1990, 22, 316-24.

Buford, BN; Koch, AJ. Glycine-arginine-alpha-ketoisocaproic acid improves performance of repeated cycling sprints. *Med Sci Sports Exerc,* 2004, 36(4), 583-87.

Cheng, W; Phillips, B; Abumrad, N. Beta-hydroxy-beta-methylbutyrate increases fatty acid oxidation by muscle cells. *FASEB Journal,* 1997, 11:381.

Cheng, W; Phillips, B; Abumrad, N. Effect of HMB on fuel utilization; membrane stability and creatine kinase content of cultured muscle cells. *FASEB Journal,* 1988, 12: 950.

Clark, RH; Feleke, G; Din, M; Yasmin, T; Singh, G; Khan, FA; Rathmacher JA. Nutritional treatment for acquired immunodeficiency virus-associated wasting using beta-hydroxy beta-methylbutyrate, glutamine, and arginine: a randomized, double-blind, placebo-controlled study. *J Parenter Enteral Nutr,* 2000,24:133-9.

Clarkson, PM; Hubal, MJ. Exercise-induced muscle damage in humans. *Am J Phys Med Rehabil,* 2002, 81,52-69.

Deutz, NE; Pereira, SL; Hays, NP; Oliver, JS; Edens, NK; Evans, CM; Wolfe, RR. Effect of β-hydroxy-β-methylbutyrate (HMB) on lean body mass during 10 days of bed rest in older adults. *Clin Nutr,* 2013, 32(5), 704-12.

Fielding, RA; Evans, WJ; Hughes, VA; Moldawer, LL; Bistrian, BR. The effects of high intensity exercise on muscle and plasma levels of alpha-ketoisocaproic acid. *Eur J Appl Physiol Occup Physiol,* 1986,55(5):482-5.

Flakoll, P; Sharp, R; Baier, S; Levenhagen, D; Carr, C; Nissen S. Effect of betahydroxy betamethylbutyrate, arginine, and lysine supplementation on strength, functionality, body composition, and protein metabolism in elderly women. *Nutrition,* 2004,20:445-51.

Fuller Jr, JC; Sharp, RL; Angus, HF; Baier, SM; Rathmacher, JA. Free acid gel form of β-hydroxy-β-methylbutyrate (HMB) improves HMB clearance from plasma in human subjects compared with the calcium HMB salt. *Br J Nutr,* 2011, 105, 367-72.

Gallagher, PM; Carrithers, JA; Godard, MP; Schulze, KE; Trappe, SW. Beta-hydroxy-beta-methylbutyrate ingestion, Part I: effects on strength and fat free mass. *Med Sci Sports Exerc,* 2000,32(12):2109-15.

Gallagher, PM; Carrithers, JA; Godard, MP; Schulze, KE; Trappe, SW. Beta-hydroxy-beta-methylbutyrate ingestion; part II: effects on hematology; hepatic and renal function. *Med Sci Sports Exerc,* 2000b, 32, 2116-19.

Gerlinger-Romero, F; Guimarães-Ferreira, L; Giannocco, G; Nunes, MT. Chronic supplementation of beta-hydroxy-beta methylbutyrate (HMβ) increases the activity of the GH/IGF-I axis and induces hyperinsulinemia in rats. *Growth Horm IGF Res,* 2011,21(2):57-62.

Hider, RC; Fern, EB; London, DR. Relationship between intracellular amino acids and protein synthesis in the extensor digitorum longus muscle of rats. *Biochem J,* 1969, 114, 171-78.

Hoffman, JR; Cooper, J; Wendell, M; Im, J; Kang, J. Effects of beta-hydroxy beta-methylbutyrate on power performance and indices of muscle damage and stress during high-intensity training. *J Strength Cond Res,* 2004,18:747-52.

Hooper, DR; Flanagan, SD; Looney, DP; White, MT; Volek, JS; Maresh, CM. Influence of HMB supplementation and resistance training on cytokine responses to resistance exercise. *J Am Coll Nutr,* 2014, 33(4), 247-55.

Hunga, W; Liub, TH; Chenc, CY; Chang, CK. Effect of [beta]-hydroxy-[beta]-methylbutyrate Supplementation During Energy Restriction in Female Judo Athletes. *J Exerc Sci Fitness,* 2010, 8, 50-53.

Jacob, MW; Peter, JF; Bill, C; Gabriel, JW; Nelo, Z; Lem, T; Colin, W; Douglas, SK; Jeffrey, RS; Jay, RH; Tim, NZ; Hector, LL; Richard, BK; Abbie, ESR; Jose, A; Wilson, JM; Kim, JS; Lee, SR; Rathmacher, JA; Dalmau, B; Kingsley, JD; Koch, H; Manninen, AH; Saadat, R; Panton, LB. Acute and timing effects of betahydroxy-beta-methylbutyrate (HMB) on indirect markers of skeletal muscle damage. Nutr Metab, 2009, 6(6). International Society of Sports Nutrition Position Stand: beta-hydroxy-beta-methylbutyrate (HMB). *J Int Soc Sports Nutr,* 2013; 10:6.

Jówko, E; Ostaszewski, P; Jank, M; Sacharuk, J; Zieniewicz, A; Wilczak, J; Nissen, S. Creatine and beta-hydroxy-beta-methylbutyrate (HMB) additively increase lean body mass and muscle strength during a weight-training program. *Nutrition,* 2001,17:558-66.

Kaminski, TW; Godfrey, MD; Braith, RW; Stevens, BR. Measurement of acute dynamic anaerobic muscle fatigue using a novel fatigue resistance index. *Isokinet Exerc Sci,* 2000, 8, 95-101.

Kornasio, R; Riederer, I; Butler-Browne, G; Mouly, V; Uni; Z; Halevy, O. Beta-hydroxy-beta-methylbutyrate (HMB) stimulates myogenic cell proliferation; differentiation and survival via the MAPK/ERK and PI3K/Akt pathways. *Biochim biophysica acta,* 2009, 23, 836-46.

Kraemer, WJ; Hatfield, DL; Comstock, BA; Fragala, MS; Davitt, PM; Cortis, C; Wilson, JM; Lee, EC; Newton, RU; Dunn-Lewis, C; Häkkinen, K; Szivak, TK; Kreider, RB; Ferreira, M; Wilson, M; Almada, AL. Effects of calcium betahydroxy-beta-methylbutyrate (HMB) supplementation during resistance training on markers of catabolism; body composition and strength. *Int J Sports Med,* 1999, 20, 503–09.

Kreider, RB; Ferreira, M; Wilson, M; Almada, AL. Effects of calcium beta-hydroxy-beta-methylbutyrate (HMB) supplementation during resistance-training on markers of catabolism, body composition and strength. *Int J Sports Med,* 1999, 20(8):503-9.

Lamboley, CR; Royer, D; Dionne, IJ. Effects of beta-hydroxybeta-methylbutyrate on aerobic-performance components and body composition in college students. *Int J Sport Nutr Exerc Metab,* 2007,17:56-69.

Lowery, RP; Joy, JM; Rathmacher, JA; Baier, SM; Fuller, J; Shelley, MC 2nd; Jäeger, R; Purpura, M; Wilson, SM; Wilson, JM. Interaction of Beta-Hydroxy-Beta-Methylbutyrate Free Acid (HMB-FA) and Adenosine Triphosphate (ATP) on Muscle Mass; Strength; and Power in Resistance Trained Individuals. *J Strength Cond Res,* 2014.

Matthews, DE; Harkin, R; Battezzati, A; Brillon, DJ. Splanchnic bed utilization of enteral alpha-ketoisocaproate in humans. *Metabolism,* 1999,48(12):1555-63.

McKenzie, S; Phillips, SM; Carter, SL; Lowther, S; Gibala, MJ; Tarnopolsky, MA. Endurance exercise training attenuates leucine oxidation and BCOAD activation during exercise in humans. *Am J Physiol Endocrinol Metab,* 2000,278(4):E580-7.

Molinero, O; Marquez, S. Use of nutritional supplements in sports: risks, knowledge, and behavioural-related factors. *Nutr Hosp,* 2009, 24:128–134.

Nissen, S; Sharp, R; Ray, M; Rathmacher, JA; Rice, D; Fuller, JC Jr; Connelly, AS; Abumrad, N. Effect of leucine metabolite beta-hydroxy-beta-methylbutyrate on muscle metabolism during resistance-exercise training. *J Appl Physiol* (1985), 1996, 81(5):2095-104.

Nissen, S; Sharp, RL; Panton, L; Vukovich, M; Trappe, S; Fuller, JC Jr. Beta-hydroxy-beta-methylbutyrate (HMB) supplementation in humans is safe and may decrease cardiovascular risk factors. *J Nutr,* 2000, 130,1937-45.

Nissen, SL; Abumrad, NN. Nutritional role of the leucine metabolite B-hydroxy B-methylbutyrate (HMB). *J Nutr Biochem,* 1997, 8, 300-11.

Noh, KK; Chung, KW; Choi, YJ; Park, MH; Jang, EJ; Park, CH; Yoon, C; Kim, ND; Kim, MK; Chung, HY. β-Hydroxy β-methylbutyrate improves dexamethasone-induced muscle atrophy by modulating the muscle degradation pathway in SD rat. *PLoS One,* 2014, 9(7), 1029-47.

O'Connor, DM; Crowe, MJ. Effects of six weeks of betahydroxy-beta-methylbutyrate (HMB) and HMB/creatine supplementation on strength, power, and anthropometry of highly trained athletes. *J Strength Cond Res,* 2007,21:419-23.

Paddon-Jones, D; Keech, A; Jenkins, D. Short-term beta-hydroxy-beta-methylbutyrate supplementation does not reduce symptoms of eccentric muscle damage. *Int J Sport Nutr Exerc Metab,* 2001,11(4):442-50.

Panton, LB; Rathmacher, JA; Baier, S; Nissen, S. Nutritional supplementation of the leucine metabolite beta-hydroxy-beta-methylbutyrate (hmb) during resistance training. *Nutrition,* 2000,16:734 9.

Park, BS; Henning, PC; Grant, SC; Lee, WJ; Lee, SR; Arjmandi, BH; Kim, JS. HMB attenuates muscle loss during sustained energy deficit induced by calorie restriction and endurance exercise. *Metabolism,* 2013, 62(12), 1718-29.

Pinheiro, CH; Gerlinger-Romero, F; Guimarães-Ferreira, L; de Souza-Jr, AL; Vitzel, KF; Nachbar, RT; Nunes, MT; Curi R. Metabolic and functional effects of beta-hydroxy-beta-methylbutyrate (HMB) supplementation in skeletal muscle. *Eur J Appl Physiol,* 2012,112(7):2531-7.

Portal, S; Zadik, Z; Rabinowitz, J; Pilz-Burstein, R; Adler-Portal, D; Meckel, Y; Cooper, DM; Eliakim, A; Nemet, D. The effect of HMB supplementation on body composition, fitness, hormonal and inflammatory mediators in elite adolescent volleyball players: a prospective randomized, double-blind, placebo-controlled study. *Eur J Appl Physiol,* 2011,111:2261-9.

Ransone, J; Neighbors, K; Lefavi, R; Chromiak, J. The effect of beta-hydroxy beta-methylbutyrate on muscular strength and body composition in collegiate football players. *J Strength Cond Res,* 2003,17:34-9.

Rathmacher, JA; Nissen, S; Panton, L; Clark, RH; Eubanks, MP; Barber AE; D'Olimpio, J; Abumrad, NN. Supplementation with a combination of beta-hydroxy-beta-methylbutyrate (HMB); arginine; and glutamine is safe and could improve hematological parameters. *JPEN J Parenter Enteral Nutr,* 2004, 28, 65-75.

Rowlands, DS; Thomson, JS. Effects of beta-hydroxy-beta-methylbutyrate supplementation during resistance training on strength; body composition; and muscle damage in trained and untrained young men: a meta-analysis. *J Strength Cond Res,* 2009, 23(3), 836-46.

Slater, G; Jenkins, D; Logan, P; Lee, H; Vukovich, M; Rathmacher, JA; Hahn, AG. Beta-hydroxy-beta-methylbutyrate (HMB) supplementation does not affect changes in strength or body composition during resistance training in trained men. *Int J Sport Nutr Exerc Metab,* 2001,11:384-96.

Smith, HJ; Mukerji, P; Tisdale, MJ. Attenuation of proteasome-induced proteolysis in skeletal muscle by {beta}-hydroxy-{beta}-methylbutyrate in cancer-induced muscle loss. *Cancer Research,* 2005, 65, 277-83.

Soares, JMC; Póvoas, S; Neuparth, MJ; Duarte JA. The effects of beta-hydroxy-beta-methylbutyrate (HMB) on muscle atrophy induced by immobilization. *Med Sci Sports Exerc,* 2001;33:S140.

Stevens, BR; Godfrey, MD; Kaminski, TW; Braith, RW. High intensity dynamic human muscle performance enhanced by a metabolic intervention. *Med Sci Sports Exerc,* 2000, 32(12), 2102-8.

Thomson, JS; Watson, PE; Rowlands, DS. Effects of nine weeks of beta-hydroxy-beta-methylbutyrate supplementation on strength and body composition in resistance trained men. *J Strength Cond Res*, 2009,23:827-35.

Tischler, ME; Desautels, M; Goldberg, AL. Does leucine; leucyltRNA; or some metabolite of leucine regulate protein synthesis and degradation in skeletal and cardiac muscle? *J Biol Chem,* 1982, 257, 1613-21.

Townsend, JR; Fragala, MS; Jajtner, AR; Gonzalez, AM; Wells, AJ; Mangine, GT; Robinson, EH 4th; McCormack, WP; Beyer, KS; Pruna, GJ; Boone, CH; Scanlon, TM; Bohner, JD; Stout, JR; Hoffman, JR. β-Hydroxy-β-methylbutyrate (HMB)-free acid attenuates circulating TNF-α and TNFR1 expression postresistance exercise. *J Appl Physiol,* (1985) 2013, 115(8), 1173-82.

Van Koevering, M; Nissen, S. Oxidation of leucine and alphaketoisocaproate to beta-hydroxy-beta-methylbutyrate in vivo. *Am J Physiol,* 1992, 262, 27-31.

Van Someren, KA; Edwards, AJ; Howatson, G. Supplementation with beta-hydroxy-beta-methylbutyrate (HMB) and alpha-ketoisocaproic acid (KIC) reduce signs and symptoms of exercise induced muscle damage in man. *Int J Sport Nutr Exerc Metab,* 2005,15:413-24.

Vukovich, MD; Dreifort, GD. Effect of beta-hydroxy beta-methyl-butyrate on the onset of blood lactate accumulation and V(O)(2)peak in endurance-trained cyclists. *J Strength Cond Res,* 2001,15:491-7.

Vukovich, MD; Dreifort, GD. Effect of beta-hydroxy beta-methylbutyrate on the onset of blood lactate accumulation and V(O)(2) peak in endurance-trained cyclists. *J Strength Cond Res,* 2001, 15(4), 491-97.

Vukovich, MD; Slater, G; Macchi, MB; Turner, MJ; Fallon, K; Boston, T; Rathmacher, J. Beta-hydroxy-beta-methylbutyrate (HMB) kinetics and the influence of glucose ingestion in humans. *J Nutr Biochem,* 2001, 12, 631-39.

Vukovich, MD; Stubbs, NB; Bohlken, RM. Body composition in 70-year-old adults responds to dietary beta-hydroxy-beta- methylbutyrate similarly to thatof young adults. *J Nutr,* 2001,131:2049-52.

Wax, B; Hilton, L; Vickers, B; Gilliland, K; Conrad, M. Effects of glycine-arginine-α-ketoisocaproic acid supplementation in college-age trained females during multi-bouts of resistance exercise. *J Diet Suppl,* 2013b, 10(1), 6-16.

Wax, B; Kavazis, AN; Brown, SP; Hilton, L. Effects of supplemental GAKIC ingestion on resistance training performance in trained men. *Res Q Exerc Sport,* 2013a, 84(2), 245-51.

Wilson, GJ; Wilson, JM; Manninen, AH. Effects of beta-hydroxy-beta-methylbutyrate (HMB) on exercise performance and body composition across varying levels of age; sex; and training experience: a review. Nutr *Metab (Lond),* 2008, 5:1.

Wilson, JM; Kim, JS; Lee, SR; Rathmacher, JA; Dalmau, B; Kingsley, JD; Koch, H; Manninen, AH; Saadat, R; Panton, LB. Acute and timing effects of beta-hydroxy-beta-methylbutyrate (HMB) on indirect markers of skeletal muscle damage. *Nutr Metab (Lond).* 2009 Feb 4;6:6.

Wilson, JM; Lowery, RP; Joy, JM; Andersen, JC; Wilson, SM; Stout, JR; Duncan, N; Fuller, JC; Baier, SM; Naimo, MA; Rathmacher, J. The effects of 12 weeks of beta-hydroxy-beta-methylbutyrate free acid supplementation on muscle mass; strength; and power in resistance-trained individuals: a randomized; double-blind; placebo-controlled study. *Eur J Appl Physiol,* 2014, 114(6), 1217-27.

Wilson, JM; Lowery, RP; Joy, JM; Walters, JA; Baier, SM; Fuller, JC; Stout, JR; Norton, LE; Sikorski, EM; Wilson, SM; Duncan, NM, Zanchi, NE; Rathmacher, J. β-Hydroxy-β-methylbutyrate free acid reduces markers of exercise-induced muscle damage and improves recovery in resistance-trained men. *Br J Nutr,* 2013, 110(3), 538-44.

Yarrow, JF; Parr, JJ; White, LJ; Borsa, PA; Stevens, BR. The effects of short-term alpha-ketoisocaproic acid supplementation on exercise performance: a randomized controlled trial. *J Int Soc Sports Nutr,* 2007, 4:2.

Yonamine, CY; Teixeira, SS; Campello, RS; Gerlinger-Romero, F; Rodrigues, CF Jr; Guimarães-Ferreira, L; Machado, UF; Nunes, MT. Beta hydroxy beta methylbutyrate supplementation impairs peripheral insulin sensitivity in healthy sedentary Wistar rats. *Acta Physiol (Oxf),* 2014,212(1):62-74.

Zanchi, NE; Gerlinger-Romero, F; Guimarães-Ferreira, L; de Siqueira Filho, MA; Felitti, V; Lira, FS; Seelaender, M; Lancha, AH Jr. HMB supplementation: clinical and athletic performance-related effects and mechanisms of action. *Amino Acids,* 2011, 40(4):1015-25.

INDEX

A

B

C

D

E

F

G

H

O

P

W

X

Y

Z

A First Course in Quantum Mechanics

Based on years of teaching experience, this book provides a clearly structured and accessible introduction to the fundamental principles of quantum mechanics, tailored for undergraduate students.

The text builds a strong conceptual and mathematical foundation while promoting intuitive understanding and critical thinking. Students progress from basic concepts such as wave mechanics to more advanced topics within matrix mechanics. Complex concepts are presented in a diligent and understandable way, achieving a balance between conceptual insight and mathematical rigor without oversimplifying the subject.

Thoroughly class-tested, this book is designed to support both classroom instruction and independent study. It will serve as a great tool for undergraduates studying introductory and intermediate quantum mechanics, as well as modern physics and quantum systems for engineers.

A First Course in Quantum Mechanics

Nonrelativistic Theory

Ahmad Amer

CRC Press
Taylor & Francis Group
Boca Raton London New York

CRC Press is an imprint of the
Taylor & Francis Group, an **informa** business

First edition published 2026
by CRC Press
2385 NW Executive Center Drive, Suite 320, Boca Raton FL 33431

and by CRC Press
4 Park Square, Milton Park, Abingdon, Oxon, OX14 4RN

CRC Press is an imprint of Taylor & Francis Group, LLC

ISBN: 978-1-041-14097-9 (hbk)
ISBN: 978-1-041-14979-8 (pbk)
ISBN: 978-1-003-67710-9 (ebk)

DOI: 10.1201/9781003677109

Typeset in CMR10 font
by KnowledgeWorks Global Ltd.

Publisher's note: This book has been prepared from camera-ready copy provided by the authors.

To the memory of Ahmad As Sayed
student and friend

1986-2013

Ahmad As Sayed was a dedicated researcher and assistant lecturer in the Mathematics Department at the Faculty of Science, Cairo University. He earned his B.Sc. in Pure and Applied Mathematics in 2008 and his M.Sc. in Applied Mathematics in 2013, both from Cairo University. Known for his deep commitment to theoretical physics and mathematical modeling, Ahmad made significant contributions to the study of particle physics at the intersection of mathematics and high-energy theory.

In addition to his teaching and mentoring, Ahmad contributed to the field through several important publications in peer-reviewed journals, including "Higgs Vacuum Stability in the $B-L$ Extended Standard Model," "Right-Handed Sneutrino-Antisneutrino Oscillations in a TeV-Scale Supersymmetric $B-L$ Model," and "Higgs Mass Corrections in the SUSY $B-L$ Model with Inverse Seesaw." His research continues to inform and inspire ongoing work in related areas.

Ahmad will be remembered not only for his scholarly achievements but also for his humility, generosity, and the sincere dedication he brought to both science and education.

Contents

Preface

This book is based on lectures given by the author for several years at Cairo university to undergraduate students studying for a pure and applied mathematics degree.

The text is intended to be as clear, simple and systematic as possible. Topics and applications given here are treated mathematically with detail and rigor. The rigor here is observed not for the sake of rigor itself, but to make the fundamental, conceptual and physical ideas clearer and easier to grasp, comprehend and understand. Indexation and notations in this book, specially in chapter 8, are chosen attentively to make things easier to follow.

Although they are supposed to be known by the reader, there are some preliminaries and definitions in both physics and mathematics on which the text relies, they are presented briefly in a condensed manner in the first chapter of this book for reference.

Prerequisites needed to study this book are: First courses of Functional Analysis, Differential Equations, Mathematical Analysis, Linear Algebra, Statistics and Classical and Modern Physics.

In chapter 2 a prelude of the historical development of quantum mechanics is summarized concisely to make the axioms of quantum mechanics plausible. The uninterested reader can skip this chapter and begin directly with the axioms of quantum mechanics in chapter 3. (Although sections 2.5–2.7 are recommended to read.)

In chapters 4 and 5 the one-dimensional motion and the motion in central fields are considered. In these two chapters, some standard applications are chosen to illustrate *Schrödinger* theory and the new physics predicted by it in these applications.

More about the formal mathematical structure of quantum mechanics and more abstract and general formalism are given in chapters 8 and 9.

Linear harmonic oscillator and the angular momentum (including spin and *Pauli* equation) are demonstrated in the new formalism in chapters 10 and 11.

Chapter 12 is devoted to approximation methods and applications.

In the last chapter a discussion of the concept of identity of particles in quantum mechanics, the symmetric and antisymmetric states and the different types of statistics are given.

The author appreciates the useful discussions of the asymptotic behavior of the solutions of the associated *Legendre* equation with professors Mourad E. M. Ismail, university of Central Florida and Zainab S. I. Mansour, Cairo University.

The author is grateful to A. Moursy, W. Abdallah and M. Ashry (A. As Sayed also took part in this in the beginning) for typing the text including the figures. W. Abdallah and M. Ashry precised some dates and numerical values of some constants in the text. Also, the author is grateful to W. Abdallah for making the book index.

Ahmad Amer
Giza, Egypt

Author

Prof. Ahmad Amer is a theoretical physicist and educator specializing in quantum electrodynamics and mathematical physics. He earned his Ph.D. in mathematical and physical sciences from the Faculty of Physics, Moscow State University, where his research focused on quantum field theory and the quark model.

With more than 20 years of teaching experience at the undergraduate and graduate levels, Prof. Amer has guided students through the complexities of quantum theory using a clear and pedagogically sound approach. His work emphasizes both the mathematical structure and physical intuition behind quantum mechanics.

He has published several research articles in peer-reviewed journals and contributed to curriculum development in modern physics. *A First Course in Quantum Mechanics: Nonrelativistic Theory* is his first book designed to bridge the gap between abstract formalism and hands-on problem solving for students encountering quantum mechanics for the first time.

Currently, Prof. Amer serves as a full professor at the Mathematics Department, Faculty of Science, Cairo University, Egypt, where he continues to teach, mentor students, and develop educational resources in theoretical physics.

Part I

Wave Mechanics

1

Some Definitions and Preliminaries

In this chapter, the definition of a field in physics is given. The plane monochromatic waves are very briefly presented. The relations between the frequency ω, the wavelength λ, the velocity of the wave u, and the wave vector $\vec{k}$ are given. Fields which can be considered locally as plane monochromatic waves and the limits of geometrical optics are discussed. Exponential, cosine, and sine *Fourier* transforms as well as the *Dirac* δ-function are given. Moreover, two examples as applications of the *Dirac* δ-function and its derivatives are discussed.

1.1 Field

In physics, a field is any physical quantity the value of which, in general, differs from position to position and from time to time. The field may be scalar, vector, tensor, or otherwise. The field, or one of its components, is represented mathematically by a function of four variables t, x, y, and z.
Examples: In 3-dimensional space, the heat temperature at the different points of a medium and also the mass and the electric charge densities of a medium (mass and charge distributions) are scalar fields. The displacement and velocity of the particles of a fluid are two vector fields. In the 4-dimensional space-time, the electromagnetic potential is a vector field and the electromagnetic field intensity is a tensor field.

1.2 Plane Wave

Let g be a function describing a field (or a component of a field).
If

$$g(t, \vec{r}) = f(t - \frac{\vec{n} \cdot \vec{r}}{u}), \tag{1.1}$$

DOI: 10.1201/9781003677109-1

then the value of the field is the same for all moments and positions satisfying

$$t - \frac{\vec{n} \cdot \vec{r}}{u} = \text{const.}, \tag{1.2}$$

or

$$\vec{n} \cdot \vec{r} = ut + \text{const.} \tag{1.3}$$

This is an equation of a plane (for each t) with the normal in the direction of $\vec{n}$, or a propagating plane in the direction of its normal $\vec{n}$ with velocity u. In this case the field is said to be a plane wave.

1.3 Plane Monochromatic Wave

If the field is a simple periodic function of time, i.e., if it is in the form

$$g = \text{Re}\ a\ e^{-i\omega t}, \tag{1.4}$$

where a is a complex function of the coordinates only, the field is said to be a monochromatic wave, ω is called the frequency, and the period of this function is clearly $\frac{2\pi}{\omega}$.

If, moreover, the field is a plane wave, it will be called a plane monochromatic wave, and it will be clearly also a simple periodic function of $\vec{n} \cdot \vec{r}$ but with another period λ which is called the wavelength and

$$g = \text{Re}\ a\ e^{-i(\omega t - \vec{k} \cdot \vec{r})}, \tag{1.5}$$

a in this case is just a complex constant which is called the complex amplitude, and $\vec{k}$ is called the wave vector, and its magnitude k is the wave number. Clearly

$$\lambda = \frac{2\pi}{k}, \quad u = \frac{\omega}{k}, \tag{1.6}$$

or

$$k = \frac{\omega}{u}, \quad \vec{k} = k\vec{n}. \tag{1.7}$$

The function

$$\psi = -\omega t + \vec{k} \cdot \vec{r} + \alpha \ (\alpha \text{ is a real constant}) \tag{1.8}$$

is called the phase.

$$\psi = \text{const.}, \tag{1.9}$$

is called the wave surface or the wave front which (at each t) is the surface at all of whose points the values of the field are the same.

For ψ in the form (1.8), the wave front is a plane (for each t). In general, the case is not like this. If g can be factorized in the form

$$g = \text{Re}\ a\ e^{i\psi}, \tag{1.10}$$

where ψ and a are functions of $(t, \vec{r})$; ψ is real and a in general is complex. Let ψ be differentiable at $(t_0, \vec{r_0})$. In a narrow enough neighborhood of $(t_0, \vec{r_0})$, we have

$$g = \text{Re } f \; e^{i[(t-t_0)\frac{\partial \psi_0}{\partial t_0} + (\vec{r}-\vec{r_0})\cdot\vec{\nabla}\psi_0]}, \text{ where} \tag{1.11}$$

$$f = a \; e^{i\psi_0}, \quad \psi_0 = \psi(t_0, \vec{r_0}). \tag{1.12}$$

If the neighborhood is narrow enough to consider a as constant in it, g can be considered locally as a plane monochromatic wave with frequency $\omega = -\frac{\partial \psi_0}{\partial t_0}$ and wave vector $\vec{k} = \vec{\nabla}\psi_0$. Therefore, $\lambda = \frac{2\pi}{|\vec{\nabla}\psi_0|}$, i.e., if the changes of a are negligible over distances $\sim \lambda$ (the extension of the neighborhood in any space direction can not be less than one wavelength), g can be considered in each small region of space (and each small interval of time) as a plane wave, i.e., g is considered as a wave with varying amplitude and varying direction of propagation. This is the geometrical optics approximation, and more accurate it is the smaller the wavelength is. It is exact in the limit $\lambda \to 0$ (the limit of geometrical optics) (see Fig. 1.1). For more details, see e.g., [1, § 53].

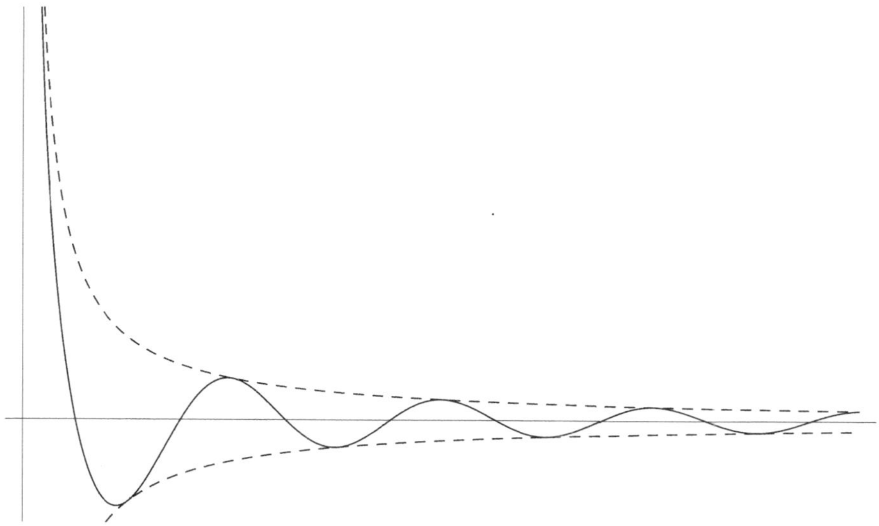

FIGURE 1.1: *The solid line represents the function* $\frac{\cos kx}{x}$ *(for* $k = \frac{1}{4}$*), and the dashed ones represent the amplitude* $\pm\frac{1}{x}$*. Near* $x = 0$*, the amplitude changes rapidly through a phase change* $\Delta(kx) = 2\pi$*, while far from the origin, the amplitude* $\frac{1}{x}$ *changes slowly through a change* $\Delta x = \frac{2\pi}{k} = \lambda$*; hence, the field can be considered in each small region on the* x*-axis as a plane wave.*

1.4 *Fourier* Integrals

1.4.1 Exponential *Fourier* Transforms

Let $f(t)$ be a periodic function with period T, then it can be expanded as a *Fourier* series:

$$f(t) = \sum_{n=-\infty}^{\infty} C_n e^{-i\omega_n t}, \quad \omega_n = n\Omega, \quad \Omega = \frac{2\pi}{T}, \tag{1.13}$$

where

$$C_n = \frac{1}{T}\int_{-\frac{T}{2}}^{\frac{T}{2}} f(t)e^{i\omega_n t}dt. \tag{1.14}$$

i.e.,

$$f(t) = \sum_{n=-\infty}^{\infty} \frac{1}{T}\int_{-\frac{T}{2}}^{\frac{T}{2}} f(u)e^{i\omega_n (u-t)}du. \tag{1.15}$$

Now,

$$\frac{1}{T} = \frac{\Omega}{2\pi} = \frac{\Delta\omega_n}{2\pi} \quad \text{(equipartition)}, \tag{1.16}$$

therefore

$$f(t) = \frac{1}{2\pi}\sum_{n=-\infty}^{\infty} \Delta\omega_n \int_{-\frac{T}{2}}^{\frac{T}{2}} f(u)e^{i\omega_n (u-t)}du. \tag{1.17}$$

In the limit $\Delta\omega_n \to 0$ $(T \to \infty)$, the integral will be taken between $\pm\infty$ and the sum transforms into an integral:

$$f(t) = \frac{1}{2\pi}\int_{-\infty}^{\infty} d\omega \int_{-\infty}^{\infty} f(u)e^{i\omega(u-t)}du = \frac{1}{\sqrt{2\pi}}\int_{-\infty}^{\infty} \tilde{f}(\omega)e^{-i\omega t}d\omega, \tag{1.18}$$

where

$$\tilde{f}(\omega) = \frac{1}{\sqrt{2\pi}}\int_{-\infty}^{\infty} f(t)e^{i\omega t}dt. \tag{1.19}$$

Notice that, since the period T tends to ∞, the condition of periodicity on the function f to be represented as a *Fourier* integral is removed.

1.4.2 *Fourier* Cosine and Sine Transforms

$$f(x) = \frac{1}{\sqrt{2\pi}}\int_{-\infty}^{\infty} \tilde{f}(k)e^{ikx}dk, \tag{1.20}$$

where

$$\tilde{f}(k) = \frac{1}{\sqrt{2\pi}}\int_{-\infty}^{\infty} f(x)e^{-ikx}dx. \tag{1.21}$$

Let f be an even function. Therefore,

$$\tilde{f}(k) = \sqrt{\frac{2}{\pi}} \int_0^\infty f(x) \cos kx dx, \tag{1.22}$$

$$\begin{aligned} f(x) &= \frac{1}{\pi} \int_{-\infty}^{\infty} \left(\int_0^\infty f(\xi) \cos k\xi d\xi \right) e^{ikx} dk \\ &= \frac{1}{\pi} \int_0^\infty d\xi f(\xi) \int_{-\infty}^{\infty} \cos k\xi e^{ikx} dk \\ &= \frac{2}{\pi} \int_0^\infty \int_0^\infty f(\xi) \cos k\xi \cos kx d\xi dk \\ &= \sqrt{\frac{2}{\pi}} \int_0^\infty \tilde{f}_c(k) \cos kx dk, \end{aligned} \tag{1.23}$$

where the *Fourier* cosine transform $\tilde{f}_c$ is

$$\tilde{f}_c(k) = \sqrt{\frac{2}{\pi}} \int_0^\infty f(x) \cos kx dx. \tag{1.24}$$

The definitions (1.23), (1.24) require $\tilde{f}_c(k)$ and $f(x)$ to be defined only for positive values of k and x, respectively, therefore, the condition that f is even is removed (the question about the parity of the function here, has no place!). Similarly, the *Fourier* sine transform pair can be defined starting with an odd function, a condition which will be removed at the end:

$$f(x) = \sqrt{\frac{2}{\pi}} \int_0^\infty \tilde{f}_s(k) \sin kx dk, \tag{1.25}$$

$$\tilde{f}_s(k) = \sqrt{\frac{2}{\pi}} \int_0^\infty f(x) \sin kx dx. \tag{1.26}$$

1.5 *Dirac* δ-Function (Distribution)

Let $\omega_n = n\Omega$, $\Delta\omega_n = \Omega = \frac{2\pi}{T}$. Then

$$\frac{1}{T} \int_{-\frac{T}{2}}^{\frac{T}{2}} g(\omega_n + \eta) e^{i\omega_n t} dt = g(\eta) \delta_{n,0}, \tag{1.27}$$

therefore,

$$\int_{-\frac{T}{2}}^{\frac{T}{2}} \frac{1}{2\pi} \sum_{n=-\infty}^{\infty} g(\omega_n + \eta) e^{i\omega_n t} \Delta\omega_n dt = g(\eta). \tag{1.28}$$

In the limit as $\Delta\omega_n = \Omega \to 0$ $(T \to \infty)$, we obtain

$$\int_{-\infty}^{\infty} \frac{1}{2\pi} \int_{-\infty}^{\infty} g(\omega+\eta) e^{i\omega t} d\omega dt = g(\eta), \tag{1.29}$$

or

$$\int_{-\infty}^{\infty} \frac{1}{2\pi} \int_{-\infty}^{\infty} g(\omega) e^{i(\omega-\eta)t} d\omega dt = g(\eta). \tag{1.30}$$

The order of integration here is quite important: the integration must be carried out first by ω, then by t; otherwise, if it is carried out vice versa, the integral $\frac{1}{2\pi}\int_{-\infty}^{\infty} e^{i(\omega-\eta)t} dt$ would not converge. If we introduce the smeared integral for $\alpha > 0$:

$$\delta(\omega-\eta,\alpha) = \frac{1}{2\pi} \int_{-\infty}^{\infty} e^{-\alpha|t|} e^{i(\omega-\eta)t} dt, \tag{1.31}$$

the order of integration in (1.30) becomes interchangeable, and in the final result we take the limit $\alpha \to 0$. Evaluating integral (1.31):

$$\delta(\omega-\eta,\alpha) = \frac{1}{\pi} \frac{\alpha}{\alpha^2+(\omega-\eta)^2}, \tag{1.32}$$

therefore,

$$\lim_{\alpha\to 0} \delta(\omega-\eta \neq 0,\alpha) = \frac{1}{\pi} \lim_{\alpha\to 0} \frac{\alpha}{\alpha^2+(\omega-\eta)^2} = 0, \tag{1.33}$$

$$\lim_{\alpha\to 0} \delta(0,\alpha) = \frac{1}{\pi} \lim_{\alpha\to 0} \frac{1}{\alpha} = \infty, \tag{1.34}$$

$$\int_{-\infty}^{\infty} \delta(\omega-\eta,\alpha) d\omega = \frac{\alpha}{\pi} \int_{-\infty}^{\infty} \frac{d\omega}{\alpha^2+(\omega-\eta)^2} = \frac{1}{\pi} \tan^{-1} \frac{\omega-\eta}{\alpha}\bigg|_{-\infty}^{\infty} = 1, \ \forall\alpha, \tag{1.35}$$

that is

$$\delta(\omega-\eta) = \begin{cases} 0, & \text{for } \omega \neq \eta, \\ \infty, & \text{for } \omega = \eta, \end{cases} \tag{1.36}$$

and

$$\int_{\omega_1}^{\omega_2} g(\omega)\delta(\omega-\eta) d\omega = \begin{cases} g(\eta), & \text{if } \eta \in (\omega_1,\omega_2), \\ 0, & \text{if } \eta \notin (\omega_1,\omega_2), \end{cases} \tag{1.37}$$

as proposed by *Dirac*[1]. $\delta(\omega-\eta)$ can be considered as the limit of $\delta(\omega-\eta,\alpha)$ as $\alpha \to 0$, i.e.,

$$\delta(\omega-\eta) = \lim_{\alpha\to 0} \delta(\omega-\eta,\alpha) = \frac{1}{2\pi} \int_{-\infty}^{\infty} e^{i(\omega-\eta)t} dt \tag{1.38}$$

[1] *P. A. M. Dirac* (1902–1984) jointly with *E. Schrödinger* (1887–1961) was awarded the *Nobel* Prize in Physics in 1933 for the discovery of new productive forms of atomic theory.

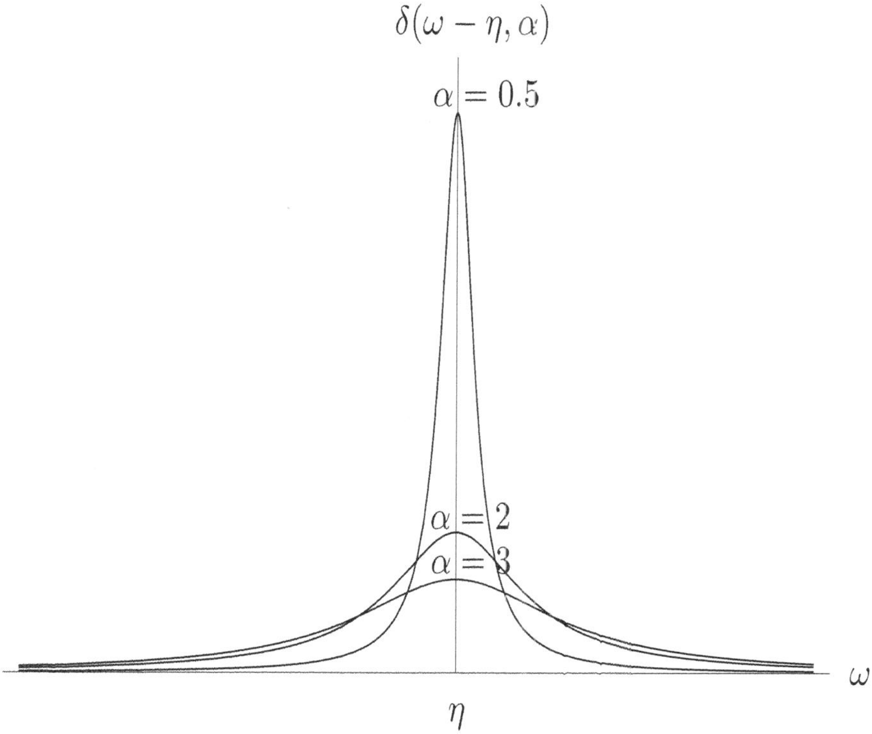

FIGURE 1.2: *The curves of this family have monotonically higher and narrower peaks at* $\omega = \eta$*, and the area under any one of them is independent of* α *and always equals* 1.

in the following, by definition, sense

$$\int_{-\infty}^{\infty} f(x)\delta(x-x_0)dx = \lim_{\alpha\to 0}\int_{-\infty}^{\infty} f(x)\delta(x-x_0,\alpha)dx, \tag{1.39}$$

for any function $f(x)$, well-behaved in a neighborhood of x_0. Indeed:

Example 1.1

$$\begin{aligned}\int_a^b f(x)\delta(x-x_0,\alpha)dx &= \frac{\alpha}{\pi}\int_a^b \frac{f(x)}{(x-x_0)^2+\alpha^2}dx \\ &= \frac{1}{\pi}f(\xi)\left(\tan^{-1}\frac{b-x_0}{\alpha} - \tan^{-1}\frac{a-x_0}{\alpha}\right), \quad \xi\in[a,b].\end{aligned} \tag{1.40}$$

(the Mean Value Theorem of integration).

1. $x_0 \notin (a,b)$*, then* $(b-x_0)(a-x_0) > 0$*, and*

$$\lim_{\alpha\to 0} \tan^{-1}\frac{b-x_0}{\alpha} = \lim_{\alpha\to 0}\tan^{-1}\frac{a-x_0}{\alpha}, \tag{1.41}$$

therefore

$$\lim_{\alpha\to 0}\int_a^b f(x)\delta(x-x_0,\alpha)dx = 0,\ \forall x_0 \notin (a,b). \tag{1.42}$$

2. $x_0 \in (a,b)$*, then* $(\beta > 0)$ *:*

$$\frac{\alpha}{\pi}\int_a^b \frac{f(x)}{(x-x_0)^2+\alpha^2}dx \tag{1.43}$$

$$\begin{aligned}
&= \frac{\alpha}{\pi}\left(\int_a^{x_0-\beta} + \int_{x_0-\beta}^{x_0+\beta} + \int_{x_0+\beta}^{b}\right)\frac{f(x)}{(x-x_0)^2+\alpha^2}dx \\
&= \frac{f(\xi_1)}{\pi}\left(\tan^{-1}\frac{x_0-a}{\alpha} - \tan^{-1}\frac{\beta}{\alpha}\right) \\
&+ \frac{f(\xi_2)}{\pi}\left(\tan^{-1}\frac{b-x_0}{\alpha} - \tan^{-1}\frac{\beta}{\alpha}\right) \\
&+ \frac{2}{\pi}f(\xi)\tan^{-1}\frac{\beta}{\alpha},\ \forall\beta,
\end{aligned} \tag{1.44}$$

where $\xi_1 \in (a, x_0-\beta)$*,* $\xi_2 \in (x_0+\beta, b)$*, and* $\xi \in (x_0-\beta, x_0+\beta)$*. Take* $\beta = \sqrt{\alpha}$ *and then take the limit as* $\alpha \to 0$*, to obtain*

$$\lim_{\alpha\to 0}\int_a^b f(x)\delta(x-x_0,\alpha)dx = f(x_0),\ \forall x_0 \in (a,b). \tag{1.45}$$

To sum up:

$$\int_a^b f(x)\delta(x-x_0)dx = \lim_{\alpha\to 0}\int_a^b f(x)\delta(x-x_0,\alpha)dx = \begin{cases} f(x_0), & \forall x_0 \in (a,b), \\ 0, & \forall x_0 \notin (a,b), \end{cases} \tag{1.46}$$

and this is just what we mean by the equation:

$$\delta(x-x_0) = \lim_{\alpha\to 0}\delta(x-x_0,\alpha). \tag{1.47}$$

Example 1.2

$$\delta(\omega-\eta) = \frac{1}{2\pi}\int_{-\infty}^{\infty} e^{i(\omega-\eta)t}dt = \lim_{T\to\infty}\delta_T(\omega-\eta), \tag{1.48}$$

where

$$\delta_T(\omega-\eta) = \frac{1}{2\pi}\int_{-T}^{T} e^{i(\omega-\eta)t}dt = \frac{1}{\pi}\frac{\sin(\omega-\eta)T}{\omega-\eta}. \tag{1.49}$$

Therefore,

$$\int_{\omega_1}^{\omega_2} f(\omega)\delta_T(\omega-\eta)d\omega = \frac{1}{\pi}\int_{\omega_1}^{\omega_2} f(\omega)\frac{\sin(\omega-\eta)T}{\omega-\eta}d\omega : \tag{1.50}$$

1. $\eta \notin (\omega_1, \omega_2)$, *hence* $\omega - \eta \neq 0$ *in the interval of the above integral and* $\frac{f(\omega)}{\omega-\eta}$ *therefore satisfies the conditions of the mean value theorem. Let* $\Omega \in (\omega_1, \omega_2)$ *be the point of this mean value theorem, therefore*

$$\int_{\omega_1}^{\omega_2} f(\omega)\delta_T(\omega-\eta)d\omega = \frac{f(\Omega)}{\pi(\Omega-\eta)T}(\cos(\omega_1-\eta)T - \cos(\omega_2-\eta)T), \tag{1.51}$$

which tends to zero as $T \to \infty$.

2. $\eta \in (\omega_1, \omega_2)$:

$$\frac{1}{\pi}\int_{\omega_1}^{\omega_2} f(\omega)\frac{\sin(\omega-\eta)T}{\omega-\eta}d\omega$$
$$= \frac{1}{\pi}\left(\int_{\omega_1}^{\eta-\epsilon} + \int_{\eta-\epsilon}^{\eta+\epsilon} + \int_{\eta+\epsilon}^{\omega_2}\right) f(\omega)\frac{\sin(\omega-\eta)T}{\omega-\eta}d\omega.$$

$\omega \neq \eta$ *in the intervals of the first and third integrals in the R.H.S., therefore, the mean value theorem can be applied to the function* $\frac{f(\omega)}{\omega-\eta}$ *there. The mean value theorem can not be applied to this function in the interval* $(\eta-\epsilon, \eta+\epsilon)$, *because* $\omega = \eta$ *in this interval, so we apply the theorem to the function* $f(\omega)$ *alone to have:*

$$\begin{aligned}\frac{1}{\pi}\int_{\omega_1}^{\omega_2} f(\omega)\frac{\sin(\omega-\eta)T}{\omega-\eta}d\omega &= \frac{f(\Omega_1)}{\pi}\frac{\cos\epsilon T - \cos(\eta-\omega_1)T}{(\eta-\Omega_1)T} \\ &+ \frac{f(\Omega_2)}{\pi}\frac{\cos\epsilon T - \cos(\omega_2-\eta)T}{(\Omega_2-\eta)T} \\ &+ f(\Omega)\frac{1}{\pi}\int_{-\epsilon T}^{\epsilon T}\frac{\sin\xi}{\xi}d\xi \quad \forall\ \epsilon > 0,\end{aligned} \tag{1.52}$$

where

$$\Omega_1 \in (\omega_1, \eta-\epsilon), \quad \Omega_2 \in (\eta+\epsilon, \omega_2), \quad \Omega \in (\eta-\epsilon, \eta+\epsilon).$$

Therefore:

$$(\eta-\Omega_1)T > \epsilon T, \quad (\Omega_2-\eta)T > \epsilon T \quad \forall\ T > 0, \text{ and } \eta-\epsilon < \Omega < \eta+\epsilon.$$

Take $\epsilon = \frac{1}{\sqrt{T}}$. *Therefore, as* $T \to \infty$, ϵ *tends to zero,* $(\eta-\Omega_1)T$ *and* $(\Omega_2-\eta)T$ *tend to* ∞ *and* Ω *tends to* η *and*

$$\lim_{T\to\infty}\int_{\omega_1}^{\omega_2} f(\omega)\delta_T(\omega-\eta)d\omega = \lim_{T\to\infty} f(\Omega)\lim_{T\to\infty}\frac{1}{\pi}\int_{-\sqrt{T}}^{\sqrt{T}}\frac{\sin\xi}{\xi}d\xi = f(\eta)$$

(f is continuous at $\omega = \eta$ and $\int_{-\infty}^{\infty} \frac{\sin\xi}{\xi} d\xi = \pi$).

To sum up:

$$\begin{aligned} \int_{\omega_1}^{\omega_2} f(\omega)\delta(\omega - \eta)d\omega &= \lim_{T\to\infty} \int_{\omega_1}^{\omega_2} f(\omega)\delta_T(\omega - \eta)d\omega \\ &= \begin{cases} f(\eta), & \forall\eta \in (\omega_1, \omega_2), \\ 0, & \forall\eta \notin (\omega_1, \omega_2), \end{cases} \end{aligned} \tag{1.53}$$

and this is just what we mean by the equation:

$$\delta(\omega - \eta) = \lim_{T\to\infty} \delta_T(\omega - \eta) = \lim_{T\to\infty} \frac{1}{2\pi} \int_{-T}^{T} e^{i(\omega-\eta)t} dt = \frac{1}{2\pi} \int_{-\infty}^{\infty} e^{i(\omega-\eta)t} dt. \tag{1.54}$$

Exercise 1.1 *In the same way demonstrated in the above two examples, show that*

1.

$$\delta(x - x_0) = \lim_{\varepsilon\to 0} \delta_\varepsilon(x - x_0), \tag{1.55}$$

where $\varepsilon > 0$,

$$\delta_\varepsilon(x - x_0) = \begin{cases} \frac{1}{2\varepsilon}, & \forall x \in (x_0 - \varepsilon, x_0 + \varepsilon), \\ 0, & \forall x \notin (x_0 - \varepsilon, x_0 + \varepsilon). \end{cases} \tag{1.56}$$

Sketch this family of functions, and show that the area under any curve of this family is independent of ε and equals 1.

2.

$$\delta(x) = \lim_{\alpha\to 0} \delta(x, \alpha), \tag{1.57}$$

where $\alpha > 0$,

$$\delta(x, \alpha) = \frac{1}{\alpha\sqrt{2\pi}} e^{-\frac{x^2}{2\alpha^2}}, \quad \textit{(Gaussian distribution)}. \tag{1.58}$$

Sketch this family of functions, and show that the area under any curve of this family is independent of α and equals 1.

3. Find the Fourier transform $\tilde{f}(\omega)$ for the function $f(t)$ given by Eq. (1.13), and substitute for $\tilde{f}(\omega)$ in the Fourier integral of $f(t)$ to reobtain Eq. (1.13).

It can be shown that:

$$\begin{aligned} \delta(x) &= \delta(-x), \quad (\delta \text{ is an even distribution}), \\ \delta(ax) &= \frac{\delta(x)}{|a|}, \; a \neq 0, \\ \delta\{\varphi(x)\} &= \sum_k \frac{\delta(x - x_k)}{|\varphi'(x_k)|}, \; \varphi'(x_k) \neq 0, \end{aligned} \tag{1.59}$$

and the summation is taken over all the zeros x_k of $\varphi(x)$ provided that φ' does not vanish there, also

$$\begin{aligned} x\delta(x) &= 0, \\ f(x)\delta(x-x_0) &= f(x_0)\delta(x-x_0), \\ \int \delta(x-y)\delta(y-x_0)dy &= \delta(x-x_0). \end{aligned} \tag{1.60}$$

All these equations must be understood in the sense that the integral on any interval of a well-behaved function multiplied by either side of any of these equations is the same, and the equality can be proved formally by manipulating the integrals as they were integrals in the ordinary sense. Rigorous proofs for (1.59) are achieved by representing the δ-functions as limits of the smeared δ-functions and proving the equalities before taking the limits. After the limits are taken, the proof is achieved. The first two equations of (1.60) are immediately proved by integrating over x, and the last one by multiplying by a well-behaved function $f(x)$ and integrating over x.

Examples of δ-Distribution:

One can think of $e\delta(x-x_0)$ as the line charge density of a point charge e located at $x = x_0$, and $m\delta(x-x_0)$ is the line mass density of a particle of mass m at $x = x_0$. $\delta(x-x_0)$ can be thought also of as the probability density of a random variable x which certainly takes the value x_0 and does not take absolutely any other values. It is, therefore, obvious that the probability density of a discrete random variable x, which acquires the value x_i with probability proportional to p_i is

$$f(x) = \frac{\sum p_i\delta(x-x_i)}{\sum p_i}. \tag{1.61}$$

Derivatives of the δ-Function:

The nth derivative $\delta^{(n)}(x)$ is defined by

$$\int_{-\infty}^{\infty} f(x)\delta^{(n)}(x)dx = (-1)^n f^{(n)}(0), \tag{1.62}$$

assuming the existence of $f^{(n)}(0)$, and $\delta^{(n)}(x)$ can be considered as the limit of the nth derivative of (the smeared δ-function) $\delta(x,\alpha)$. It can be proved that:

$$\begin{aligned} \delta^{(n)}(x) &= (-1)^n\delta^{(n)}(-x), \\ x^{n+1}\delta^{(n)}(x) &= 0, \\ \int \delta^{(m)}(x-y)\delta^{(n)}(y-a)dy &= \delta^{(n+m)}(x-a), \\ \delta^{(n)}(x) &= \frac{i^n}{2\pi}\int_{-\infty}^{\infty} k^n e^{ikx}dk. \end{aligned} \tag{1.63}$$

The definition of the δ-function can be generalized to three dimensions:

$$\delta(\vec{r}-\vec{r}_0) = \delta(x-x_0)\delta(y-y_0)\delta(z-z_0) = \frac{1}{(2\pi)^3}\int e^{i\vec{k}\cdot(\vec{r}-\vec{r}_0)}d^3k. \tag{1.64}$$

Example 1.3 *The electrostatic potential of a point charge e at the origin is $\varphi = \frac{e}{r}$. This potential φ satisfies Poisson's equation:*

$$\nabla^2 \varphi = -4\pi\rho, \tag{1.65}$$

here $\rho = e\delta(\vec{r})$, $\varphi = \frac{e}{r}$, therefore

$$\nabla^2 \frac{1}{r} = -4\pi\delta(\vec{r}). \tag{1.66}$$

For more details see [2, 3, 4].

1.6 Localization of a Wave

Let

$$g = f\ e^{-i(\omega_0 t - \vec{k}_0 \cdot \vec{r})}, \tag{1.67}$$

if f does not depend neither on the time t nor on the position $\vec{r}$, then g is a plane monochromatic wave. The intensity of the wave is proportional to $|f|^2$, i.e., the plane monochromatic wave has the same intensity everywhere and at all times. Such a wave is infinite in extent in all directions in space and exists over the whole range of time from $-\infty$ to ∞. But if g is not monochromatic, then

$$g = f(t)\ e^{-i\omega_0 t}. \tag{1.68}$$

If g is expanded as a superposition of monochromatic waves, i.e., if it is represented as a *Fourier* integral

$$g(t) = \frac{1}{\sqrt{2\pi}} \int_{-\infty}^{\infty} \tilde{g}(\omega)\ e^{-i\omega t}\ d\omega, \tag{1.69}$$

the intensity of the component with frequency ω in this superposition is proportional to $|\tilde{g}(\omega)|^2$ and the *Fourier* transform $\tilde{g}(\omega)$ is:

$$\tilde{g}(\omega') = \frac{1}{\sqrt{2\pi}} \int_{-\infty}^{\infty} f(t)\ e^{i(\omega' - \omega_0)t}\ dt. \tag{1.70}$$

If $f(t)$ were constant, then it could be taken out of the integral, and $\tilde{g}(\omega') = \sqrt{2\pi} f \delta(\omega' - \omega_0)$, and $g(t)$ again is just one monochromatic component with frequency ω_0. The two factors of the integrand in Eq. (1.70) are both functions of time. The second one is periodic with period $\frac{2\pi}{|\omega' - \omega_0|}$. For large enough $|\omega' - \omega_0|$, the periodic factor is varying very fast and $f(t)$ can be taken out of the integral, and $\tilde{g}(\omega' \gg \omega_0) \simeq \sqrt{2\pi} f \delta(\omega' - \omega_0 \neq 0) = 0$. But for ω' near enough to ω_0, $f(t)$ however slow cannot be taken out of the integral and $\tilde{g}(\omega')$ will be appreciably different from zero.

Let τ be the order of magnitude of the time interval after which the change of f can not be neglected. Therefore, $\tilde{g}(\omega')$ will be almost zero (negligible) $\forall \frac{2\pi}{|\omega'-\omega_0|} \ll \tau$ and $\tilde{g}(\omega')$ will not vanish $\forall \frac{2\pi}{|\omega'-\omega_0|} \gtrsim \tau$.

Let ω be the maximum value of ω' for which $\tilde{g}(\omega')$ is appreciably different from zero (i.e., for which f can not be taken out of the integral), then

$$\frac{2\pi}{\Delta\omega} \sim \tau; \quad \text{where } \Delta\omega = \omega - \omega_0, \tag{1.71}$$

i.e., $g(t)$ as a superposition of monochromatic waves will contain components with frequencies ranging from $\omega_0 - \Delta\omega$ to $\omega_0 + \Delta\omega$.

If the wave is radiated during a finite time interval of the order of magnitude of Δt, i.e., if Δt is of the order of magnitude of the wave duration in time (at each point of space), then

$$\Delta t \gtrsim \tau \sim \frac{2\pi}{\Delta\omega}, \tag{1.72}$$

i.e.,

$$\Delta\omega\ \Delta t \gtrsim 2\pi. \tag{1.73}$$

If we write $g = f(\vec{r})\ e^{i\vec{k}_0\cdot\vec{r}}$, to study the field as a function of the coordinates, we similarly obtain

$$\Delta k_x\ \Delta x \gtrsim 2\pi, \quad \Delta k_y\ \Delta y \gtrsim 2\pi, \quad \Delta k_z\ \Delta z \gtrsim 2\pi, \tag{1.74}$$

where $\Delta x, \Delta y$, and Δz are the order of magnitude of the extensions of the wave in space.

1.7 Superposition of Two Plane Monochromatic Waves

Let the two waves f_1 and f_2 be of different phases but with the same frequency and the same wave vector. Let $f_1 = a\cos\psi$, $f_2 = b\cos(\psi+\alpha)$, where

$$\psi = \omega t - \vec{k}\cdot\vec{r}, \tag{1.75}$$

then

$$g = f_1 + f_2 = c\ \cos(\psi+\gamma) = (a + b\cos\alpha)\cos\psi - b\sin\alpha\sin\psi, \tag{1.76}$$

therefore

$$c\ \cos\gamma = a + b\cos\alpha, \quad c\ \sin\gamma = b\sin\alpha, \tag{1.77}$$

and

$$c = \sqrt{a^2 + b^2 + 2ab\cos\alpha}, \tag{1.78}$$

therefore,

$$|a - b| \leqslant c \leqslant a + b. \tag{1.79}$$

The two extreme cases ($\alpha = \pi, 0$) correspond to the destructive and constructive superpositions. As a result of the destructive superposition of two waves with equal amplitudes ($\alpha = \pi,\ b = a$), the two waves disappear completely.

2

Historical Prelude

To expose a new theory in a logical order, one begins with the axioms. Then, some fundamental conclusions are inferred, and some basic theorems are proved. Consequent to this, the due definitions are given. This would be very economic if the axioms of the quantum theory were not so far disconnected from the intuition and common sense coming from the everyday experience. These axioms are implausible for a beginner. Therefore, a view on the developments that led to the quantum theory is in order.

2.1 Black Body Radiation

Max Planck, namely here, wrote the first letter—the lucky "h"—in the quantum theory. A black body is the body which absorbs completely all the radiation incident on it. The black body radiation is the radiation in equilibrium with an isothermal enclosure; that is, the radiation inside a cavity completely surrounded by opaque walls heated to a constant temperature. A small hole made in the wall of this cavity behaves like an ideal black body, since the probability for any ray incident on such a hole to reemerge is negligible, and it is extremely unlikely that this hole will reflect any rays incident on it; hence, any radiation coming from the hole is only the radiation from the cavity. The walls of the cavity always emit and absorb electromagnetic radiation. Kirchhoff in 1859 showed that when thermal equilibrium is reached, the two rates of emission and absorption are equal, and the spectral distribution of the radiation will depend only on the temperature of the cavity,[1] and will not depend neither on the volume, the shape of the cavity nor on the material which the walls of the cavity are made of.[2] The point here is to find the spectral distribution of this radiation as a function of the (constant) temperature of the body (the walls of the cavity). This distribution is obtained on the basis of results

[1]This resembles in some sense the vapor in equilibrium with the liquid in an isothermal enclosure. When the thermal equilibrium is reached, the two rates of evaporation and condensation are equal, and a constant vapor pressure which depends only on the temperature T is obtained.

[2]Otherwise, a thermal equilibrium would not be reached in a cavity, the walls of which are of different materials.

DOI: 10.1201/9781003677109-2

from different branches of classical physics. From thermodynamics, *Stefan-Boltzmann*'s law is deduced (1879). This law states that the energy density u of the radiation at any point in the cavity is proportional to T^4, where T is the temperature of the walls of the cavity. From thermodynamics and electrodynamics *Wien*[3] obtained the displacement law (1893), which states that the spectral distribution $\rho(\omega)$ of the energy density must be a function of the form

$$\rho(\omega) = \omega^3 f(\frac{\omega}{T}). \tag{2.1}$$

Since the spectral density $\rho(\omega)$ must be independent of the material of the walls and should depend only on their temperature, it is determined by choosing the simplest model of the walls, which is approximated as a set of harmonic oscillators.

From electrodynamics and mechanics, it was finally deduced that[4]

$$\rho(\omega) = \frac{\omega^2}{\pi^2 c^3}\overline{E}, \tag{2.2}$$

where $\overline{E}$ is the time average of the total energy of the oscillator.

The time average $\overline{E}$ of the energy of the oscillator is equal to its statistical average (see e.g., [7, § 1]). Using *Maxwell*'s distribution for the energy of the particles:

$$N(E) = Ae^{-E/kT} = Ae^{-\alpha E}, \qquad \alpha = \frac{1}{kT}, \tag{2.3}$$

where T is the absolute temperature of the cavity, k is the *Boltzmann*'s constant and A is a normalization constant, therefore

$$\overline{E} = \frac{\int_0^\infty Ee^{-\alpha E}dE}{\int_0^\infty e^{-\alpha E}dE} = -\frac{\partial}{\partial\alpha}\ln\int_0^\infty e^{-\alpha E}dE = \frac{d}{d\alpha}\ln\alpha = \frac{1}{\alpha} = kT. \tag{2.4}$$

Substituting this value of $\overline{E}$ in Eq. (2.2), we obtain

$$\rho(\omega) = \frac{\omega^2}{\pi^2 c^3}kT. \tag{2.5}$$

This is the *Rayleigh-Jeans*[5] distribution, which agrees with *Wien*'s displacement law and which though agrees with experiment at small enough frequencies, it completely fails for high frequencies. ρ is proportional to ω^2; this means that the higher the frequency, the more intense the corresponding component of radiation will be. i.e., if you open an oven, you will be faced with a torrent

---

[3] *W. Wien* (1864–1928) was awarded the *Nobel* prize in Physics in 1911 for his discoveries regarding the laws governing the radiation of heat.

[4] For details, including the deduction of *Stefan-Boltzmann*'s law and *Wien*'s law, see [5, 6].

[5] *L. Rayleigh* (1842–1919) was awarded the *Nobel* prize in Physics in 1904 for his investigations of the densities of the most important gases and for his discovery of argon in connection with these studies.

of γ-rays. This never happens; moreover, the total energy u of the radiation per unit volume will be infinite:

$$u = \int_0^\infty \rho(\omega)d\omega = \frac{kT}{\pi^2 c^3}\int_0^\infty \omega^2 d\omega = \infty, \qquad (2.6)$$

a result which was called the ultraviolet catastrophe by *Ehrenfest.*

Here *Max Planck*[6] (1900) observed that if the energy of the micro oscillator assumes only integral multiples of a certain indivisible quantum (atom of energy) ε

$$E = n\varepsilon, \quad n = 0, 1, 2, \ldots, \qquad (2.7)$$

then, the integral in Eq. (2.4) transforms into the sum:

$$\sum_{n=0}^{\infty} \varepsilon e^{-\alpha\varepsilon n} = \frac{\varepsilon}{1 - e^{-\alpha\varepsilon}}, \qquad (2.8)$$

hence

$$\overline{E} = -\frac{\partial}{\partial\alpha}\ln\frac{\varepsilon}{1 - e^{-\alpha\varepsilon}} = \frac{\varepsilon e^{-\alpha\varepsilon}}{1 - e^{-\alpha\varepsilon}} = \frac{\varepsilon}{e^{\alpha\varepsilon} - 1}, \qquad (2.9)$$

therefore

$$\rho(\omega) = \frac{\omega^2\varepsilon}{\pi^2 c^3}\frac{1}{e^{\frac{\varepsilon}{kT}} - 1}. \qquad (2.10)$$

To satisfy *Wien*'s law (Eq. (2.1)), *Planck* was led to assume that the energy quantum ε is proportional to ω, and it came to his mind to denote the constant of proportionality by $\hbar$:[7]

$$\varepsilon = \hbar\omega, \quad E_n = n\hbar\omega, \quad n = 0, 1, 2, \ldots, \qquad (2.11)$$

therefore:[8]

$$\rho(\omega) = \frac{\hbar\omega^3}{\pi^2 c^3}\frac{1}{e^{\frac{\hbar\omega}{kT}} - 1}. \qquad (2.12)$$

Here the miracle comes and the exponential of ω in the denominator guarantees the convergence of the integral for u, and thus the ultraviolet divergence for the radiation energy density is removed. Moreover, *Planck*'s distribution (Eq. (2.12)) is in excellent agreement with experiment. When ρ is plotted against ω, a family of curves (T being a parameter) is obtained, one curve

---

[6] *M. Planck* (1858–1947) awarded the *Nobel* prize in Physics in 1918 in recognition of the services he rendered to the advancement of physics by his discovery of energy quanta.

[7] At that time (1900) physicists used the frequency ν in units of cycle/sec while ω is in units of rad/sec, i.e., $\nu = \frac{\omega}{2\pi}$, and *Planck* in fact wrote $\varepsilon = h\nu$. Since $h\nu = \frac{h}{2\pi}\omega$, therefore $\hbar = \frac{h}{2\pi}$ and both are called *Planck*'s constant, $h = 6.626 \times 10^{-34}$ m^2 kg/s, $\hbar = 1.05 \times 10^{-34}$ m^2 kg/s. Notice that $\hbar$ has the units of action, since ω has the units of rad/time. The notation $\hbar$ is due to *Dirac.*

[8] A rigorous proof of *Planck*'s distribution can be found in [6]. There, methods of quantum statistical mechanics are applied (*Bose-Einstein* statistics).

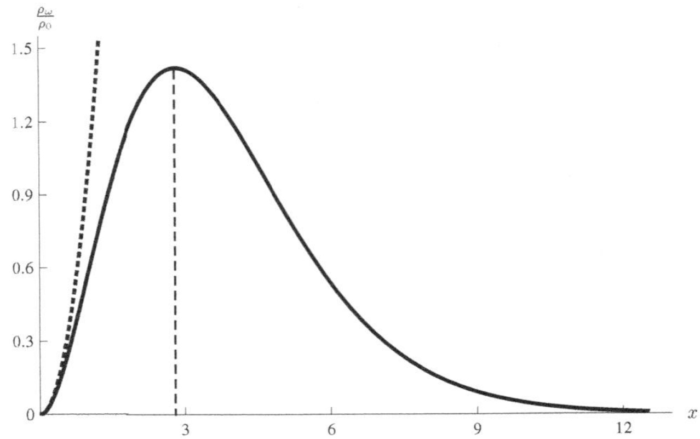

FIGURE 2.1: *Radiation spectrum of an ideal black body. The heavy dotted line indicates the Rayleigh-Jeans curve* $\rho_\omega = \rho_0 x^2$*, and the solid line the Planck* $\rho_\omega = \rho_0 x^3/(e^x - 1)$*, which is the same as the experimental curve. Here* $\rho_0 = (kT)^3/\pi^2\hbar^2 c^3$*,* $\omega = \omega_0 x$*, and* $\omega_0 = kT/\hbar$*.*

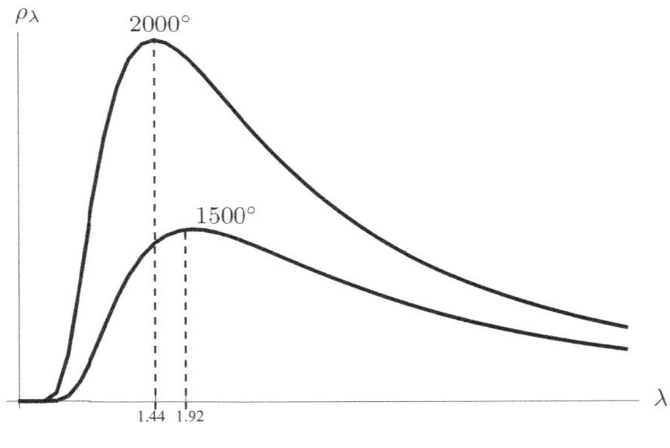

FIGURE 2.2: *Wein's displacement law curves of spectral energy distribution as a function of temperature for an ideal black body* $\lambda_m T \simeq 0.29$ *cm. deg.*

for each constant value of T, and thus the spectral distribution of the radiation can determine the temperature of its source (e.g., Figs. 2.1,2.2). *Planck*'s spectral distribution of the black body radiation from the sun corresponds to 6000 K (the temperature of the surface of the sun), the maximum intensity of this distribution lies in the green region of the spectrum. In 1964, *A. Penzias* and *R. W. Wilson*[9] made their first measurements clearly showing the pres-

[9] *A. Penzias* (1933) and *R. W. Wilson* (1936) were jointly awarded the half of *Nobel* prize in Physics in 1978 for their discovery of cosmic microwave background radiation.

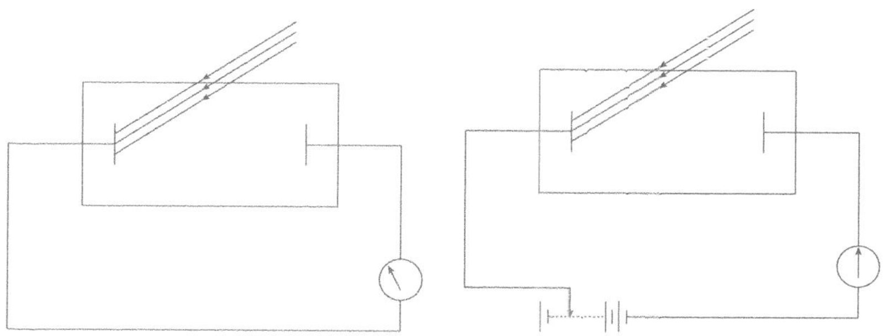

FIGURE 2.3: *Photoelectric effect.*

ence of the cosmic microwave background (CMB).[10] The CMB is the isotropic radiation left over from the time of the Big Bang; its spectrum corresponds to a radiation of a black body of 2.7 K. For this, *A. Penzias* and *R. W. Wilson* won the *Nobel* prize.

2.2 Photoelectric Effect

To explain the discreteness of the energy of an oscillator as hypothesized by *Planck*, *Einstein*[11] made the hypothesis that this energy is discrete because the electromagnetic radiation consists of discrete particles which are called photons (atoms of light), each of which carries an amount of energy $\hbar\omega$. In 1905, on this basis, he gave a theory for the photoelectric effect discovered by *H. Hertz* in 1887. When light (ultraviolet and radiation of higher frequencies) falls on a metal surface, electrons are ejected from the surface of the metal. If a counter potential difference is applied and increased gradually until it just stops the electric current in the circuit, the potential difference in this case determines the velocity of the ejected electrons in its absence (Fig. 2.3). It was found that the velocity of the ejected electrons is proportional to the frequency ω of the incident radiation and does not depend on its intensity, and the emission of photoelectrons begins only when the frequency of the radiation exceeds a certain threshold characteristic of each metal. The current intensity was found to be proportional to both the frequency and the intensity of the

[10]The CMB was first predicted in 1948 by *R. Alpher* and *R. Herman* and rediscovered by *Y. Zel'dovich* in the early 1960s and independently predicted by *R. Dicke* at the same time.

[11]*A. Einstein* (1879–1955) was awarded the *Nobel* prize in Physics in 1921 for his services to Theoretical Physics, and especially for his discovery of the law of the photoelectric effect.

radiation. *Einstein* proposed the simple equation

$$\frac{m}{2}v^2 = \hbar\omega - W, \tag{2.13}$$

where W is the work function of an electron in the metal. This is the law of conservation of energy: the light quantum (the photon) incident on the metal is absorbed by one of the electrons of its surface, giving all its energy $\hbar\omega$ to this electron, which spends an energy equal to the work W necessary to extract the electron from the metal, and the difference $\hbar\omega - W$ will be the kinetic energy of the emitted electron. The threshold of the radiation frequency to begin electron emission is clearly $\frac{W}{\hbar}$. The increase in the intensity of radiation means the increase of the number of incident photons and consequently the increase of the number of the emitted electrons which increases the current.

2.3 *Compton* Effect

An electromagnetic wave has a momentum density $\vec{d}$ (see e.g., [1, § 33])

$$\vec{d} = \frac{\vec{E}\wedge\vec{H}}{4\pi c} = \frac{E^2}{4\pi c}\vec{k}^0 = \frac{u}{c}\vec{k}^0, \tag{2.14}$$

where u is the energy density and $\vec{k}^0$ is a unit vector in the direction of propagation of the wave. Therefore, according to *Einstein*'s quantum theory of electromagnetic radiation, a photon carrying energy equal to $\hbar\omega$ must carry a momentum $\hbar\vec{k}$, where

$$\vec{k} = \frac{\omega}{c}\vec{k}^0, \tag{2.15}$$

is the wave vector. The relation between the energy $\hbar\omega$ and the momentum $\hbar\vec{k}$ is that for a relativistic particle with zero mass, which is characteristic of particles moving with the velocity of light. *Compton*[12] (1922–1923) found experimentally that the frequency of the X-rays changes by scattering by free electrons. *Compton* effect is inexplainable in the classical theory. According to which, the electron will oscillate with the frequency of the incident electromagnetic wave, hence it will radiate waves with the same frequency. Thus, the frequency of the wave is not altered by scattering. But according to *Einstein*'s quantum theory of radiation, this scattering is viewed as a collision between an incident photon with momentum $\hbar\vec{k}$ and an electron at rest. As a result of collision, the photon is scattered with momentum $\hbar\vec{k}'$ and the electron flies with momentum $\vec{p}\,'$ (Fig. 2.4). Using the laws of energy and momentum

[12]*A. Compton* (1892–1962) was awarded the half of *Nobel* prize in Physics in 1927 for his discovery of the effect named after him.

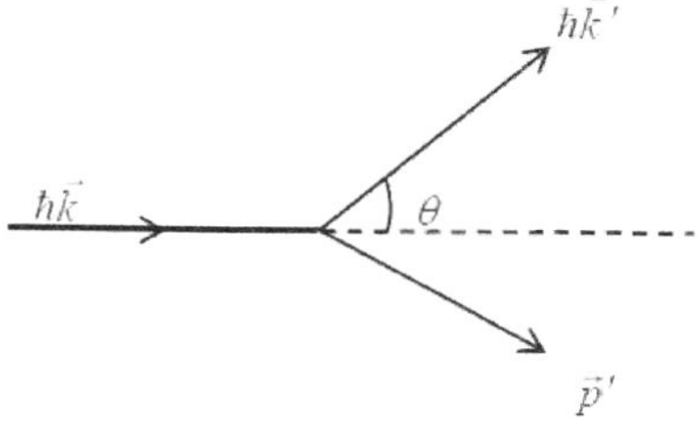

FIGURE 2.4: *Compton scattering.*

conservation *Compton*'s formula is obtained

$$\Delta\lambda = \lambda' - \lambda = 2\lambda_c \sin^2 \frac{\theta}{2}, \tag{2.16}$$

(the wavelength $\lambda = \frac{2\pi}{k} = \frac{2\pi c}{\omega}$), θ is the scattering angle and $\lambda_c = \frac{2\pi\hbar}{mc}$ is the *Compton* wavelength of the electron (m is the electron mass).

This result $\Delta\lambda = 2\lambda_c \sin^2 \frac{\theta}{2}$ was experimentally confirmed. This was particularly interesting because it confirmed not only that the photon energy is $\hbar\omega$ (as was verified by the photoelectric effect) but also that the photon momentum is $\hbar\vec{k}$.

Exercise 2.1 *Obtain Compton's formula.*

2.4 The Atom

2.4.1 *Democritus* Atom

The "Atom" first appeared in ancient Greek philosophy in connection with the question: To what extent matter is divisible? Some preferred this division process to have an ultimate term the "atom" which (by definition) is indivisible (*Democritus* in 460 B.C.), others decided that this process can continue infinitely. Through centuries (about 2500 years) there were no scientific evidences to overweight one of these two positions to the other. To adhere to one of these two views was a matter of faith, belief, or even taste rather than a scientific knowledge.

2.4.2 *Dalton*'s Atom

Recently (little more than two centuries ago) scientific facts accumulated, and the phenomenological laws in chemistry such as the laws of constant ratios, multiple ratios, valence, equivalent weights, ..., etc. began to overweight the "atom" point of view. *Dalton* (1808) explained the chemical reactions as the combination of a certain number of atoms of a chemical element with other certain numbers of atoms of other chemical elements to form a molecule of the chemical compound. The "atom" point of view was more confirmed by the Brownian motion and by the kinetic theory of gases. The study of electric discharge in rarefied gasses (gasses under low pressure of few millimeters of mercury)[13] was conclusive for the atomic theory for the structure of matter to win out, and matter was understood to consist ultimately of atoms and molecules.

2.4.3 *J. J. Thomson*'s Atom

At these early times physicists always associated the positive charge of the atom with its mass because they noticed that positively charged ions are all heavy, while among the negative ions there were the heavy and the light. The mass of the lightest negative particle, which was later called the electron, is about 2000 times less than the mass of the lightest positive ion, the positive hydrogen ion which was later called the proton.

Physicists tried to uncover the structure of the atom. *J. J. Thomson*[14] proposed a static model for it, where he distributed uniformly the positive charge (with the mass) throughout the entire volume of the atom (being a sphere), and he located the electrons as negative point charges inside this sphere. This was called the "cake with raisin" model. For a hydrogen atom there is only one electron; in the ground state, it is located at the center of the atom. At the center, the electric field of the positive charge is zero, and the electron is at rest. When the electron is displaced anyhow from the center, the atom will be in an excited state, and the electron will be acted on by a force towards the center and proportional to the distance from it. The electron will perform simple harmonic motion and thus emits electromagnetic radiation at frequencies equal to the electron frequency and its harmonics. This prediction of the model disagrees with the well-established *Balmer*'s empirical formula:

$$\omega_{nn'} = R\left(\frac{1}{n'^2} - \frac{1}{n^2}\right), \tag{2.17}$$

where R is the *Rydberg* constant, n' and n are positive integers. This formula determines the frequencies of the bright-line spectra of the hydrogen atom.

[13] The weather pressure is 760 millimeters of mercury.

[14] *J. J. Thomson* (1856–1940) was awarded the *Nobel* prize in Physics in 1906 in recognition of the great merits of his theoretical and experimental investigations on the conduction of electricity by gases.

Setting $n' = 1,\ n = 2, 3, 4, \ldots$, we obtain the *Lyman* sequence, which lies in the ultraviolet part of the spectrum. For $n' = 2,\ n = 3, 4, 5, \ldots$, we have the *Balmer* sequence, which lies in the visible part of the spectrum. The *Paschen* sequence corresponds to $n' = 3,\ n = 4, 5, 6, \ldots$, and it is in the infrared region of the spectrum.

2.4.4 *Rutherford*'s Atom

Here comes *Rutherford*[15] who was thinking that we shall not be able to uncover the structure of the atom without striking it with a projectile which penetrates through it. The available at that time projectiles were not suitable. The heavy ions were heavy enough compared with the atomic weight but slow, while projectiles fast enough to penetrate the atom, were light.

Becquerel[16] in 1896 (by chance) discovered radio activity, that is uranium spontaneously emits radiation. The physical identity of *Becquerel*'s radiation was unknown. It was useless to strike the atom (unknown structure) with this radiation (unknown identity). You will never discover an unknown with the aid of another unknown. *Rutherford* put this radiation in a magnetic field. The radiation split into three components which he called α, β, γ-rays. This reflects how he knew nothing about their identities. *Rutherford* and his collaborators put Democritus the α-ray under experimental investigation and proved that it is a beam of particles with mass equal to that of the helium atom and a positive charge which is twice the elementary charge. Now it is known that α-particles are the nuclei of helium atoms. Moreover, the emitted α particles were fast enough, and *Rutherford* obtained the desired tool to investigate the atomic structure. In 1911–1913 *Rutherford* performed his famous experiment (one of the most pioneering experiments in physics) by passing α-particles through metal foils. An α-particle is invisible by bare eyes and uncatchable by hands. There was a need in a device to count them. *Geiger*, one of the students and collaborators of *Rutherford*, did the job by inventing the counter known after his name. Thus, *Rutherford* and his co-workers could study the scattering of α-particles by an atom. They found that most of the incident α-particles are deflected through small angles (2°–3°). Beside this they also detected individual deflections of α-particles through very large angles up to 180° (from 8000 particles, one is deflected through an angle greater than 90°). In spite of this very small number, *Rutherford* and his collaborators were very astonished by these large deflections. One of his students said that, to imagine how were we astonished, imagine a cannon shooting heavy shots, and even one out of 8000 of these heavy shots is reflected back by air without meeting any target

---

[15] *E. Rutherford* (1871–1937) was awarded the *Nobel* prize in Chemistry in 1908 for his investigations into the disintegration of the elements, and the chemistry of radioactive substances.

[16] *A. Becquerel* (1852–1908) was awarded the half of *Nobel* prize in Physics 1903 in recognition of the extraordinary services he has rendered by his discovery of spontaneous radioactivity.

to reflect it. From the analysis of this experiment, *Rutherford* established that atom is fairly transparent to α-particles but it reflects very strongly a very few number of them. So he suggested that the atom is a "hollow" structure with a very strong electric field inside it which can not be produced unless the positive charge (with the mass) is concentrated in a very small volume which he called "nucleus". The radius of the atomic nucleus is of order of 10^{-13} cm. *Rutherford* proposed a dynamic model for the atom, the planetary model: A positively charged nucleus having almost the entire mass of the atom is concentrated at the center and negatively charged electrons move in closed orbits about it like planets around the sun (the sun constitutes almost the entire mass of the solar system).

Rutherford did not only qualitatively analyze his experiment to propose his model, but also he calculated theoretically the scattering of charged particles by a *Coulomb* field and obtained the *Rutherford*'s formula for the differential cross-section [8, 5] which is in excellent agreement with experiment. Nevertheless, *Blackett*[17], another student of *Rutherford*, found that the number of α-particles flying very near to the nucleus (order of 10^{-12}–10^{-13} cm apart) and which must be scattered at large angles is counted to be noticeably less than what is predicted by *Rutherford*'s formula. *Blackett* inferred from this that at such small distances there is a new fundamental force and the interaction between the α-particles and the nucleus is a very strong mutual attraction which was called later the "strong interaction." The strong interaction drops off rapidly with the distance from the nucleus with a range of order of 10^{-13} cm.

In *Rutherford* model of a hydrogen-like atom (an electron with charge $-e$ moving around a nucleus of charge Ze), the electron moves in the central potential

$$V = \frac{-Ze^2}{r}, \tag{2.18}$$

with the effective potential

$$V_{\text{eff}} = \frac{-Ze^2}{r} + \frac{L^2}{2mr^2}, \tag{2.19}$$

where L is the constant angular momentum. If a is the radius of the circular orbit then $V'_{\text{eff}}(a) = 0$ and in this case $E = (V_{\text{eff}})_{\text{min}} = V_{\text{eff}}(a)$. Therefore,

$$\frac{Ze^2}{a^2} - \frac{L^2}{ma^3} = 0, \tag{2.20}$$

that is

$$a = \frac{L^2}{mZe^2}, \tag{2.21}$$

[17] *P. Blackett* (1897–1974) was awarded the *Nobel* prize in Physics in 1948 for his development of the *Wilson* cloud chamber method, and his discoveries therewith in the fields of nuclear physics and cosmic radiation.

and

$$E = -\frac{mZ^2e^4}{L^2} + \frac{mZ^2e^4}{2L^2} = -\frac{mZ^2e^4}{2L^2}, \tag{2.22}$$

also

$$\dot{\phi} = \frac{L}{ma^2} = \frac{mZ^2e^4}{L^3} = \omega. \tag{2.23}$$

For a periodic coordinate *Ehrenfest* defines the adiabatic integral I_i as (see e.g. [8, § 49])

$$I_i = \oint p_i dq_i, \tag{2.24}$$

which according to *Ehrenfest*, is invariant under slow (adiabatic) changes of the parameters of the system (e.g., the charge), and the frequency ω_i of the periodic coordinate q_i is

$$\omega_i = 2\pi \frac{\partial E}{\partial I_i}. \tag{2.25}$$

In our case we have only one coordinate ϕ and it is periodic with period 2π, its canonically conjugate momentum is $p_\phi = L = \text{const.}$ Therefore

$$I = \oint p_\phi d\phi = \oint L d\phi = 2\pi L. \tag{2.26}$$

Notice that (Eq. (2.23)):

$$\omega = \dot{\phi} = \frac{\partial E}{\partial L} = 2\pi \frac{\partial E}{\partial I}, \tag{2.27}$$

and the atom will emit radiation at the mechanical frequency ω and at its harmonics $n\omega$

$$\omega_n = n\omega = 2\pi n \frac{\partial E}{\partial I}. \tag{2.28}$$

Here, the dynamic model arises a big conflict with the classical theory. Electrons moving in closed orbits are accelerated, hence they must, according to classical electrodynamics, radiate, thus the electron will be loosing energy. This loss of energy causes the electron to go nearer and nearer to the nucleus until it falls into it. Therefore, the dynamic model gives an unstable atom according to classical electrodynamics.

2.4.5 *Bohr*'s Atom

Bohr[18] (1913), to avoid this conflict of *Rutherford*'s dynamic model of the atom, made two postulates:

[18] *N. Bohr* (1885–1962) was awarded the *Nobel* prize in Physics in 1922 for his services in the investigation of the structure of atoms and of the radiation emanating from them.

1. The atom has stationary states characterized by the quantization of the adiabatic invariants:

$$I_i = \oint p_i dq_i = hn_i, \tag{2.29}$$

the electron in these states does not radiate although its motion is accelerated.

2. When the electron goes from a stationary state with energy E_n (the initial state) to a state with energy $E_{n'} < E_n$ (the final state), the atom radiates a quantum of energy $\hbar\omega_{nn'}$, where

$$\omega_{nn'} = \frac{E_n - E_{n'}}{\hbar}. \tag{2.30}$$

Since $\Delta I = 2\pi(n - n')\hbar$, $\Delta E = E_n - E_{n'}$, then

$$\omega_{nn'} = 2\pi(n - n')\frac{\Delta E}{\Delta I} = 2\pi k\frac{\Delta E}{\Delta I}, \tag{2.31}$$

which is the same as Eq. (2.28) with the ratio of the finite differences instead of the derivative.

Eq. (2.26) for a hydrogen-like atom with the quantization condition of the first postulate (Eq. (2.29)) gives

$$I = \oint Ld\phi = 2\pi L = hn, \tag{2.32}$$

i.e.,

$$L = \hbar n. \tag{2.33}$$

Substituting this value for L in Eqs. (2.21), (2.22) we obtain the radii and energies for an electron in the different circular orbits in a hydrogen-like *Bohr*'s atom:

$$a_n = \frac{\hbar^2}{mZe^2}n^2, \tag{2.34}$$

$$E_n = -\frac{mZ^2e^4}{2\hbar^2 n^2}. \tag{2.35}$$

The ground state (lowest energy) corresponds to $n = 1$.

In the ground state:

$$a_1 = \frac{\hbar^2}{mZe^2} = \frac{1}{Z}a_0, \tag{2.36}$$

where

$$a_0 = \frac{\hbar^2}{me^2} \simeq 0.529 \times 10^{-8}\ \text{cm} \tag{2.37}$$

is the radius of the first circular *Bohr*'s orbit in the hydrogen atom ($Z = 1$) or briefly, the *Bohr*'s radius.

With the second postulate (the frequency postulate), we find

$$\omega_{nn'} = \frac{E_n - E_{n'}}{\hbar} = \frac{m Z^2 e^4}{2\hbar^3}\left(\frac{1}{n'^2} - \frac{1}{n^2}\right). \tag{2.38}$$

For hydrogen (Z=1), we obtain *Balmer*'s formula (Eq. (2.17)) with

$$R = \frac{me^4}{2\hbar^3}, \tag{2.39}$$

this relates the *Rydberg* constant R to *Planck*'s constant $\hbar$, and gives for R a value in good agreement with its experimental value, this is one of the greatest achievements of the theory.

Bohr's theory had also a direct experimental confirmation, namely to its first postulate of discrete stationary states from the experiments of *J. Franck* and *G. Hertz*[19] (1914), who showed that when an electron beam (an electric current) passes through mercury vapor with electron energies < 4.9 eV, the collision of electrons with mercury atoms do not influence the current intensity. When the electron energy reaches 4.9 eV the current drops suddenly. With further increase of the electron energy, periodic sharp falls of the current are observed. This effect is clear in the light of *Bohr*'s theory. If the energy difference between the first excited state and the ground state of the mercury atom is 4.9 eV, the energy of an electron in the beam is not sufficient to raise an atom to an excited state if this electron energy is less than 4.9 eV, and the collision will be elastic, hence the current does not change. When the electron energy $E \geq 4.9$ eV, the electron in the beam may give a part (namely 4.9 eV) of its energy to an atom and therefore, the current will decrease. If 9.8 eV$< E <$ 14.7 eV, the transfer of energy to the atoms can occur twice; 4.9 eV is given up in a first collision and another 4.9 eV is given up in the second (see Fig. 2.5).

In spite of this success, *Bohr*'s theory has inherent defects which did not allow its further development.

2.5 Wave Properties of Matter Particles—*De Broglie* Waves

It was mentioned in 2.2 and 2.3 that light has particle and wave properties. *Louis De Broglie*[20] in (1924–1925) suggested that electrons and other matter

[19] *G. Hertz* (1887–1975) jointely with *J. Franck* (1882–1964) was awarded the *Nobel* prize in Physics in 1925 for their discovery of the laws governing the impact of an electron upon an atom.

[20] *L. De Broglie* (1892–1987) was awarded the *Nobel* prize in Physics in 1929 for his discovery of the wave nature of electrons.

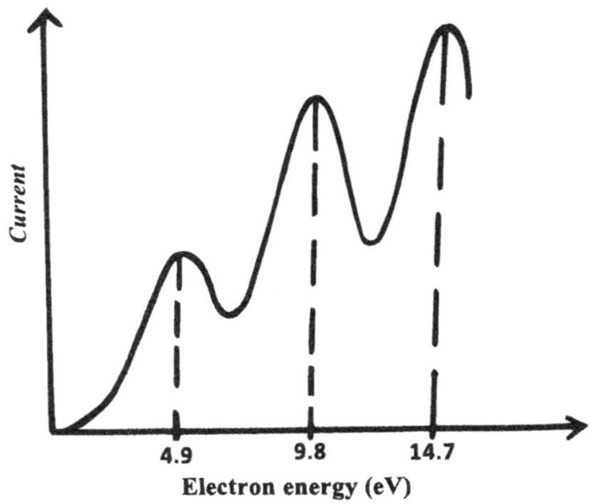

FIGURE 2.5: *The dependence of current on electron energy in the Franck-Hertz experiment.*

particles have also wave and particle properties. The energy ϵ and momentum $\vec{p}$ of the particle (the photon) associated with the light wave of frequency ω and wave vector $\vec{k}$ are

$$\epsilon = \hbar\omega, \quad \vec{p} = \hbar\vec{k}. \tag{2.40}$$

De Broglie suggested that the frequency ω and the wave vector $\vec{k}$ of the wave associated with a matter particle have the same relations to the particle energy E and momentum $\vec{p}$ as in Eq. (2.40), i.e.,

$$E = \hbar\omega, \quad \vec{p} = \hbar\vec{k}. \tag{2.41}$$

That is *De Broglie* generalized the wave-particle duality to be common to all particles as to light.

De Broglie hypothesis explains at once *Bohr*'s postulate of stationary states (see Eq. (2.29)). Indeed, the wavelength λ of the wave associated with an electron of momentum p is (Eq. (2.41))

$$\lambda = \frac{2\pi\hbar}{p}. \tag{2.42}$$

If the electron circulates with this momentum in a circle of radius a, the circumference $2\pi a$ must be an integral multiple of λ, i.e.,

$$2\pi a = n\lambda, \tag{2.43}$$

otherwise the associated wave function $e^{-i(\omega t - \vec{k}\cdot\vec{r})}$ would have different values at one and the same point on the circle (the orbit) at one and the same instant t. Since

$$p_\phi = L = ap, \tag{2.44}$$

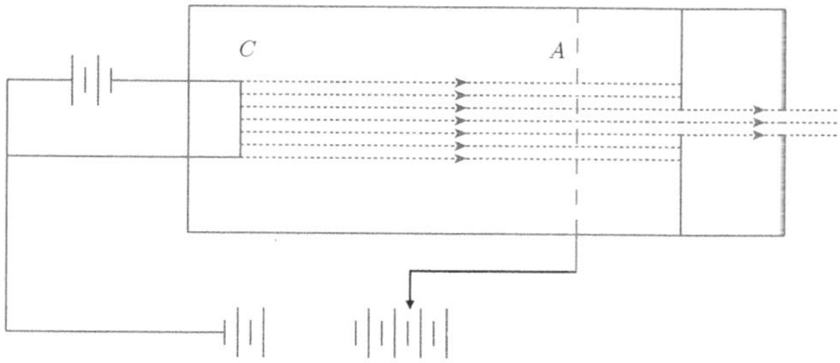

FIGURE 2.6: *An electron gun that produces a monochromatic beam of electrons. C is the cathode, and A is the anode (the secondary battery is for heating the cathode to have thermionic emission of electrons).*

therefore, from Eqs. (2.41) and (2.43) we have

$$I = \oint p_\phi d\phi = 2\pi L = 2\pi a p = n\lambda p = 2\pi\hbar n. \tag{2.45}$$

which is the *Bohr*'s condition for stationary orbits (Eq. (2.29)).

Experimental Proof of the *De Broglie* Waves

2.5.1 Electron Gun

About three years later *De Broglie* suggestion received a direct experimental confirmation. To study the wave properties of electrons and to prove that they have a specific wavelength, there is a need in a monochromatic electron beam (that is electrons moving with the same velocity). A source of such a beam is the electron gun Fig. 2.6 which emits electrons with velocity v, when $v \ll c$:

$$\frac{1}{2}mv^2 = \frac{e\Phi}{300}, \tag{2.46}$$

Φ is the accelerating potential (in volts) of the grid A relative to the cathode C. The wavelength λ of the associated with the electron wave is

$$\lambda = \frac{2\pi\hbar}{mv} = \frac{2\pi\hbar\sqrt{150}}{\sqrt{me\Phi}} = \frac{1.2\times10^{-7}}{\sqrt{\Phi}}\ \text{cm}. \tag{2.47}$$

This equation shows that the lower the potential, the longer the wavelength is and thus the more apparent the wave properties will be. But a certain energy

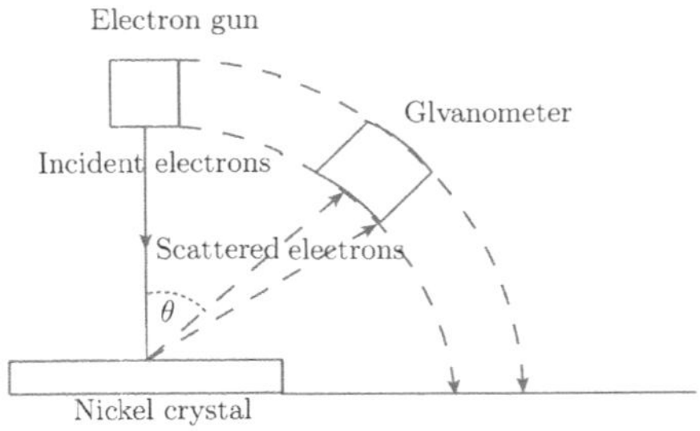

FIGURE 2.7: *Diagram of the Davisson-Germer experiment on electron diffraction.*

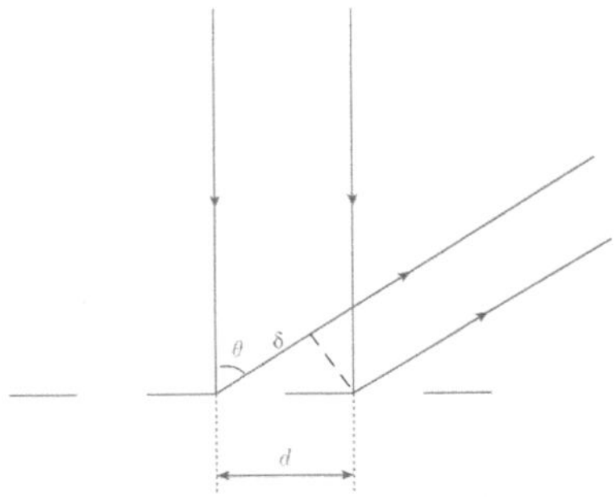

FIGURE 2.8: *Electron Scattering by a two-dimensional diffraction grating.*

(the work function) is expended in the ejection of the electron from the surface of the cathode. This work function is of order of several eV's and results in a spread in the electron velocities. The smallest potential Φ at which the beam is relatively monochromatic is 15–20 volts which gives a value for $\lambda \sim 10^{-8}$ cm which is the same as that of soft X-rays.

2.5.2 *Davisson* and *Germer* Experiment

C. Davisson[21] and *L. Germer* in 1927 were the first to observe experimentally the electron waves. Since the wavelength of the associated with the electron wave is of the same order as that of soft X-rays, they proceeded in the same way as for soft X-rays and used a crystal with a lattice constant of the order of 10^{-8} cm for the diffraction grating Fig. 2.7. The electron beam falls perpendicularly on the surface of the crystal, where the electrons are scattered at different angles by the surface lattice. The penetration of the electrons into the crystal is neglected and the position of the diffraction maxima are determined by the condition that the path difference δ is equal to an integral number of wavelengths. This is because a path difference δ corresponds to a phase difference $2\pi\dfrac{\delta}{\lambda}$ (see Eqs. (1.6-1.8)). Therefore,

$$\delta = d \sin\theta_n = n\lambda, \qquad (n \text{ is a positive integer}) \tag{2.48}$$

where d is the two-dimensional lattice constant and θ_n is the scattering angle Fig. 2.8. The diffraction maxima were detected by a galvanometer to have their angular positions as given by Eq. (2.48).

[21] *C. Davisson* (1881–1958) jointly with *G. Thomson* (1892–1975) was awarded the *Nobel* prize in Physics 1937 for their experimental discovery of the diffraction of electrons by crystals.

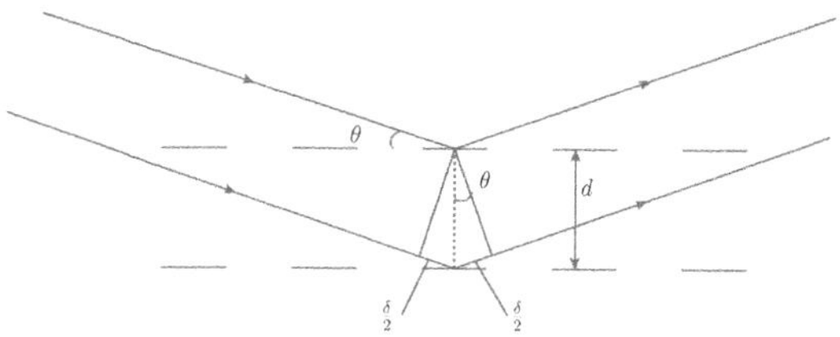

FIGURE 2.9: *Reflection from a three-dimensional crystal lattice.*

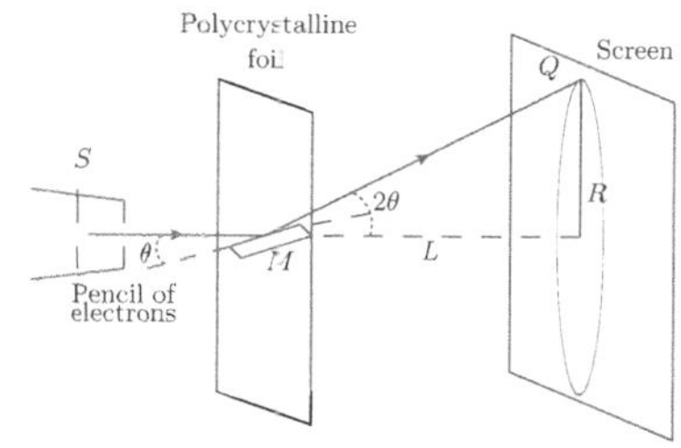

FIGURE 2.10: *Diffraction of electron waves in a polycrystalline substance.*

2.5.3 *G. Thomson* Experiment

G. Thomson[22] (1927–1928) obtained diffraction patterns of electrons (of energy of about ten thousand of eV's which corresponds to wavelength of order of 10^{-9} cm as that of hard X-rays) by passing them through metal foils. The electron beam is incident on a polycrystalline thin foil where it meets single crystals and is reflected from them. The path difference δ of two rays reflected from two neighboring layers (Fig. 2.9) is

$$\delta = 2d \sin\theta, \tag{2.49}$$

where d is the lattice constant of the three-dimensional crystal and θ is the angle between the ray and the lattice plane. The single crystals of the foil are oriented at random, and a ray can leave the foil at any angle with respect to the direction of incidence. Now when

$$2d \sin\theta = n\lambda, \quad (n \text{ is a positive integer}) \tag{2.50}$$

a diffraction maximum occurs, and a bright spot is found on the screen (Fig. 2.10). The apparatus in the experiment is axially symmetric, and the bright spots form diffraction rings with radii R_n:

$$\tan 2\theta = \frac{R}{L}, \tag{2.51}$$

where L is the distance from the foil to the screen. In these experiments θ is small, therefore

$$R_n = n\frac{L}{d}\lambda = n\frac{L}{d}\,\frac{1.2 \times 10^{-7}}{\sqrt{\Phi}}. \tag{2.52}$$

This relation is completely confirmed by the experiment.

[22] *G. Thomson* (1892–1975) jointly with *C. Davisson* (1881–1958) was awarded the *Nobel* prize in Physics 1937 for their experimental discovery of the diffraction of electrons by crystals.

The analogy between light waves and electron waves has led to the development of a new branch of physics, electron optics, the application of which was the invention of the electron microscope. Electric and magnetic lenses can be made for electron waves. The electric lens is a capacitor which has an aperture in the middle of its plate, while the magnetic lens is a magnetic coil. The resolving power of the electron microscope is much higher than that of optical ones. An optical microscope can give a magnification of about two thousand times, while an electron microscope can magnify up to two million times. Besides the electron microscope, proton microscopes are also used; their resolving power is even higher. Diffraction of neutrons is very effective for investigating the structure of matter. Neutrons are electrically neutral particles; hence, even for small energies when their *De Broglie* wavelength is relatively large, they can freely pass through matter.

2.5.4 Physical Identity of *De Broglie* Waves

Up till now, we have been speaking about the frequency and the wavelength of the *De Broglie* wave. What about its nature? *G. Thomson*'s experiment gives the answer. The nature of the obtained in this experiment diffraction pattern is independent of the intensity of the incident beam of electrons, i.e., on weakening the incident beam intensity, nothing will change, except that the contrast of the diffraction pattern will be also weakened. If the intensity of the incident beam is gradually decreased, the diffraction pattern will fade gradually, but it will still be there, until there is only one electron at a time passing through the apparatus, the diffraction pattern will then be destroyed and we shall have instead, random points on the screen. After a long enough time of exposure, the **same** diffraction rings (with radii satisfying Eq. (2.52)) will appear again, or equally well, if transparencies are exposed, each for a short time, and then compared, we shall see random points on them, but when we put them in large enough numbers on each other, we shall have again the **same** diffraction rings. We infer from this that the wave nature is an inherent property of each individual particle, and the position of each particle is a random variable with a probability distribution related, somehow, to this wave.

2.6 *Heisenberg* Uncertainty Principle

Since the inequalities (1.74) hold for any wave, they hold for the waves associated with the motion of a particle. Multiplying each side of these inequalities by $\hbar$ we obtain

$$\Delta x\, \Delta p_x \gtrsim h, \quad \Delta y\, \Delta p_y \gtrsim h, \quad \Delta z\, \Delta p_z \gtrsim h, \tag{2.53}$$

which are the essential content of the uncertainty principle introduced by *Heisenberg*[23] (1927), and which can be derived in a more precise form from the quantum theory. A fourth uncertainty relation for the energy can be deduced from inequality (1.73):

$$\Delta t \, \Delta E \gtrsim h. \tag{2.54}$$

These relations mean that it is impossible for a particle to have exactly a definite position x and a definite momentum p_x simultaneously, the more certain p_x, the more indefinite x will be. The extreme case where p_x is exactly definite ($\Delta p_x = 0$), x is absolutely indefinite ($\Delta x \to \infty$), and vice versa. The same holds for (y, p_y), (z, p_z), and (t, E), see [9, chapter 4§ 6] and [10, § 1].

One concludes from this that a quantum particle has no definite path. A classical particle moves on a definite path. Each point on a path determines exactly and simultaneously the position of the particle and the direction of its momentum (which is the direction of the tangent to the path at this point). The existence of such a path for a quantum particle contradicts the uncertainty principle. In other words, while for a classical particle, the three positions $\vec{r}(t)$, $\vec{r}(t+\Delta t)$, and $\vec{r}(t+2\Delta t)$ get collinear for small enough Δt (along the tangent to the path at $\vec{r}(t)$), these three positions of a quantum particle, although they come nearer and nearer to each other, do not get collinear for any value of Δt, however small it may be.

2.7 *Bohr*'s Correspondence Principle

Niels Bohr (1923), taking account of the experience that the classical theory had proved itself correct within certain circumstances, put a restriction on a new theory when constructed:

> *"A new theory must have the classical theory as a certain limit of it."*

[23] *W. Heisenberg* (1901–1976) was awarded the *Nobel* prize in Physics in 1932 for the creation of quantum mechanics, the application of which has, inter alia, led to the discovery of the allotropic forms of hydrogen.

3

Axiomatic Basis of Quantum Mechanics

An elementary form of the basic axioms of quantum mechanics is given. Elements of the algebra of state functions in a vector space and the linear operators in this vector space are discussed. The physical meaning of the state function is inferred and as a result of this meaning, certain boundary conditions of the state function are imposed. In addition, a theorem on the normalization constant, the time rate of change of a physical variable and the *Ehrenfest*'s theorem are given. The probability density and probability current density are defined and the equation of continuity for the probability is deduced. The uncertainty relations are proved and the conservation of energy is discussed. The momentum eigenvalues and eigenstates are determined as well as the normalization of the state function in a box and to the *Dirac* δ-function are given. The state function of a free particle is found. The passage to the limiting case of classical mechanics is performed and discussed.

3.1 Introduction

In classical mechanics, the mechanical state of a mechanical system at a time t is, by definition, the aggregate of its coordinates and momenta at this time. It is assumed that the mechanical state at any moment determines completely the mechanical state at any other moment[1] (using the equations of motion).

Any dynamical variable of the system is represented mathematically by a (real valued) function of its mechanical state and the time. Thus, once the equations of motion with the given initial conditions (the mechanical state at a certain moment) are solved, i.e., once the mechanical state of the system is found as a function of time, everything about the system is known.

In quantum Mechanics, it is impossible to determine two canonically conjugate (in the Hamiltonian sense, see § 2.6) variables simultaneously thus, one can not speak about functions of coordinates and momenta (pairs of canonically conjugate variables), hence, the approach of classical Mechanics, is inapplicable.

[1] This is called the *Laplacian* determinism.

DOI: 10.1201/9781003677109-3

3.2 An Elementary Form of the Basic Axioms of Quantum Mechanics

The basic axioms of quantum mechanics (they can be thought of as the rules of the game) are as follows:

1. The state of a physical system is described by a (in general complex valued) function of its coordinates[2] and time. This function is an element of a certain vector space and must satisfy certain conditions which will be investigated later.
2. **The superposition principle:** Any linear combination of any state functions of a physical system is also a possible state function of this system, that is, if ψ_1, ψ_2, ... are state functions of a physical system, then $\sum_a c_a\psi_a$ is a possible state of this system. Therefore, the space of the state functions must be a linear space.
3. A physical variable is represented mathematically by a linear operator in the space of the state functions.
4. A measurement of a physical variable can yield **exclusively only** one of the eigenvalues of the operator representing it mathematically. Since these measured values are always real, the eigenvalues of an operator representing a physical variable must be real and this operator must then be Hermitian.
5. The mean value of a large number of measurements on an ensemble of identical systems in the same state ψ, of a physical variable represented by the operator A is

$$\langle A\rangle = (\psi, A\psi). \tag{3.1}$$

6. The state function at any time t is determined by its initial value by the *Schrödinger*[3] equation (1926):

$$i\hbar\frac{\partial\psi}{\partial t} = H\psi, \tag{3.2}$$

where H is the Hamiltonian operator. The linearity of *Schrödinger* equation guarantees it not to contradict with the superposition principle, that is, if the time development of ψ_a is given by $\psi_a(t)$, then $\sum_a c_a\psi_a$ is also a possible time development of a state, where the c_a's are constants.

---

[2]This means that we have chosen the coordinate representation (see § 8.5.1).

[3]*E. Schrödinger* (1887–1961), jointly with *P. A. M. Dirac* (1902–1984), was awarded the *Nobel* prize in Physics in 1933 for the discovery of new productive forms of atomic theory.

Appendix

1. The state functions form a vector space. The underlying field is the field of complex numbers. The addition and multiplication by a number, are defined as:

$$\begin{aligned} (f \pm g)(\vec{r}) &= f(\vec{r}) \pm g(\vec{r}), \\ (\alpha f)(\vec{r}) &= \alpha f(\vec{r}). \end{aligned}$$

The scalar product of two elements f, g is denoted by (f, g) and is defined to be a complex number having the following properties

(a) $(f, g) = (g, f)^* \rightarrow (f, f)$ is real.
(b) $(f, f) \geq 0 \;\; \forall f$ and it is zero only if $f = 0$.
(c) Antilinear, that is:

$$\begin{aligned} (f, \alpha_1 g_1 + \alpha_2 g_2) &= \alpha_1 (f, g_1) + \alpha_2 (f, g_2), \\ (\alpha_1 f_1 + \alpha_2 f_2, g) &= \alpha_1^* (f_1, g) + \alpha_2^* (f_2, g), \end{aligned}$$

f, g are any two states and α_1, α_2 are any two complex numbers. (Notice that this is consistent with property 1a).

In the *Schrödinger* wave theory the scalar product of two states f, g is defined as

$$(f, g) = \int_D f^* g d\tau,$$

where $d\tau$ is the integration element and D is the domain of the state functions.

With this definition, the scalar product has the above-mentioned three properties, and $\langle A \rangle$ of Eq. (3.1) will be

$$\langle A \rangle = (\psi, A\psi) = \int_D \psi^* A \psi d\tau. \tag{3.3}$$

2. (a) The sum and product of two operators A and B are defined as:

$$\begin{aligned} (A \pm B) f &= Af \pm Bf, \; \forall f, \\ (AB) f &= A(Bf), \; \forall f. \end{aligned}$$

(b) The null and unit operators $\mathcal{O}$ and $\mathcal{I}$ are

$$\begin{aligned} \mathcal{O} f &= 0, \; \forall f, \\ \mathcal{I} f &= f, \; \forall f. \end{aligned}$$

(c) The inverse, if exists, of an operator A is denoted by A^{-1} and must satisfy by definition:

$$A^{-1}A = AA^{-1} = \mathcal{I}.$$

(d) The Hermitian adjoint operator of an operator A is denoted by $A^\dagger$ and is defined as:

$$(f, A^\dagger g) = (Af, g), \ \forall f, g,$$

it is clear that $(A^\dagger)^\dagger = A$, indeed,

$$(f, (A^\dagger)^\dagger g) = (A^\dagger f, g) = (g, A^\dagger f)^* = (Ag, f)^* = (f, Ag), \ \forall f, g.$$

$$(AB)^\dagger = B^\dagger A^\dagger,$$

indeed:

$$(f, (AB)^\dagger g) = (ABf, g) = (Bf, A^\dagger g) = (f, B^\dagger A^\dagger g), \forall f, g.$$

Corollary 3.1 $(ABCD\dots)^\dagger = \dots D^\dagger C^\dagger B^\dagger A^\dagger$.

(e) If $A^\dagger = A$, A is said to be Hermitian, and if $A^\dagger = A^{-1}$, A is said to be unitary, and if $A^\dagger = -A$, A is said to be antihermitian.

(f) The Hermitian adjoint of any complex number α being a multiplication operator is just its complex conjugate, indeed:

$$(f, \alpha^* g) = \alpha^*(f, g) = (\alpha f, g), \ \forall f, g,$$

therefore, all real numbers are Hermitian (they are also self conjugate).

(g) The expectation values of Hermitian operators are real, and therefore, their eigenvalues also are real. Indeed:

$$\langle A\rangle^* = (\psi, A\psi)^* = (A\psi, \psi) = (\psi, A\psi) = \langle A\rangle.$$

(h) The eigenfunctions corresponding to different eigenvalues of a Hermitian operator are orthogonal:
Let A be a Hermitian operator and u_a, u_b are two of its eigenfunctions corresponding to its eigenvalues a, b, that is:

$$Au_a = au_a, \quad Au_b = bu_b.$$

Therefore

$$(u_a, Au_b) = b(u_a, u_b) = (Au_a, u_b) = a(u_a, u_b),$$

hence

$$(b - a)(u_a, u_b) = 0,$$

therefore

$$(u_a, u_b) = 0, \ \forall a \neq b.$$

3. (a) The energy operator $\hat{E}$ is

$$\hat{E} = i\hbar\frac{\partial}{\partial t}.$$

(b) In the coordinate representation:

i. The coordinates are multiplication operators.

ii. The momentum canonically conjugate to a Cartesian coordinate is represented by the partial differential operator with respect to this Cartesian coordinate times $\frac{\hbar}{i}$, e.g.,

$$\hat{p}_x = \frac{\hbar}{i}\frac{\partial}{\partial x},$$

whenever p_x is the canonical conjugate of x, and

$$\hat{f}(\vec{r}_j, \vec{p}_j, t) = f(\vec{r}_j, \frac{\hbar}{i}\frac{\partial}{\partial \vec{r}_j}, t),$$

whenever the components of $\vec{p}_j$ are the canonical conjugate momenta of the Cartesian coordinates x_j, y_j, z_j.

This choice of the operators representing $E, \vec{p}$ is closely connected with the following: for a plane monochromatic wave $f = e^{-\frac{i}{\hbar}(Et - \vec{p}\cdot\vec{r})}$ (see § 1.3):

$$i\hbar\frac{\partial f}{\partial t} = Ef, \tag{3.4}$$

$$\frac{\hbar}{i}\vec{\nabla} f = \vec{p}f, \quad \text{(see § 8.5)} \tag{3.5}$$

4. **Theorem 3.2** $\vec{\nabla}$ *is antihermitian when it operates on functions in a domain D, if either:*

(a) D is unbounded and the functions go fast enough to zero at great distances, or

(b) D is bounded and either:

i. The functions vanish on the surface S bounding D, or

ii. The functions satisfy periodicity conditions on S.

Proof We have

$$\begin{aligned}(f, \vec{\nabla} g) = \int_D f^* \vec{\nabla} g d^3 r &= -\int_D (\vec{\nabla} f^*) g d^3 r + \int_D \vec{\nabla}(f^* g) d^3 r \\ &= \int_D (-\vec{\nabla} f)^* g d^3 r + \oint f^* g d\vec{S}.\end{aligned}$$

Under either one of the conditions (4a), (4(b)i), (4(b)ii), the surface integral is zero, and

$$(f, \vec{\nabla} g) = (-\vec{\nabla} f, g),$$

that is

$$\vec{\nabla}^\dagger = -\vec{\nabla}. \quad (\vec{\nabla} \text{ is antihermitian}).$$

Corollary 3.3

(a) The momentum operator $\hat{\vec{p}} = \frac{\hbar}{i}\vec{\nabla}$ *is Hermitian.*

(b) The Laplacian operator ∇^2 *is Hermitian:*

$$(\nabla^2)^\dagger = (\vec{\nabla}^\dagger)^2 = (-\vec{\nabla})^2 = \nabla^2.$$

(c) The kinetic energy operator $T = \frac{p^2}{2m} = -\frac{\hbar^2}{2m}\nabla^2$ *is Hermitian.*

Example 3.1 *Prove, by integration that* ∇^2 *is Hermitian:*

$$f^*\nabla^2 g - (\nabla^2 f)^* g = f^*\nabla^2 g - (\nabla^2 f^*) g = \text{div}\{f^*\vec{\nabla} g - (\vec{\nabla} f^*) g\}, \tag{3.6}$$

therefore,

$$(f, \nabla^2 g) = (\nabla^2 f, g) + \oint \vec{B} \cdot d\vec{S}, \tag{3.7}$$

where $\vec{B} = f^*\vec{\nabla} g - (\vec{\nabla} f^*) g$. The above surface integral is zero under either one of the conditions of the above theorem, and

$$(f, \nabla^2 g) = (\nabla^2 f, g). \tag{3.8}$$

Exercise 3.1

1. Find the Hamiltonian operator for a particle with potential energy $U(\vec{r})$*, hence write down Schrödinger equation for the state* ψ *of this particle.*

2. As in 1, if a magnetic field with vector potential $\vec{A}(\vec{r})$ *is added, the particle has a charge e. (The Lagrange function in this case is* $L = \frac{1}{2}mv^2 + \frac{e}{c}\vec{A} \cdot \vec{v} - U$*.)*

3. For the case of the particle in 1, obtain Schrödinger equation in spherical coordinates.

4. Find the fundamental commutators: $[p_\alpha, x_\beta]$, $[p_\alpha, p_\beta]$, $[x_\alpha, x_\beta]$.

5. Prove that $[\vec{p}, f(\vec{r})] = \frac{\hbar}{i}\vec{\nabla} f$.

6. Knowing that $\vec{r}$ and $\vec{p}$ are Hermitian, prove that $\vec{r}\wedge\vec{p}$ is Hermitian, while $\vec{r}\cdot\vec{p}$ is not.

7. Prove that:

(a) $[A, B] = -[B, A]$
(b) $[A + B, C] = [A, C] + [B, C]$
(c) $[A, B + C] = [A, B] + [A, C]$
(d) $[AB, C] = A[B, C] + [A, C]B$
(e) $[A, BC] = [A, B]C + B[A, C]$

8. Prove that:

(a) $[L_x, L_y] = i\hbar L_z$ *and by cyclic permutations find the two other commutators.*
(b) $[L^2, L_\alpha] = 0,\ \alpha = x, y, z.$
(c) $[L_z, L_\pm] = \pm\hbar L_\pm$, *where* $L_\pm = L_x \pm iL_y$.

9. Find $L^2, L_\alpha, L_\pm$ in the spherical coordinates.

3.3 The Physical Interpretation of the State Function—*Born*'s Interpretation

1. From axiom 5:

$$\langle\vec{r}\rangle = \int \psi^* \vec{r}\psi d^3r = \int \vec{r}|\psi(\vec{r}, t)|^2 d^3r. \tag{3.9}$$

Thus, the direct meaning of the state function ψ is that $|\psi(\vec{r}, t)|^2$ is the probability density for finding the particle at the position $\vec{r}$. $\psi(\vec{r}, t)$ itself is called the probability amplitude.[4] This imposes on the state function the following conditions:

(a) Single valuedness (the state function, as a function of spatial space points, must have only one value at each point).[5]
(b) Continuity.

[4]This probabilistic interpretation of the state function is due to *Max Born* (1926). *M. Born* (1882–1970) was awarded the half of *Nobel* prize in Physics in 1954 for his fundamental research in quantum mechanics, especially for his statistical interpretation of the wavefunction.

[5]One may think that this condition must be put on $|\psi|^2$ and not necessarily on ψ itself, but since the superposition principle holds here, that is any superposition (any linear combination) of any state functions ψ_α is also a state function, the condition must be put on ψ_α themselves to guarantee that $|\sum_\alpha c_\alpha\psi_\alpha|$ is single valued.

(c) To have continuous first partial derivatives. This is for the probability current density to be continuous (see below).

To be normalizable, that is

(d) To be finite everywhere (if not square integrable).

(e) The integral of $\psi^*\psi$ over its domain in coordinate space must be time-independent. But this is guaranteed by the *Schrödinger* equation (see § 3.4).

2. **The A-representation.** The operator A representing the physical variable A is Hermitian. The set of its eigenfunctions ϕ_a is therefore a complete set, and can be taken to be an orthonormal set. Any state function ψ can be represented as a linear combination of these eigenfunctions:

$$\psi = \sum c_a \phi_a. \tag{3.10}$$

To determine the coefficient c_a, we write

$$(\phi_a, \psi) = (\phi_a, \sum c_b \phi_b) = \sum_b c_b(\phi_a, \phi_b) = \sum_b c_b \delta_{ab} = c_a. \tag{3.11}$$

$$\begin{aligned} \langle A \rangle &= (\psi, A\psi) = (\psi, A \sum c_a \phi_a) = (\psi, \sum a c_a \phi_a) \\ &= \sum a c_a (\psi, \phi_a) = \sum a c_a c_a^* = \sum |c_a|^2 a. \end{aligned} \tag{3.12}$$

Thus, $|c_a|^2$ is the probability for the variable A to have the value a when the physical system is in the state ψ, where c_a is the coefficient of ϕ_a in the expansion of ψ. c_a itself is called the probability amplitude. This means that, if the system is (definitely) in the state ψ, it is partly in each state ϕ_a of a complete orthonormal set $\{\phi_a\}$, with probability amplitude $c_a = (\phi_a, \psi)$, that is with probability $|c_a|^2 = |(\phi_a, \psi)|^2$.

If ψ is one of the eigenfunctions of A corresponding to one of its eigenvalues a, say, i.e., if

$$\psi = \phi_a, \tag{3.13}$$

then

$$c_b = \delta_{ba}, \quad |c_b|^2 = \delta_{ba},$$

i.e., **any** measurement of A on the system in the state ϕ_a will yield **certainly** the value a, and

$$\langle A \rangle = (\psi, A\psi) = (\phi_a, A\phi_a) = a(\phi_a, \phi_a) = a. \tag{3.14}$$

In this case, the standard deviation (the uncertainty) ΔA of A is zero. Indeed:

$$\begin{aligned} (\Delta A)^2 &= \langle (A - \langle A \rangle)^2 \rangle = \langle (A^2 - 2\langle A \rangle A + \langle A \rangle^2) \rangle = \langle (A^2 - \langle A \rangle^2) \rangle \\ &= \langle A^2 \rangle - \langle A \rangle^2, \end{aligned}$$

$\langle A^2 \rangle = (\phi_a, A^2\phi_a) = a^2$, and $\Delta A = 0$.

In the above formula for $\langle A \rangle$, ψ is assumed to be normalized. In this case

$$1 = (\psi, \psi) = (\sum c_a\phi_a, \psi) = \sum c_a^*(\phi_a, \psi) = \sum_a |c_a|^2. \tag{3.15}$$

Thus, it is clear that, once the state function $\psi(\vec{r}, t_0)$ of a particle is given, not only the probability distribution of the coordinates of the particle $|\psi(\vec{r}, t_0)|^2$ is determined, but also the probability distribution $|c_a|^2$ of any of its dynamical variables at t_0 is determined. The time development of the probability amplitudes is given by *Schrödinger* equation. That is, while the dynamical variables of a classical particle are functions of time and the mechanical state $(\vec{r}, \vec{p})$ of the particle, the dynamical variables of a quantum particle are random variables taking their values at any time according to certain probability distributions determined by the state function of the quantum particle at this time.

3. **The closure property**. A state function ψ is expanded in terms of a complete orthonormal set ϕ_a:

$$\psi(\vec{r}) = \sum_a c_a\phi_a(\vec{r}), \quad c_a = \int \phi_a^*(\vec{r})\psi(\vec{r})d^3r, \tag{3.16}$$

therefore

$$\begin{aligned}\psi(\vec{r}) &= \sum_a \int \psi(\vec{r}\,')\phi_a^*(\vec{r}\,')\phi_a(\vec{r})d^3r' = \int \psi(\vec{r}\,') \sum_a \phi_a^*(\vec{r}\,')\phi_a(\vec{r})d^3r' \\ &= \int \psi(\vec{r}\,')\delta(\vec{r}\,' - \vec{r})d^3r', \ \forall\psi \end{aligned} \tag{3.17}$$

therefore $\sum_a \phi_a^*(\vec{r}\,')\phi_a(\vec{r})$ must equal $\delta(\vec{r}\,' - \vec{r})$, i.e.,

$$\sum_a \phi_a^*(\vec{r}\,')\phi_a(\vec{r}) = \delta(\vec{r}\,' - \vec{r}), \tag{3.18}$$

which is an expression of the condition of completeness of ϕ_a, and it is called the closure property.

3.4 The Normalization Integral Is Time-Independent

The normalization integral is

$$\int \psi^*(t, \vec{r})\psi(t, \vec{r})d^3r = (\psi, \psi), \tag{3.19}$$

$$\frac{d}{dt}(\psi, \psi) = (\psi, \frac{\partial \psi}{\partial t}) + (\frac{\partial \psi}{\partial t}, \psi), \tag{3.20}$$

substituting for the partial derivatives from *Schrödinger* equation (3.2), we have

$$\frac{d}{dt}(\psi, \psi) = (\psi, \frac{1}{i\hbar}H\psi) + (\frac{1}{i\hbar}H\psi, \psi) = \frac{1}{i\hbar}\{(\psi, H\psi) - (H\psi, \psi)\} = 0, \tag{3.21}$$

(since the Hamiltonian operator represents a dynamical variable, it must then be Hermitian), i.e., the normalization integral $\int \psi^* \psi d^3r$ is constant (time-independent).

3.5 The Time Rate of Change of a Physical Variable

Denote by A the operator representing this physical variable, and by ψ the function describing the state of the system:

$$\frac{d}{dt}\langle A \rangle = \frac{d}{dt}(\psi, A\psi) = (\psi, \frac{\partial A}{\partial t}\psi) + (\frac{\partial \psi}{\partial t}, A\psi) + (\psi, A\frac{\partial \psi}{\partial t}). \tag{3.22}$$

Substituting $\frac{1}{i\hbar}H\psi$ for $\frac{\partial \psi}{\partial t}$ (from *Schrödinger* equation), we have

$$\frac{d}{dt}\langle A \rangle = \langle \frac{\partial A}{\partial t} \rangle + (\psi, \frac{1}{i\hbar}AH\psi) + (\frac{1}{i\hbar}H\psi, A\psi), \tag{3.23}$$

since H represents a dynamical variable, it must then be Hermitian:

$$\frac{d}{dt}\langle A \rangle = \langle \frac{\partial A}{\partial t} \rangle + (\psi, \frac{1}{i\hbar}[A, H]\psi), \tag{3.24}$$

i.e.,

$$\frac{d}{dt}\langle A \rangle = \langle \frac{\partial A}{\partial t} + \frac{1}{i\hbar}[A, H] \rangle. \tag{3.25}$$

That is, the time rate of change of the expectation value of a physical variable is the sum of the expectation value of the partial time derivative of the operator representing this variable and the expectation value of the commutator of this operator with the Hamiltonian divided by $i\hbar$. Therefore, if the operator representing a physical variable commutes with the Hamiltonian of the system and does not depend explicitly on time, the expectation value of this variable will be constant in time and the physical variable in this case is said to be a constant of motion.

3.6 *Ehrenfest*'s Theorem

This theorem gives the time rate of change of the expectation values $\langle\vec{r}\rangle$ and $\langle\vec{p}\rangle$ of the position and the momentum operators. Applying Eq. (3.25), and since $\vec{r}$ does not depend explicitly on time, we have

$$\frac{d}{dt}\langle\vec{r}\rangle = (\psi, \frac{1}{i\hbar}[\vec{r}, H]\psi). \tag{3.26}$$

Let

$$H = \frac{p^2}{2m} + U(\vec{r}). \tag{3.27}$$

Therefore

$$[\vec{r}, H] = \frac{1}{2m}[\vec{r}, p^2], \tag{3.28}$$

$$[x, p^2] = [x, p_x^2] = [x, p_x]p_x + p_x[x, p_x] = 2i\hbar p_x, \tag{3.29}$$

therefore

$$[\vec{r}, H] = \frac{i\hbar}{m}\vec{p}, \tag{3.30}$$

$$\frac{d}{dt}\langle\vec{r}\rangle = \frac{1}{m}(\psi, \vec{p}\psi) = \frac{1}{m}\langle\vec{p}\rangle, \tag{3.31}$$

and:

$$m\frac{d^2}{dt^2}\langle\vec{r}\rangle = \frac{d}{dt}\langle\vec{p}\rangle = \frac{1}{i\hbar}\langle[\vec{p}, H]\rangle. \tag{3.32}$$

$$[\vec{p}, H] = [\vec{p}, U(\vec{r})] = \frac{\hbar}{i}\vec{\nabla}U, \tag{3.33}$$

therefore

$$m\frac{d^2}{dt^2}\langle\vec{r}\rangle = \frac{d}{dt}\langle\vec{p}\rangle = -\langle\vec{\nabla}U\rangle = \langle\vec{F}(\vec{r})\rangle. \tag{3.34}$$

Ehrenfest's theorem is an example of the correspondence principle. Here the expectation values replace the values of the variables in the analogous classical equations.

3.7 Probability Density and Probability Current Density

Since

$$\frac{\partial}{\partial t}\psi^*\psi = \psi^*\frac{\partial\psi}{\partial t} + \frac{\partial\psi^*}{\partial t}\psi = \frac{1}{i\hbar}\{\psi^* H\psi - (H\psi)^*\psi\}. \tag{3.35}$$

Let

$$H = \frac{p^2}{2m} + U(\vec{r}), \tag{3.36}$$

therefore

$$\begin{aligned}\frac{\partial}{\partial t}\psi^*\psi &= -\frac{\hbar}{2mi}\{\psi^*\nabla^2\psi - (\nabla^2\psi^*)\psi\} \\ &= -\frac{\hbar}{2mi}\vec{\nabla}\cdot\{\psi^*\vec{\nabla}\psi - (\vec{\nabla}\psi^*)\psi\} = -\text{div}\vec{j}, \qquad (3.37)\end{aligned}$$

where

$$\vec{j} = \frac{\hbar}{2mi}\{\psi^*\vec{\nabla}\psi - (\vec{\nabla}\psi^*)\psi\}. \qquad (3.38)$$

If the probability density $\psi^*\psi$ is denoted by ρ, then

$$\frac{\partial\rho}{\partial t} + \text{div}\vec{j} = 0, \qquad (3.39)$$

which is the equation of continuity for the probability, $\vec{j}$ is the probability current density.

Exercise 3.2

1. *Show that the integral of the probability current density over a closed surface is the probability flux through this surface.*

Notice: *The probability flux through a closed surface is the time rate of decreasing of the total probability to find the particle anywhere in the domain bounded by this surface.*

2. *Prove that*

$$\int \vec{j}(\vec{r},t)d^3r = \frac{\langle\vec{p}\rangle}{m}. \qquad (3.40)$$

3.8 The Uncertainty Relations

Let the system be in a (normalized) state u and A, B be two Hermitian operators:

$$\langle A^2\rangle = (u, A^2u) = (Au, Au), \text{ also } \langle B^2\rangle = (Bu, Bu).$$

Applying *Schwartz* inequality:

$$(Au, Au)(Bu, Bu) \geq |(Au, Bu)|^2.$$

Now

$$(Au, Bu) = (u, ABu) = \frac{1}{2}(u, (F + iC)u),$$

where

$F = AB + BA, \quad iC = [A, B]$. (Notice that $AB + BA + [A, B] = 2AB$.)

Exercise 3.3 *Prove that F, C are Hermitian.*

Therefore

$$(Au, Bu) = \frac{1}{2}\left(\langle F\rangle + i\langle C\rangle\right),$$

$\langle F\rangle, \langle C\rangle$ are real since F, C are Hermitian, therefore

$$|(Au, Bu)|^2 = \frac{1}{4}\left(\langle F\rangle^2 + \langle C\rangle^2\right) \geq \frac{1}{4}\langle C\rangle^2.$$

Therefore

$$\langle A^2\rangle\langle B^2\rangle \geq \frac{1}{4}\langle C\rangle^2. \tag{3.41}$$

The uncertainty (the standard deviation) ΔA of A is $\Delta A = \sqrt{\langle (A - \langle A\rangle)^2\rangle}$. Replace A, B in Eq. (3.41) by $A - \langle A\rangle, B - \langle B\rangle$ which are also Hermitian since A, B are so and $\langle A\rangle, \langle B\rangle$ are real, then:

$$\langle (A - \langle A\rangle)^2\rangle\langle (B - \langle B\rangle)^2\rangle \geq \frac{1}{4}\langle C\rangle^2, \quad \text{i.e.} \tag{3.42}$$

$$\Delta A \Delta B \geq \frac{1}{2}|\langle C\rangle|. \tag{3.43}$$

Example 3.2 *1. $[x, p_x] = i\hbar$, therefore*

$$\Delta x \Delta p_x \geq \frac{\hbar}{2}. \tag{3.44}$$

2. $[L_x, L_y] = i\hbar L_z$, therefore

$$\Delta L_x \Delta L_y \geq \frac{\hbar}{2}|\langle L_z\rangle|. \tag{3.45}$$

3.9 Separation of Time in *Schrödinger* Equation

Schrödinger equation is

$$i\hbar\frac{\partial\psi}{\partial t} = H(t, \vec{r}, \frac{\hbar}{i}\vec{\nabla})\psi. \tag{3.46}$$

If H does not depend explicitly on time,

$$i\hbar\frac{\partial\psi}{\partial t} = H(\vec{r}, \frac{\hbar}{i}\vec{\nabla})\psi. \tag{3.47}$$

Consider a particular solution in the form of a product

$$\psi(\vec{r}, t) = u(\vec{r})f(t), \tag{3.48}$$

therefore

$$\frac{i\hbar}{f}\frac{df}{dt} = \frac{1}{u}H(\vec{r}, \frac{\hbar}{i}\vec{\nabla})u = E = \text{const.} \tag{3.49}$$

(since the L.H.S. depends on t only and the other side depends on $\vec{r}$ only ($t, \vec{r}$ here are independent variables)),

$$\frac{i\hbar}{f}\frac{df}{dt} = E, \tag{3.50}$$

gives

$$f(t) = Ce^{-i\frac{E}{\hbar}t}. \tag{3.51}$$

For u, we have[6]

$$Hu_E(\vec{r}) = Eu_E(\vec{r}), \tag{3.52}$$

i.e., u_E is an eigenfunction of H corresponding to the eigenvalue E. It is also called the total energy (H being the Hamiltonian operator) eigenfunction corresponding to the energy eigenvalue E (see below). For ψ, we have

$$\psi(\vec{r}, t) = e^{-i\frac{E}{\hbar}t}u_E(\vec{r}). \tag{3.53}$$

C is taken to be 1, since u_E is determined from the eigenvalue equation up to a multiplication factor. It is clear that ψ is normalized if u is normalized.

3.10 Significance of the Separation Constant— Conservation of Energy

If the system is in the state ψ of Eq. (3.53), the state is an energy eigenfunction, since

$$\hat{E}\psi = i\hbar\frac{\partial}{\partial t}\psi = E\psi, \tag{3.54}$$

$$H\psi = E\psi, \tag{3.55}$$

hence any measurement of the energy will yield certainly the value E which is constant and the variance

$$(\Delta E)^2 = \langle(\hat{E} - \langle\hat{E}\rangle)^2\rangle = \langle\hat{E}^2\rangle - \langle\hat{E}\rangle^2 = 0. \tag{3.56}$$

In such a case we say that the system is in a pure energy state.

$$|\psi|^2 = |u(\vec{r})|^2, \tag{3.57}$$

[6] This equation is called the time-independent *Schrödinger* equation, it is also called the energy eigenvalue equation.

that is $|\psi|^2$ is time-independent and the state is said to be stationary. The general solution of *Schrödinger* equation is

$$\psi(\vec{r},t) = \sum_{E,\lambda} C_{E\lambda} e^{-i\frac{E}{\hbar}t} u_{E\lambda}(\vec{r}), \tag{3.58}$$

$C_{E\lambda}$ are constants, where $Hu_{E\lambda} = Eu_{E\lambda}$, and $\sum_\lambda |C_{E\lambda}|^2$ is the probability for the energy to be equal to E in this state ψ ($u_{E\lambda}$ are assumed to be orthonormal). Also

$$(\psi(\vec{r},t),\psi(\vec{r},t)) = \sum_{E',E,\lambda,\lambda'} C^*_{E'\lambda'} C_{E\lambda} e^{\frac{i}{\hbar}(E'-E)t} (u_{E'\lambda'}, u_{E\lambda}) = \sum_{E,\lambda} |C_{E\lambda}|^2, \tag{3.59}$$

and the normalization condition for ψ will read

$$\sum_{E,\lambda} |C_{E\lambda}|^2 = 1. \tag{3.60}$$

3.11 The Momentum Eigenvalues and Eigenfunctions

We have

$$-i\hbar\vec{\nabla}\phi_{\vec{p}} = \vec{p}\phi_{\vec{p}}, \tag{3.61}$$

or

$$-i\hbar\frac{\partial\phi_{\vec{p}}}{\partial x_\alpha} = p_\alpha\phi_{\vec{p}}, \quad \alpha = x,y,z. \tag{3.62}$$

A separation of the variables can be achieved if we search for a solution in the form $\phi_{\vec{p}} = f_1(x)f_2(y)f_3(z)$. Thus

$$\frac{1}{f_\alpha}\frac{df_\alpha}{dx_\alpha} = \frac{i}{\hbar}p_\alpha, \tag{3.63}$$

$$f_\alpha = Ce^{i\frac{p_\alpha}{\hbar}x_\alpha}, \tag{3.64}$$

and

$$\phi_{\vec{p}} = Ae^{\frac{i}{\hbar}\vec{p}\cdot\vec{r}}. \tag{3.65}$$

1. The variables x, y, z range from $-\infty$ to ∞ without returning to any previous position, hence, there can be no question regarding single valuedness.
2. $\phi_{\vec{p}}$ for any A is not square integrable, so it is necessary to impose the milder condition of being finite, and this is satisfied for all real values of $\vec{p}$. That is all real values of $\vec{p}$ are possible eigenvalues of $\hat{\vec{p}}$ and they form a continuous spectrum, and we write $\phi(\vec{p})$ instead of $\phi_{\vec{p}}$, that is

$$\phi(\vec{p}) = Ae^{\frac{i}{\hbar}\vec{p}\cdot\vec{r}}. \tag{3.66}$$

3.12 Normalization of the State Function

1. *Schrödinger* equation and the eigenvalue equations of linear operators are linear and homogeneous; therefore, a solution of any such equation multiplied by a constant will still be a solution of this equation. If the normalization integral converges and is not equal to 1 (from its definition, it is non-negative), denote the unnormalized solution by ϕ and write

$$\int \phi^* \phi \, d^3r = N, \quad \psi = \frac{1}{\sqrt{N}}\phi, \tag{3.67}$$

then ψ is the normalized solution.

2. **Normalization in a box (*Born* normalization).** An example of the nonexistence of the normalization integral is that for the momentum eigenfunctions. Consider a free particle in one dimension (x, say). Its momentum eigenfunction is

$$\phi(p) = Ae^{ikx}, \quad p = \hbar k. \tag{3.68}$$

Born periodicity length L is introduced and an artificial periodicity condition is imposed:

$$\phi(x + L) = \phi(x). \tag{3.69}$$

Therefore,

$$e^{ikL} = 1 \to k_n = \frac{2\pi}{L}n, \quad p_n = \frac{2\pi\hbar}{L}n, \quad n = 0, \pm 1, \pm 2, \ldots, \tag{3.70}$$

$$\Delta p_n = \frac{2\pi\hbar}{L} \to 0 \text{ as } L \to \infty, \tag{3.71}$$

(thus the artificial periodicity condition is removed, and the spectrum of $\vec{p}$ is again continuous), the particle is supposed to be in the interval $[-\frac{L}{2}, \frac{L}{2}]$, therefore

$$\int_{-L/2}^{L/2} \phi^* \phi dx = A^2 L \to A = \frac{1}{\sqrt{L}} \tag{3.72}$$

$$\phi_n = \frac{1}{\sqrt{L}} e^{i\frac{2\pi}{L}nx}, \tag{3.73}$$

and

$$(\phi_{n'}, \phi_n) = \frac{1}{L}\int_{-L/2}^{L/2} e^{i\frac{2\pi}{L}(n-n')x} dx = \frac{\sin(n-n')\pi}{(n-n')\pi} = \delta_{nn'}. \tag{3.74}$$

This can be easily generalized to three dimensions to obtain

$$\phi(\vec{p},\vec{r}) = \frac{1}{L^{3/2}}e^{i\vec{k}\cdot\vec{r}}, \quad \vec{p} = \hbar\vec{k}, \tag{3.75}$$

$$k_\alpha = \frac{2\pi}{L}n_\alpha, \quad \alpha = x, y, z, \quad n_\alpha = 0, \pm 1, \pm 2, \dots, \tag{3.76}$$

$$\int \phi^*_{n'_x n'_y n'_z}\phi_{n_x n_y n_z}dxdydz = \delta_{n'_x n_x}\delta_{n'_y n_y}\delta_{n'_z n_z}. \tag{3.77}$$

The domain of integration is the normalization cube centered at the origin.

At the end of the calculations, the limit is taken as $L \to \infty$.

Example 3.3 *The number of different states in the element of volume d^3k in the wave vector-space is*

$$\left(\frac{L}{2\pi}\right)^3 d^3k : \tag{3.78}$$

Indeed, since $(\Delta k_\alpha)_{\min} = \frac{2\pi}{L}$ $\forall\alpha = x, y, z$, the sequence of points $\frac{2n+1}{L}\pi$ $(n = 0, \pm 1, \pm 2, \dots)$ partitions the axis k_α into intervals each of length $\frac{2\pi}{L}$. In each of these intervals there is only one state with a given value of k_α thus the wave vector-space is divided into cubic cells each of volume $(\frac{2\pi}{L})^3$. In each one of these cells there is only one state with given (k_x, k_y, k_z). Therefore, the number of different states in the element of volume Δ^3k is $\Delta k_x\Delta k_y\Delta k_z/(\frac{2\pi}{L})^3$.

3. **Normalization to δ-function (*Dirac* normalization).** For the momentum eigenfunctions of free particles in one dimension (x), *Dirac* put the normalization condition to be:

$$(\phi(p'), \phi(p)) = \delta(p - p'), \tag{3.79}$$

as a generalization of the *Kronicker* $\delta_{n'n}$ in Eq. (3.74) for discrete spectra, hence,

$$\begin{aligned}\delta(p'-p) = (\phi(p'), \phi(p)) &= \int_{-\infty}^{\infty}\phi^*(p')\phi(p)dx = A^2\int_{-\infty}^{\infty}e^{i\frac{p-p'}{\hbar}x}dx \\ &= A^2\hbar\int_{-\infty}^{\infty}e^{i(p-p')\xi}d\xi = 2\pi\hbar A^2\delta(p-p').\end{aligned} \tag{3.80}$$

Therefore, $A = \dfrac{1}{\sqrt{2\pi\hbar}}$,

$$\phi(p, x) = \frac{1}{\sqrt{2\pi\hbar}}e^{i\frac{p}{\hbar}x}. \tag{3.81}$$

In three dimensions:

$$\phi(\vec{p},\vec{r}) = \frac{1}{(2\pi\hbar)^{3/2}} e^{\frac{i}{\hbar}\vec{p}\cdot\vec{r}}. \tag{3.82}$$

Notice that, for any scalar product of states, the transformation from the artificially discrete spectrum to the continuous one is effected formally by

$$\frac{1}{L^3}\sum_{n_1,n_2,n_3} = \frac{1}{(2\pi)^3}\sum_{n_1,n_2,n_3}\Delta^3 k \to \frac{1}{(2\pi)^3}\int d^3k = \frac{1}{(2\pi\hbar)^3}\int d^3p. \tag{3.83}$$

3.13 The State Function of a Free Particle

For a free particle, the Hamiltonian operator is

$$\hat{H} = \frac{\hat{p}^2}{2m}, \tag{3.84}$$

therefore the time-independent *Schrödinger* equation for a free particle is

$$\hat{H}u = \hat{E}u.$$

Apply $\hat{\vec{p}}$ twice on the function $e^{\frac{i}{\hbar}\vec{p}\cdot\vec{r}}$ to have

$$\frac{\hat{\vec{p}}^{\,2}}{2m} e^{\frac{i}{\hbar}\vec{p}\cdot\vec{r}} = \frac{p^2}{2m} e^{\frac{i}{\hbar}\vec{p}\cdot\vec{r}},$$

and

$$\phi_{\vec{p}}(\vec{r}) = A\, e^{\frac{i}{\hbar}\vec{p}\cdot\vec{r}} \tag{3.85}$$

is an eigenfunction of $\hat{H}$ corresponding to the eigenvalue $\frac{p^2}{2m}$, the energy and momentum of a free particle are continuous. The complete state function of a free particle with momentum $\vec{p}$ is

$$\psi_{\vec{p}}(\vec{r},t) = A\, e^{\frac{i}{\hbar}(-Et+\vec{p}\cdot\vec{r})}. \tag{3.86}$$

This result could be obtained in another way absolutely independent of *Schrödinger* equation and the proposition that $\hat{\vec{p}} = \frac{\hbar}{i}\vec{\nabla}$. It can be looked at as a direct mathematical expression of *De Broglie* hypothesis (see § 2.5), according to which a free particle with definite energy E and momentum $\vec{p}$ is described by a wave with definite frequency $\frac{E}{\hbar}$ and definite wave vector $\vec{k} = \frac{\vec{p}}{\hbar}$. This is the plane wave given by Eq. (3.86).

The general solution of the time-dependent *Schrödinger* equation for a free particle is obtained by substituting $\phi_{\vec{p}}(\vec{r})$ from Eq. (3.85) in Eq. (3.58) and replacing the summation by a triple integral:

$$\psi(\vec{r},t) = \frac{1}{(2\pi\hbar)^{3/2}}\int C(\vec{p}) e^{-\frac{i}{\hbar}(Et-\vec{p}\cdot\vec{r})} d^3p, \tag{3.87}$$

where $E = \frac{p^2}{2m}$ and $(2\pi\hbar)^{-3/2}$ was substituted for A (normalization to δ-function).

If normalization in a box is adopted, A is substituted by $L^{-3/2}$,

$$\psi(\vec{r}, t) = \frac{1}{L^{3/2}} \sum_{n_1, n_2, n_3} C_{n_1 n_2 n_3} e^{-\frac{i}{\hbar}(Et - \vec{p}\cdot\vec{r})}, \tag{3.88}$$

$$p_\alpha = \frac{2\pi\hbar}{L} n_\alpha, \quad E = \frac{h^2}{2mL^2}(n_1^2 + n_2^2 + n_3^2). \tag{3.89}$$

Notice that

$$\Delta p_\alpha = \frac{h}{L} \to 0 \text{ as } L \to \infty, \tag{3.90}$$

$$\Delta E = \frac{h^2}{2m} \frac{2(n_1 + n_2 + n_3) + 3}{L^2} \to 0 \text{ as } L \to \infty. \tag{3.91}$$

(Δn_α has been taken to be 1 for all α). That is, the spectra of $\hat{\vec{p}}$ and $\hat{E}$ are continuous ones as the normalization length tends to ∞.

3.14 The Passage to the Limiting Case of Classical Mechanics

According to *Bohr*'s correspondence principle, a new theory must have the classical theory as a certain limit of it. For example, relativistic mechanics (special theory) reduces automatically to classical mechanics in the limit $\frac{v}{c} \to 0$, e.g., in this limit, the *Lorentz* transformation tends to the *Galileo* one:

$$x = \frac{x' + vt'}{\sqrt{1 - \frac{v^2}{c^2}}} \to x' + vt', \qquad t = \frac{t' + \frac{v}{c^2}x'}{\sqrt{1 - \frac{v^2}{c^2}}} \to t'.$$

Also, for the momentum and energy:

$$\vec{p} = \frac{m\vec{v}}{\sqrt{1 - \frac{v^2}{c^2}}} \to m\vec{v},$$

$$E = \frac{mc^2}{\sqrt{1 - \frac{v^2}{c^2}}} = c\sqrt{p^2 + m^2c^2} \to mc^2 + \frac{p^2}{2m} = mc^2 + \frac{1}{2}mv^2.$$

The passage from quantum mechanics to classical mechanics is not that simple. Here we imitate the method of passing from wave optics to geometrical optics (see § 1.3). To pass from quantum mechanics to classical mechanics taking as the physical system, for simplicity and clearness, only one particle in an external field, write the state function ψ in the modulus and phase form:

$$\psi = a\ e^{\frac{i}{\hbar}\Sigma}, \tag{3.92}$$

where $a, \frac{1}{\hbar}\Sigma$ being the modulus and phase of a complex function, are real. Substituting Eq. (3.92) in *Schrödinger* equation with

$$H = -\frac{\hbar^2}{2m}\nabla^2 + U,$$

we obtain

$$a[\frac{\partial\Sigma}{\partial t} + \frac{1}{2m}(\vec{\nabla}\Sigma)^2 + U] - \frac{\hbar^2}{2m}\nabla^2 a - i\hbar(\frac{\partial a}{\partial t} + \frac{a}{2m}\nabla^2\Sigma + \frac{1}{m}\vec{\nabla}\Sigma\cdot\vec{\nabla}a) = 0.$$

Equating to zero the real and imaginary parts, we have

$$\frac{\partial\Sigma}{\partial t} + \frac{1}{2m}(\vec{\nabla}\Sigma)^2 + U - \frac{\hbar^2}{2ma}\nabla^2 a = 0, \tag{3.93}$$

$$\frac{\partial a}{\partial t} + \frac{a}{2m}\nabla^2\Sigma + \frac{1}{m}\vec{\nabla}\Sigma\cdot\vec{\nabla}a = 0. \tag{3.94}$$

In the classical limit ($\hbar \to 0$) Eq. (3.93) reduces to the *Hamilton-Jacobi* equation for the action S as a function of the time and coordinates, that is:

$$\lim_{\hbar\to 0}\Sigma = S, \quad \lim_{\hbar\to 0}\vec{\nabla}\Sigma = \vec{\nabla}S = \vec{p},$$

$$\lim_{\hbar\to 0}\frac{\partial\Sigma}{\partial t} = \frac{\partial S}{\partial t} = -H\ (= -E\ \text{ for constant } H),$$

and the *De Broglie* wavelength λ here is (see the comment after Eq. (1.12))

$$\lambda = \frac{2\pi\hbar}{|\vec{\nabla}\Sigma|} \to \frac{2\pi\hbar}{p} \to 0 \quad \text{in this limit.}$$

In optics this limit is called the geometrical optics limit and in quantum mechanics, it is the classical limit. Multiply Eq. (3.94) by $2a$ to obtain

$$\frac{\partial a^2}{\partial t} + \operatorname{div}(a^2\frac{\vec{\nabla}S}{m}) = 0 \quad \text{in the limit } \Sigma \to S. \tag{3.95}$$

This equation has a clear physical meaning: $a^2 = |\psi|^2$ is the probability density to find the particle at some position in space and $\frac{1}{m}\vec{\nabla}S = \vec{v}$ is the classical velocity of the particle. Hence Eq. (3.95) shows that the density "moves" according to *Hamilton-Jacobi* equation along classical paths with the classical velocity. This means that even in the classical limit we do not achieve the classical (*Laplacian*) determinism. All what we achieved is that the probability distribution for the coordinates changes with time along the classical paths according to the laws of classical mechanics.

Obtaining in the classical limit a motion along a definite path is a matter of initial conditions. We must start with a wave function appreciably different from zero only in a small region whose dimensions tend to zero with $\hbar$. Then in the quasiclassical case, this wave will move in space along a classical path of the particle.

The operators in quantum mechanics tend in the classical limit to multiplication by the corresponding classical quantities. For example, operating by $\hat{\vec{p}} = \frac{\hbar}{i}\vec{\nabla}$ on the wave function of Eq. (3.92), we have

$$\hat{\vec{p}}\psi = \frac{\hbar}{i}\vec{\nabla}a\, e^{\frac{i}{\hbar}\Sigma} = (\vec{\nabla}\Sigma + \frac{\hbar}{i}\frac{1}{a}\vec{\nabla}a)\psi,$$

$$\lim_{\hbar\to 0}\hat{\vec{p}}\psi = \lim_{\hbar\to 0}(\vec{\nabla}\Sigma + \frac{\hbar}{i}\frac{1}{a}\vec{\nabla}a)\psi = \vec{p}\psi.$$

For more on the analogy between geometrical optics and classical mechanics, see [1, § 53] and [10, § 6, § 17], see also [11, Chap. 9 § 8].

4

One-Dimensional Motion

In this chapter, some cases of one-dimensional motion are studied. For a particle in a rectangular potential well the energy spectrum is determined, so also the normalization constant, the calculations are performed in the simple case of infinitely deep potential well. The results are compared in quantum and classical mechanics. In addition, rectangular potential barrier is studied. The reflection and transmission coefficients are determined. Moreover, the tunnel effect and tunnel effect paradox are discussed and two applications, namely, the α-decay and the cold emission of electrons by metals, are given for illustration. The linear harmonic oscillator is studied, where its energy eigenvalues and eigenstates are determined, and the parity of these eigenstates is found and the zero-point energy is discussed as a result of the uncertainty relations and as a manifestation of the wave properties of the microparticles.

4.1 A Particle in a Rectangular Potential Well

The Hamiltonian operator is

$$H = \frac{p^2}{2m} + U. \tag{4.1}$$

In this example, U is given by:

$$U(x) = \begin{cases} U_0 > 0, & \forall x < 0, & \text{(region I)}, \\ 0, & \forall x \in (0, a) & \text{(region II)}, \\ U_0 > 0, & \forall x > a & \text{(region III)}. \end{cases} \tag{4.2}$$

See Fig. 4.1. The potential U and hence the Hamiltonian H does not depend on time. Therefore, the state function ψ has the form

$$\psi(t, x) = e^{-i\frac{E}{\hbar}t} u(x), \tag{4.3}$$

where $E > 0$ is the total conserved energy of the particle and u is the energy (Hamiltonian) eigenfunction, i.e., the solution of the energy (Hamiltonian) eigenvalue problem:

$$Hu = Eu. \tag{4.4}$$

DOI: 10.1201/9781003677109-4

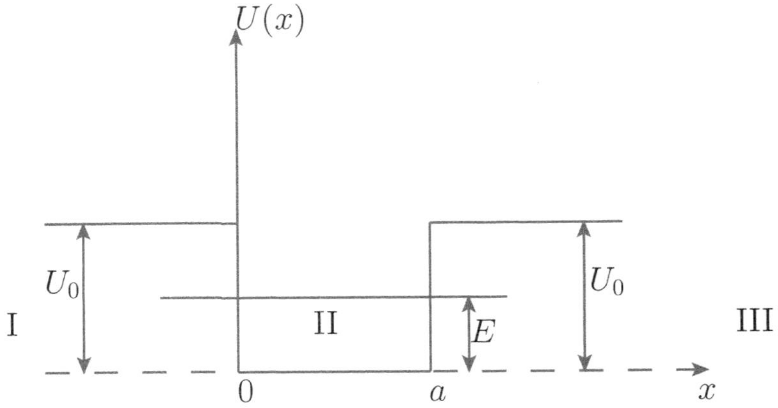

FIGURE 4.1: *The motion of a particle in a potential well.*

Here, $p = \frac{\hbar}{i}\frac{d}{dx}$, hence

$$u''_{\rm II} + k^2 u_{\rm II} = 0, \tag{4.5}$$

where $k^2 = \frac{2m}{\hbar^2}E > 0$, therefore

$$u_{\rm II} = A_{\rm II}\sin kx + B_{\rm II}\cos kx. \tag{4.6}$$

Let $E < U_0$. Then

$$u''_{\rm I,III} - \chi^2 u_{\rm I,III} = 0, \quad \chi^2 = \frac{2m}{\hbar^2}(U_0 - E) > 0, \tag{4.7}$$

therefore

$$u_{\rm I,III} = A_{\rm I,III}e^{\chi x} + B_{\rm I,III}e^{-\chi x}. \tag{4.8}$$

See Fig. 4.2. Since u must be finite everywhere, $B_{\rm I}$ and $A_{\rm III}$ must equal zero. Therefore,

$$u_{\rm I} = Ae^{\chi x}, \quad u_{\rm III} = Be^{-\chi(x-a)}. \tag{4.9}$$

For the probability density and the probability current density to be continuous everywhere, we must match $u_{\rm I}, u_{\rm II}$ and their first derivatives at $x = 0$, and $u_{\rm II}, u_{\rm III}$ and their first derivatives at $x = a$, to obtain

1. all the arbitrary constants in terms of one of them, A say, which will be determined from the normalization condition, and to obtain also
2. the energy spectrum (eigenvalues).

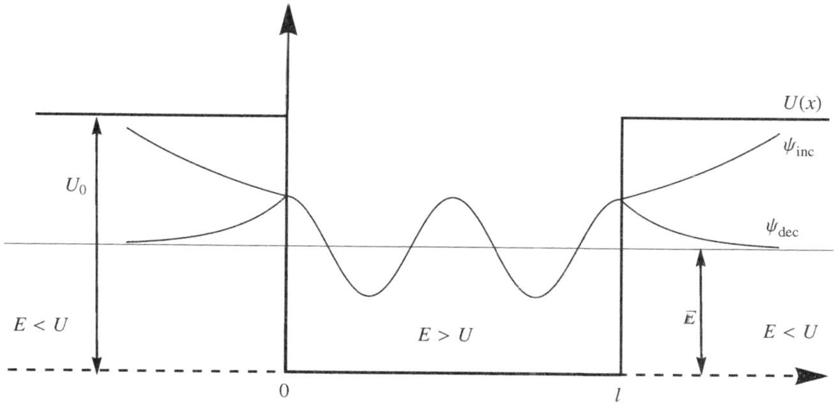

FIGURE 4.2: *Wave function for a given value of E. The energy level is taken to be the abscissa of the wave function.*

We shall perform these calculations in the simple case when U_0 is infinite, therefore χ tends also to infinity and $u_{\mathrm{I}}, u_{\mathrm{III}}$ tend to zero. Therefore, the conditions on u_{II} will be

$$u_{\mathrm{II}}(0) = 0, \quad u_{\mathrm{II}}(a) = 0, \tag{4.10}$$

this gives:

$$B_{\mathrm{II}} = 0, \quad \sin ka = 0, \tag{4.11}$$

therefore

$$k = \frac{\pi}{a} n, \quad n = 1, 2, \dots, \tag{4.12}$$

$$u_n = A_n \sin n\pi \frac{x}{a}. \tag{4.13}$$

The possible negative values of n were excluded because $u_{-n} = -u_n$, that is u_{-n} is proportional to u_n and therefore they are essentially the same, also $n = 0$ was excluded, because in this case the state (wave) function is identically zero.

$$E = \frac{\hbar^2 k^2}{2m}, \tag{4.14}$$

therefore the energy eigenvalues E_n are

$$E_n = \frac{\pi^2 \hbar^2}{2ma^2} n^2 = n^2 E_1, \tag{4.15}$$

where

$$E_1 = \frac{\pi^2 \hbar^2}{2ma^2} \tag{4.16}$$

is the ground state energy.

The constant A_n is determined by the normalization condition:

$$1 = A_n^2 \int_0^a \sin^2 n\pi \frac{x}{a} dx = A_n^2 \frac{a}{2}, \tag{4.17}$$

that is

$$A_n = \sqrt{\frac{2}{a}}, \tag{4.18}$$

and

$$u_n = \sqrt{\frac{2}{a}} \sin n\pi \frac{x}{a}. \tag{4.19}$$

A particle moving in a rectangular potential well with energy $E < U_0$, according to classical mechanics, in both cases of U_0 being finite or infinite, will be moving in the well forward and backward between the two turning points $0, a$, where it receives impulses reflecting the direction of its velocity keeping its magnitude unchanged.

According to quantum mechanics, the particle moves with a given momentum which acquires only discrete values. The probability distribution for its position is an oscillatory function inside the well and is zero, in the case of infinite potential well outside it ($x < 0$ or $x > a$). Outside the well in the case of finite U_0, the probability distribution will be exponentially decreasing as going far from the well.

4.2 Rectangular Potential Barrier

The potential in this case is

$$U(x) = \begin{cases} 0, & \forall x < 0, \\ U_0 > 0, & \forall x \in (0, a), \\ 0, & \forall x > a. \end{cases} \tag{4.20}$$

See Fig. 4.3. We consider a particle approaching this barrier from the region of negative x colliding with it to be reflected or transmitted by the barrier.

In the corresponding classical problem, the particle is always reflected whenever its energy is less than the hight of the barrier, and it is always transmitted if its energy is greater than this height.

In the quantum mechanical treatment, we have

$$u''_{\text{I,III}} + k^2 u_{\text{I,III}} = 0, \quad k^2 = \frac{2mE}{\hbar^2}, \tag{4.21}$$

where E is the total constant energy, therefore

$$\begin{aligned} u_{\text{I}} &= Ae^{ikx} + Be^{-ikx}, \\ u_{\text{III}} &= Ce^{ikx}. \end{aligned} \tag{4.22}$$

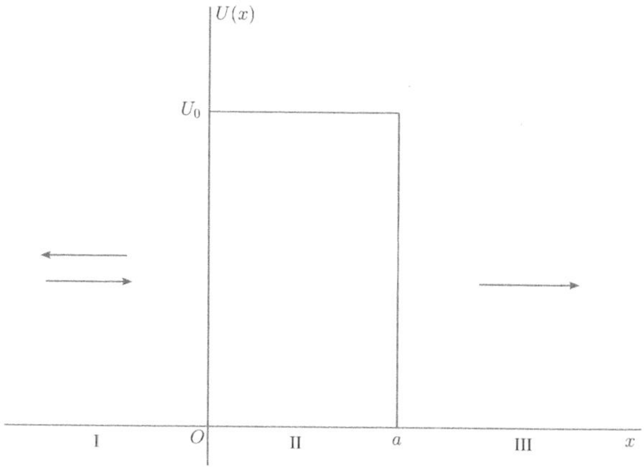

FIGURE 4.3: *One-dimensional rectangular potential barrier of height U_0 and thickness a.*

The first term in the R.H.S. of the first equation represents the incident wave, the second represents the reflected wave, while the R.H.S. of the second equation represents the transmitted wave (in region III, there are no waves propagating in the negative direction).

Substituting Eq. (4.22) in Eq. (3.38) for the probability current density, we have

$$j_{\rm I} = v(|A|^2 - |B|^2), \quad j_{\rm III} = v|C|^2, \tag{4.23}$$

where $v = \dfrac{\hbar k}{m}$ is the velocity of the particle in either of the regions I or III. Since the probability current densities do not depend on x, $v|A|^2, v|B|^2, v|C|^2$ can be interpreted as the flux density of the incident, reflected and transmitted waves in full agreement with the interpretation of A, B, C as the amplitudes of the corresponding waves.

Inside the barrier (region II), the solution depends on whether E is greater or less than U_0.

1. Let $E > U_0$. Therefore,

$$u_{\rm II} = Fe^{i\chi x} + Ge^{-i\chi x}, \quad \chi^2 = \frac{2m(E-U_0)}{\hbar^2} > 0. \tag{4.24}$$

The continuity of u and u' at $x = 0$ and $x = a$ give

$$\begin{aligned} F + G &= A + B, & Fe^{i\chi a} + Ge^{-i\chi a} &= Ce^{ika}, \\ \frac{\chi}{k}(F - G) &= A - B, & \frac{\chi}{k}(Fe^{i\chi a} - Ge^{-i\chi a}) &= Ce^{ika}, \end{aligned} \tag{4.25}$$

which give

$$\frac{B}{A} = \frac{(k^2-\chi^2)(1-e^{2i\chi a})}{(k+\chi)^2 - (k-\chi)^2 e^{2i\chi a}}, \quad \frac{C}{A} = \frac{4k\chi e^{i(\chi-k)a}}{(k+\chi)^2 - (k-\chi)^2 e^{2i\chi a}}. \tag{4.26}$$

Therefore, the reflection coefficient $R = \left|\frac{B}{A}\right|^2$ and the transmission coefficient $T = \left|\frac{C}{A}\right|^2$ will be:

$$R = \frac{(k^2 - \chi^2)^2 \sin^2 \chi a}{(k^2 - \chi^2)^2 \sin^2 \chi a + 4k^2\chi^2} = \frac{U_0^2 \sin^2 \chi a}{U_0^2 \sin^2 \chi a + 4E(E - U_0)}, \tag{4.27}$$

$$T = \frac{4k^2\chi^2}{(k^2 - \chi^2)^2 \sin^2 \chi a + 4k^2\chi^2} = \frac{4E(E - U_0)}{U_0^2 \sin^2 \chi a + 4E(E - U_0)}. \tag{4.28}$$

It is clear that $R+T = 1$ as it must be. That is the incident particle is either reflected from the barrier with probability R or transmitted through it with probability T and the sum of the two probabilities is 1.

Notice that:

(a) Since in general $R \neq 0$, the particle will be reflected from the barrier with nonvanishing probability although its energy exceeds the barrier hight ($E > U_0$). This is a quantum mechanical phenomenon called overbarrier reflection and has no classical analogue. The overbarrier reflection is a manifestation of the wave properties of microparticles. There is also a reflection from the potential well since $R \neq 0$ for the substitution $U_0 \to -U_0$ in Eq. (4.27) and in the expression for χ (Eq. (4.24)).

(b) There will be a perfect transmission when $\chi a = n\pi$ (n is an integer) that is when the barrier contains an integral number of half wavelengths. This effect is also true for transmission through a potential well ($U_0 < 0$). Phenomena of this type is known in the transmission of light through refracting thin layers.

2. **Tunnel effect.** In the case of $0 < E < U_0$, to obtain the scattering coefficients, just replace χ by $i\beta$ in Eq. (4.24), where

$$\beta^2 = \frac{2m(U_0 - E)}{\hbar^2} > 0. \tag{4.29}$$

The result for T is

$$T = \frac{4k^2\beta^2}{(k^2 + \beta^2)^2 \sinh^2 \beta a + 4k^2\beta^2} = \frac{4E(U_0 - E)}{U_0^2 \sinh^2 \beta a + 4E(U_0 - E)}, \tag{4.30}$$

which means that there is a transmission through the barrier, even though $E < U_0$. The penetration of microparticles through a potential barrier is another manifestation of their wave properties, and has an analogue in optics. This analogue is the total reflection of

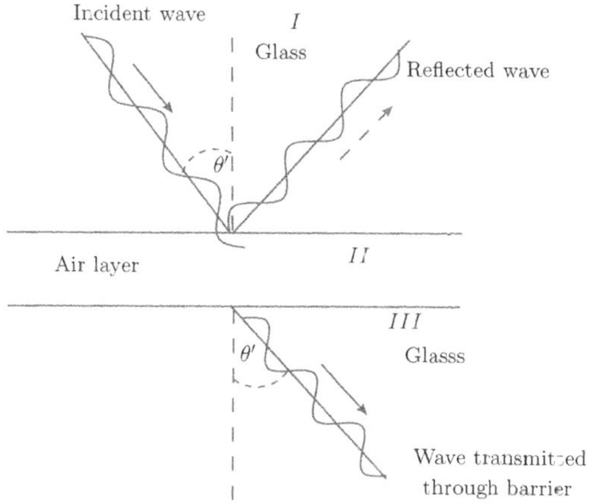

FIGURE 4.4: *Schematic diagram of the tunnel effect in optics (total internal reflection).*

light from an optically rarer medium (e.g., reflection in glass from air). If in the cases of total internal reflection the refracted wave encounters a second glass surface, that is if we have two pieces of glass separated by an air layer, then the wave penetrates through the air layer and continues propagating in the second glass piece (see Fig. 4.4). When $\beta a \gg 1$, the transmission coefficient T becomes very small and approximately is:

$$T = \frac{16E(U_0 - E)}{U_0^2} e^{-2\beta a}, \tag{4.31}$$

which decreases exponentially with the breadth of the barrier.

Tunnel effect paradox. [5, Ch. 6, § B] Here we face the tunnel effect paradox. At first glance, it seems that the microparticles inside the potential barrier are moving with imaginary momentum:

$$u_{\text{II}} = Fe^{-\beta x} + Ge^{\beta x}, \tag{4.32}$$

therefore $p = \pm\hbar\chi = \pm i\hbar\beta$, where $\beta = \dfrac{\sqrt{2m(U_0 - E)}}{\hbar}$ is real.

But it is important to notice that in this exclusively quantum mechanical effect, it is only the probability density (of a particle to be at a certain point inside the barrier) which decreases exponentially as the particle goes far inside the barrier. The momentum of the particle is real, because the probability amplitude, i.e., the wave function u_{II} in Eq. (4.32) can be represented as a *Fourier* integral, i.e., as a superposition of wave functions each of which has a real momentum, and the particle inside the barrier will have a real momentum within certain uncertainty consistent with the uncertainty relations of quantum mechanics.

Indeed, calculations using Eqs. (4.25) give

$$G = -\frac{k - i\beta}{k + i\beta}\, e^{-2\beta a}\, F, \text{ that is, } \quad |G| = e^{-2\beta a}|F|.$$

For βa large enough to neglect the factor $e^{-2\beta a}$ with respect to one, the wave function inside the barrier is

$$u = Fe^{-\beta x} = \sqrt{\frac{2}{\pi}} \int_0^\infty \tilde{u}_c(k) \cos kx \, dk, \tag{4.33}$$

where the *Fourier* cosine transform $\tilde{u}_c(k)$ is

$$\tilde{u}_c(k) = \sqrt{\frac{2}{\pi}} F \int_0^\infty e^{-\beta x} \cos kx \, dx = \sqrt{\frac{2}{\pi}}\, F\, \frac{\beta}{\beta^2 + k^2}. \tag{4.34}$$

The modulus of the amplitude $|\tilde{u}_c|$ is a monotonic decreasing function. Its maximum is at $k = 0$, and it loses half this value at $k = \beta$, that is β is the width of the peak of this amplitude. Therefore, the uncertainty in the momentum is

$$\Delta p \sim \hbar\beta. \tag{4.35}$$

The width of the barrier is a, hence $\Delta x \sim a$ and

$$\Delta x \Delta p \sim \hbar\beta a > \hbar. \tag{4.36}$$

In what follows two phenomena are discussed, for illustration. The study of these two phenomena is based on models of potential barriers which are not rectangular, therefore formula (4.30) does not apply. The transmission coefficient D through the barriers in these two examples is given by

$$D = e^{-\frac{2}{\hbar}\sqrt{2m}\int_a^b \sqrt{U(x)-E}\, dx}, \tag{4.37}$$

where a, b are the roots of the equation $U(x) = E$ (Fig. 4.5). This formula is obtained using the WKB approximation method for the exposition of which and the deduction of (4.37), see e.g. [4, 5, 10].

4.3 Alpha-Decay

Curies[1] assumed that the number dN of disintegrating nuclei at time t in a time interval dt is proportional to this interval and to the number N of the nuclei at this time

$$dN = -\lambda N dt, \tag{4.38}$$

[1] *P. Curie* (1859–1906), and *M. Curie* (1867–1934), was awarded the half of *Nobel* prize in Physics 1903 in recognition of the extraordinary services they have rendered by their joint researches on the radiation phenomena discovered by Professor Henri Becquerel.

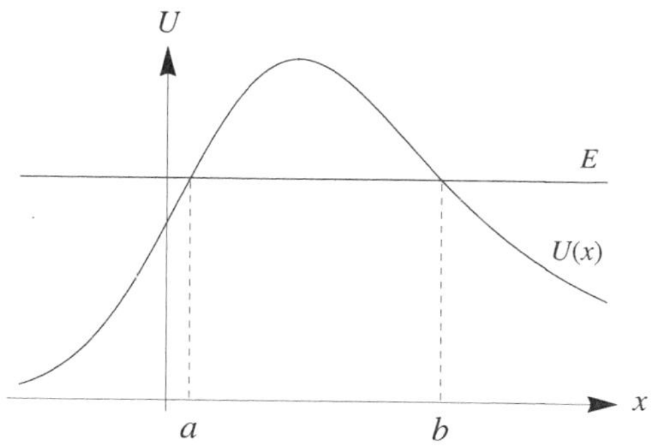

FIGURE 4.5: *The potential barrier of an arbitrary (but sufficiently smooth) shape.*

and obtained their law of radioactive decay:

$$N = N_0 \; e^{-\lambda t}.$$

It is clear that the half-lifetime $\tau_{1/2}$ after which half of the initial number of the nuclei had already decayed is related to the decay constant λ by

$$\tau_{1/2} = \frac{\ln 2}{\lambda}.$$

Gamow, *Condon*, and *Gurney* (1928) assumed that the α-particle is already formed in the α-emitter nucleus[2] and exists in the strong nuclear attractive and the *Coulomb* repulsive fields of the daughter nucleus. The potential energy of the α-particle is taken to be

$$U(r) = \begin{cases} 0 & \forall r < R, \\ \frac{2(Z-2)e_0^2}{r} & \forall r > R, \end{cases}$$

where $-e_0$ is the electron charge and $R \sim 1$ F$= 10^{-13}$ cm is the nucleus radius (Fig. 4.6).

They explained the α-emission by the tunnelling of the α-particle through the potential barrier shown in Fig. 4.6 which shows that the higher the energy of the emitted α-particle the narrower the width of the barrier to be crossed and therefore, the larger the probability for the α-particle to be emitted, that is the shorter the lifetime the emitter will have. Applying Eq. (4.37), the transmission coefficient D is:

$$D = e^{-\frac{2}{\hbar}\sqrt{2M}\int_R^{R_1}\sqrt{U(r)-E}\;dr},$$

where M is the mass of the α-particle, R, R_1 are the beginning and the end of the barrier. D is normalized to unit incident current (see Eqs. (4.23), (4.28)),

[2]The α-particle is the most stable nucleus. It has the maximum mass defect per nucleon, that is the maximum binding energy/nucleon.

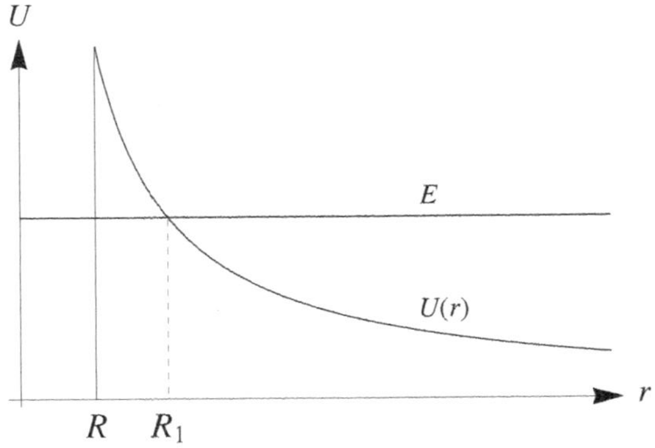

FIGURE 4.6: *The potential energy of an α-particle in the field of a radioactive nucleus.*

therefore it is the probability for an α-particle to cross the barrier as a result of one strike on its wall in each second, hence the probability for emission of one α-particle in a second is nD where n is the number of strikes per second on the wall of the barrier. Comparing with Eq. (4.38), we have

$$\lambda = nD,$$

if v, p are the velocity and momentum of the α-particle inside the nucleus, then in each time interval $\Delta t \sim \frac{R}{v} = \frac{MR}{p}$ there is one strike on the barrier wall and the number n of strikes on it in a second is $n = \frac{1}{\Delta t} \sim \frac{p}{MR}$. p is estimated from the uncertainty principle ($pR \sim \hbar$) to be $\sim \frac{\hbar}{R}$, therefore, $n \sim \frac{\hbar}{MR^2}$, that is

$$\lambda = \frac{\hbar}{MR^2} e^{-\frac{2}{\hbar}\sqrt{2M}\int_R^{R_1}\sqrt{U(r)-E}\,dr},$$

where E is the energy of the emitted α-particle. Performing the integration (use, e.g., the substitution $r = R_1 \sin^2 \xi$) and taking into account that practically $\frac{R}{R_1} \ll 1$, make the approximation $\sin\sqrt{\frac{R}{R_1}} = \sqrt{\frac{R}{R_1}}$ to finally obtain

$$\lambda = \frac{\hbar}{MR^2} A\, e^{-\frac{B}{\sqrt{E}}}, \qquad \tau_{1/2} = \frac{MR^2 \ln 2}{\hbar A}\, e^{\frac{B}{\sqrt{E}}}, \tag{4.39}$$

where

$$B = \frac{2(Z-2)\pi e_0^2}{\hbar}\sqrt{2M}, \qquad A = e^{\frac{8e_0}{\hbar}\sqrt{MR(Z-2)}}. \tag{4.40}$$

A result which agrees with experiment. It is clear from Eq. (4.39) that the higher the energy E of the emitted α-particle, the lower the half-lifetime will be. A small increase in E, e.g. from 4 MeV to 9 MeV (which are approximately the values of the lower and higher energies of the emitted α-particles by the radioactive family of uranium) decreases drastically the half-lifetime from 10^9 years to 10^{-7} seconds.

4.4 The Electron Gas Model

The high electric conductivity of metals shows that electrons move comparatively freely within the metal. Their escape from the metal into vacuum, however is difficult and requires a certain amount of energy called the work function. Thus, in the simplest model, a metal is considered as a gas of electrons moving in a potential well inside which (i.e., inside the metal) the potential is zero, while outside the metal (i.e., in vacuum) the potential is greater than zero. Since the electrons are considered inside the metal as free particles, to count the number of states with momenta in the element of volume $\Delta p_x \Delta p_y \Delta p_z$ of the momentum space centered at the point (p_x, p_y, p_z), we introduce the *Max Born* artificial periodicity length L which will be taken to tend to infinity at the end of the calculations, the momentum of an electron will be $\vec{p} = \hbar \vec{k}$, where p_α is given by (see Eq. (3.76))

$$p_\alpha = \hbar k_\alpha = \frac{2\pi\hbar}{L} n_\alpha, \qquad \alpha = x, y, z, \qquad n_\alpha = 0, \pm 1, \pm 2, \ldots .$$

The distance between two neighboring eigenvalues of p_α is $\frac{2\pi\hbar}{L}$ $\forall \alpha$. Therefore, the momentum space is divided into cubic cells of side length $\frac{2\pi\hbar}{L}$, each of which contains only one momentum eigenvalue and consequently corresponds to only one momentum eigenstate. Therefore, the number of independent quantum states of an electron with momentum in the element of volume d^3p is $\frac{L^3}{(2\pi\hbar)^3} d^3p$, according to *Pauli*[3] exclusion principle[4] (see § 13.6) each of these states can be occupied by only two electrons (the electron spin is $\frac{1}{2}$ and has only two independent spin states corresponding to the two possible spin orientations (see § 11.4.3)). Therefore, the number of electrons with momenta in the element of volume d^3p is $\frac{2L^3}{(2\pi\hbar)^3} d^3p$. If ρ_0 is the number of electrons in a unit volume of the metal, the maximum momentum of an electron at absolute zero ($T = 0$) is related to ρ_0 by

$$\begin{aligned} \rho_0 &= \frac{2}{L^3} \left(\frac{L}{2\pi\hbar} \right)^3 \sum \Delta p_x \Delta p_y \Delta p_z \\ &= \frac{2}{(2\pi\hbar)^3} \sum \Delta p_x \Delta p_y \Delta p_z \to \frac{2}{(2\pi\hbar)^3} \int_0^{p_{\max}} p^2 dp \oint d\Omega \\ &= \frac{1}{\pi^2 \hbar^3} \int_0^{p_{\max}} p^2 dp = \frac{p_{\max}^3}{3\pi^2 \hbar^3}, \end{aligned}$$

i.e.

$$p_{\max} = \hbar (3\pi^2 \rho_0)^{1/3},$$

[3] *W. Pauli* (1900–1958) was awarded the *Nobel* prize in Physics in 1945 for the discovery of the Exclusion Principle, also called the *Pauli* Principle.

[4] No more than one electron can occupy a one particle eigenstate.

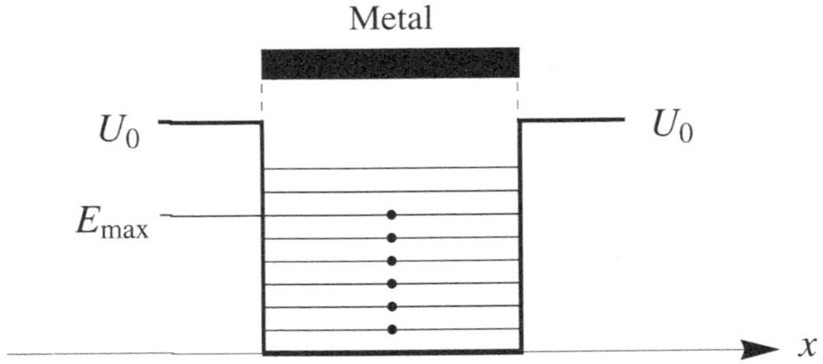

FIGURE 4.7: *The model of potential well for a metal. E_{max} is the upper boundary of the filled levels at $T = 0$ (Fermi energy). The heavy dots denote occupied levels where there are in general many electrons differing in their momenta and spin orientation and have the same energy in each level.*

and the maximum energy of an electron (at $T = 0$) is

$$E_{\rm max} = E_{\rm F} = \frac{p_{\rm max}^2}{2m} = \frac{\hbar^2}{2m}(3\pi^2 \rho_0)^{2/3}, \qquad (4.41)$$

this energy is called the *Fermi*[5] energy or the *Fermi* level. The scheme of filling the energy levels by electrons in a metal is shown in Fig. 4.7.

As a numerical example, the *Fermi* energy is calculated for silver see [5, Ch. 6, § c]. Since silver is monovalent, the outer shell of a silver atom contains only one electron, and the number of free electrons is equal to the number of atoms in the same volume. The density of silver is 10.5 g/cm^3, its atomic weight is 107.9, and the *Avogadro* number (the number of atoms in one gram-atom) is 6.02×10^{23}, therefore

$$\rho_0 = \frac{10.5}{107.9} \times 6.02 \times 10^{23} = 5.8 \times 10^{22},$$

from Eq. (4.41)

$$E_{\rm F} = 8.5 \times 10^{-12} \text{ erg} = 5.3 \text{ eV}.$$

Since the work function for silver is 3.7 eV, the depth of the potential well in silver is 9 eV. The average energy of an electron in a metal (at $T = 0$) is given

[5] *E. Fermi* (1901–1954) was awarded the *Nobel* prize in Physics 1938 for his demonstrations of the existence of new radioactive elements produced by neutron irradiation, and for his related discovery of nuclear reactions brought about by slow neutrons.

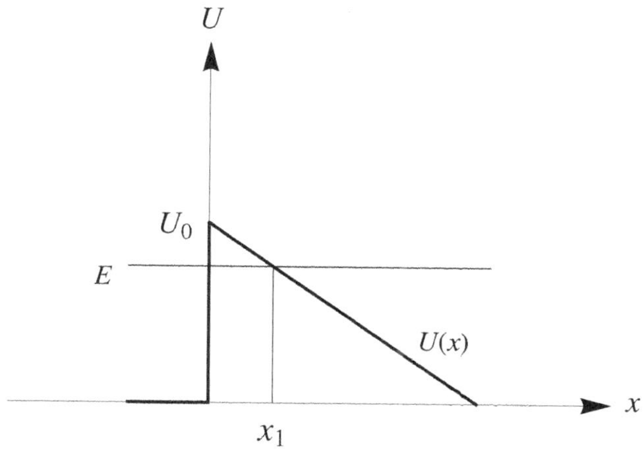

FIGURE 4.8: *The potential energy of an electron in the presence of a constant external electric field.*

by

$$E^0_{\text{avg}} = \frac{2}{\rho_0}\int \frac{p^2}{2m(2\pi\hbar)^3}d^3p = \frac{3}{5}E_{\text{F}},$$

in agreement with experiment, according to the model shown in Fig. 4.7. To extract an electron from a metal, the electron must be given an energy not less than the work function:

$$W = U_0 - E_{\text{F}}.$$

In the photoelectric emission, the electron receives an energy $\hbar\omega$ from the absorbed photon and leaves the metal with a kinetic energy

$$\frac{1}{2}mv^2 = \hbar\omega - W \quad (\textit{Einstein}\text{'s equation}).$$

In a metal at a temperature higher than the absolute zero, some of the electrons occupy energy levels higher than the *Fermi* level. If the kinetic energy of the electron gas is increased by heating the metal, some of the electrons will acquire an energy exceeding the potential barrier hight and an electric current will flow from the metal. This effect which is called the thermionic emission is used to obtain electron beams in electron tubes. Under the action of an external electrostatic field, such a current can also arise even at lower temperatures. This effect is known as the cold emission of electrons by metals.

4.5 Cold Emission of Electrons by Metals

If a constant homogeneous electric field of intensity $\mathscr{E}$ is applied along the surface of a metal in the negative x-direction, the potential energy of the electron (neglecting the electrical image force which has a negligible contribution to the final result) will be

$$U(x) = \begin{cases} U_0 - e_0\mathscr{E}x & \text{in vacuum (outside the metal),} \\ 0 & \text{inside the metal,} \end{cases}$$

where $-e_0$ is the electron charge.

It is seen (Fig. 4.8) that the external electric field outside the metal produces a potential barrier of finite width, and the cold emission is explained quantum mechanically by the penetration of the electrons through this barrier from the metal to vacuum. The transmission coefficient D in this case Eq. (4.37) is

$$D = e^{-\frac{2}{\hbar}\sqrt{2m}\int_0^{x_1}\sqrt{U(x)-E}\,dx},$$

$$U_0 - e_0\mathscr{E}x_1 = E, \quad \text{that is,} \quad x_1 = \frac{U_0 - E}{e_0\mathscr{E}},$$

therefore

$$\begin{aligned} \int_0^{x_1}\sqrt{U(x)-E}\,dx &= \int_0^{x_1}\sqrt{U_0 - E - e_0\mathscr{E}x}\,dx \\ &= \sqrt{e_0\mathscr{E}}\int_0^{x_1}\sqrt{x_1 - x}\,dx = \frac{2}{3}\sqrt{e_0\mathscr{E}}\,x_1^{3/2}, \end{aligned}$$

therefore,

$$D = e^{-\frac{4}{3}\sqrt{2m}\frac{(U_0-E)^{3/2}}{e_0\mathscr{E}\hbar}} = e^{-\frac{\mathscr{E}_0}{\mathscr{E}}},$$

where $\mathscr{E}_0 = \frac{4}{3}\sqrt{2m}\frac{(U_0-E)^{3/2}}{e_0\hbar}$ which depends on the nature of the metal (U_0) and the energy E of the free electrons in the metal.

Since D is normalized to a unit incident current on the wall of the barrier (see Eq. (4.23)), the cold emission electric current is proportional to the transmission coefficient D:

$$J = J_0 D = J_0 e^{-\frac{\mathscr{E}_0}{\mathscr{E}}}.$$

Models based on potential wells, barriers and sequences of them are very useful in studying many phenomena in crystals, conductivity, solid state and nuclear physics.

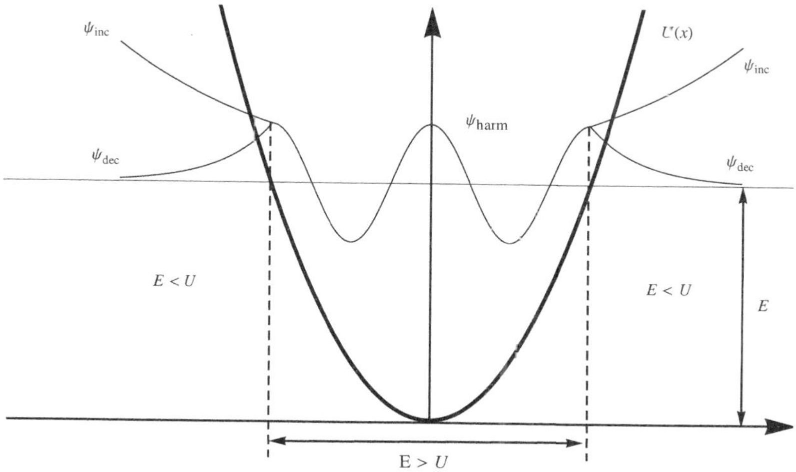

FIGURE 4.9: *Wave function of the harmonic oscillator for an arbitrary value of the energy.*

4.6 Linear Harmonic Oscillator

For a linear harmonic oscillator H is given by

$$H = \frac{p^2}{2m} + \frac{1}{2}m\omega^2 x^2. \tag{4.42}$$

The Hamiltonian does not depend on time, therefore the state function ψ can have the form

$$\psi(t, x) = e^{-i\frac{E}{\hbar}t}u(x), \tag{4.43}$$

where $E > 0$ (Fig. 4.9) is the total conserved energy of the particle and u_E is an energy eigenstate, i.e., is a solution of the time-independent *Schrödinger* equation (the energy eigenvalue problem).

4.6.1 The Energy Eigenvalues

$$Hu_E = Eu_E. \tag{4.44}$$

Therefore,

$$-\frac{\hbar^2}{2m}\frac{d^2u}{dx^2} + (\frac{1}{2}m\omega^2 x^2 - E)u = 0, \tag{4.45}$$

that is

$$\frac{\hbar}{m\omega}\frac{d^2u}{dx^2} + (\frac{2E}{\hbar\omega} - \frac{m\omega}{\hbar}x^2)u = 0, \tag{4.46}$$

put

$$\lambda = \frac{2E}{\hbar\omega}, \quad \xi = \frac{x}{x_0}, \quad x_0 = \sqrt{\frac{\hbar}{m\omega}}, \tag{4.47}$$

(x_0 has the dimension of length, so ξ is dimensionless) we have

$$u'' + (\lambda - \xi^2)u = 0, \tag{4.48}$$

where $u'' = \frac{d^2u}{d\xi^2}$.

First we find the asymptotic behavior of the state function as $\xi \to \pm\infty$, that is the constant λ is negligible compared with ξ^2, then

$$u''_\infty - \xi^2 u_\infty = 0. \tag{4.49}$$

Put $u_\infty = e^{\alpha\xi^2}$, hence

$$u''_\infty = (4\alpha^2\xi^2 + 2\alpha)e^{\alpha\xi^2} \simeq 4\alpha^2\xi^2 e^{\alpha\xi^2}, \tag{4.50}$$

therefore $4\alpha^2 = 1$, $\alpha = \pm\frac{1}{2}$, and

$$u_\infty = c_1 e^{-\frac{1}{2}\xi^2} + c_2 e^{\frac{1}{2}\xi^2}. \tag{4.51}$$

The second term in the above expression for u_∞ tends to ∞ as $\xi \to \pm\infty$, therefore, c_2 must be taken equal to zero. Coefficient c_1 can be taken equal 1, since the state function is not yet normalized. We seek a solution for the state function in the form

$$u = u_\infty w = e^{-\frac{1}{2}\xi^2} w, \tag{4.52}$$

therefore

$$u'' = (w'' - 2\xi w' + (\xi^2 - 1)w)e^{-\frac{1}{2}\xi^2}. \tag{4.53}$$

Substituting for u'', u in Eq. (4.48), we have

$$w'' - 2\xi w' + (\lambda - 1)w = 0, \tag{4.54}$$

we solve this equation in series, putting

$$w = \sum_{k=0}^{\infty} a_k \xi^k \tag{4.55}$$

and substituting for w in Eq. (4.54), we have

$$\sum_{k=0}^{\infty} \{(k+2)(k+1)a_{k+2} - (2k+1-\lambda)a_k\}\xi^k = 0, \ \forall \xi \in (-\infty, \infty), \tag{4.56}$$

therefore

$$a_{k+2} = \frac{2k+1-\lambda}{(k+2)(k+1)} a_k, \tag{4.57}$$

which relates a_k to a_{k+2}. Therefore, all the even numbered coefficients are proportional to a_0 and the odd numbered ones are proportional to a_1. If we take $a_0 = 1$ and $a_1 = 0$, the solution W_0 is an even series, while taking $a_0 = 0$ and $a_1 = 1$, the solution W_1 is an odd series. W_0 and W_1 being of different parities are linearly independent (can not be proportional to one another). Therefore, the general solution of the second-order differential equation (4.54) is $a_0 W_0 + a_1 W_1$. Beginning from $k > \frac{\lambda - 1}{2}$ all the coefficients are positive and

$$\frac{a_k}{a_{k-2}} = \frac{2k - \lambda - 3}{k(k-1)} \xrightarrow{k\to\infty} \frac{2}{k}, \tag{4.58}$$

which is the same as that for the series

$$e^{\xi^2} = \sum_{k=0,2,4,\ldots} \frac{1}{(\frac{k}{2})!}\xi^k. \tag{4.59}$$

Therefore, if the series does not terminate, we shall have for $\xi \to \pm\infty$, $w \to e^{\xi^2}$ and

$$u = e^{-\frac{1}{2}\xi^2} w \to e^{\frac{1}{2}\xi^2} \to \infty. \tag{4.60}$$

Therefore, the power series must terminate and must be reduced to a polynomial of certain degree, n, say (n is any non-negative integer). That is

$$a_n \neq 0, \quad a_{n+2} = 0, \tag{4.61}$$

hence

$$a_k = 0, \quad \forall k = n+2, n+4, \ldots \tag{4.62}$$

This can be achieved only if

$$\lambda = 2n + 1, \tag{4.63}$$

therefore (see Eq. (4.47))

$$E_n = (n + \frac{1}{2})\hbar\omega, \ n = 0, 1, 2, \ldots \tag{4.64}$$

This is the energy eigenvalues (the energy spectrum). Comparing Eq. (4.64) with Eq. (2.11) of *Planck*, we have here an additional term

$$E_0 = \frac{1}{2}\hbar\omega, \tag{4.65}$$

which is the lowest energy (the energy of the ground state) and is called the zero-point energy.

4.6.2 The Energy Eigenstates

To obtain the energy eigenstates for the harmonic oscillator, put $\lambda = 2n + 1$ in Eq. (4.57) and obtain

$$a_{k-2j} = (-1)^j \frac{k!(n-k)!!}{(k-2j)!(n-k+2j)!!2^j} a_k. \tag{4.66}$$

Take the coefficient a_n of the highest power to be 2^n (this coefficient can be chosen arbitrarily, since the normalization constant of the state function is not yet determined). Therefore,

$$a_{n-2j} = (-1)^j \frac{n!}{j!(n-2j)!} 2^{n-2j}, \tag{4.67}$$

and we have

$$w = H_n(\xi) = \sum_{j=0}^{[n/2]} (-1)^j \frac{n!}{j!(n-2j)!} (2\xi)^{n-2j}. \tag{4.68}$$

This is the *Hermite* polynomial. The *Hermite* polynomial can be written in closed form:

$$H_n(\xi) = (-1)^n e^{\xi^2} \frac{d^n}{d\xi^n} e^{-\xi^2}. \tag{4.69}$$

For this write

$$v = e^{-\xi^2}, \tag{4.70}$$

therefore

$$v' + 2\xi v = 0, \tag{4.71}$$

differentiate $n+1$ times to obtain

$$v^{(n+2)} + 2\xi v^{(n+1)} + 2(n+1)v^{(n)} = 0. \tag{4.72}$$

Put

$$v^{(n)} = e^{-\xi^2} g, \tag{4.73}$$

and obtain

$$g'' - 2\xi g' + 2ng = 0, \tag{4.74}$$

which is Eq. (4.54) with $\lambda = 2n+1$, therefore g is proportional to the *Hermite* polynomial, i.e.,

$$g = e^{\xi^2} \frac{d^n}{d\xi^n} e^{-\xi^2} = A_n H_n(\xi). \tag{4.75}$$

Equating the coefficients of ξ^n in both sides, we have $A_n = (-1)^n$, and thus Eq. (4.69) is proved. Finally we have

$$u_n = C_n e^{-\xi^2/2} H_n(\xi). \tag{4.76}$$

To determine the normalization coefficient C_n write:

$$1 = \int_{-\infty}^{\infty} u_n^* u_n dx = C_n^2 x_0 \int_{-\infty}^{\infty} e^{-\xi^2} H_n(\xi) H_n(\xi) d\xi. \tag{4.77}$$

Substituting the closed form for one of the polynomials H_n we have

$$(-1)^n C_n^2 x_0 \int_{-\infty}^{\infty} H_n(\xi) \frac{d^n}{d\xi^n} e^{-\xi^2} d\xi = 1. \tag{4.78}$$

Transferring the differentiation from the second factor to the first by integrating by parts n times, and noting that all the terms

$$H_n^{(k-1)}(\xi) \frac{d^{n-k}}{d\xi^{n-k}} e^{-\xi^2} d\xi, \tag{4.79}$$

being a polynomial multiplied by $e^{-\xi^2}$, tend to zero as $\xi \to \pm\infty$, we have

$$C_n^2 x_0 \int_{-\infty}^{\infty} e^{-\xi^2} \frac{d^n}{d\xi^n} H_n(\xi) d\xi = 1. \tag{4.80}$$

From Eq. (4.68)

$$\frac{d^n}{d\xi^n} H_n(\xi) = n! 2^n. \tag{4.81}$$

Since

$$\int_{-\infty}^{\infty} e^{-\xi^2} d\xi = \sqrt{\pi}, \tag{4.82}$$

we have

$$C_n^2 n! 2^n x_0 \sqrt{\pi} = 1, \tag{4.83}$$

that is

$$C_n = \frac{1}{\sqrt{n! 2^n x_0 \sqrt{\pi}}}, \tag{4.84}$$

and

$$u_n = \frac{1}{\sqrt{n! 2^n x_0 \sqrt{\pi}}} e^{-\xi^2/2} H_n(\xi), \qquad (\xi = \frac{x}{x_0}, \quad x_0 = \sqrt{\frac{\hbar}{m\omega}}). \tag{4.85}$$

The u_n's (with different n's) are orthogonal since they are eigenstates of the Hermitian operator H corresponding to different eigenvalues. This can also be proved by direct calculation of $(u_{n'}, u_n)$, $n > n'$. In the integral for the scalar product $(u_{n'}, u_n)$, substitute for H_n $(n > n')$ its closed form. After transferring the differentiation n times from the closed form to the polynomial, we shall have the factor

$$\frac{d^n}{d\xi^n} H_{n'} = 0, \ (n > n') \tag{4.86}$$

in the integrand, therefore

$$(u_{n'}, u_n) = 0, \ \text{for } n' \neq n. \tag{4.87}$$

4.6.3 Parity of the Eigenstates

Since H_n is an even or an odd polynomial according to n being an even or an odd integer (see Eq. (4.68)), and since $e^{-\xi^2/2}$ is an even function of x, therefore $u_n(x)$ (see Eq. (4.76)) has a definite parity equals to $(-1)^n$.

4.6.4 The Zero-Point Energy

The zero-point energy, as was mentioned above is a prediction of the wave theory. We shall show here that it is a direct result of the uncertainty relations.

Let the state of the oscillator be u_n, since it has definite parity and x is odd, then $u_n^2 x$ is odd, and

$$\langle x\rangle = (u_n, xu_n) = \int_{-\infty}^{\infty} u_n^2 x dx = 0, \tag{4.88}$$

also $p = \frac{\hbar}{i}\frac{d}{dx}$ is an odd operator, and

$$\langle p\rangle = (u_n, pu_n) = \frac{\hbar}{i}\int_{-\infty}^{\infty} u_n \frac{du_n}{dx} dx = 0. \tag{4.89}$$

Therefore,

$$(\Delta x)^2 = \langle x^2\rangle, \quad (\Delta p)^2 = \langle p^2\rangle. \tag{4.90}$$

$$E = \langle H\rangle = \frac{\langle p^2\rangle + m^2\omega^2\langle x^2\rangle}{2m}. \tag{4.91}$$

According to the uncertainty relations:

$$\langle x^2\rangle\langle p^2\rangle \geq \frac{\hbar^2}{4}, \tag{4.92}$$

therefore

$$E \geq \frac{\hbar^2}{8m\langle x^2\rangle} + \frac{1}{2}m\omega^2\langle x^2\rangle. \tag{4.93}$$

The minimal value of the R.H.S. of this inequality is obtained by equating to zero its first derivative with respect to $\langle x^2\rangle$:

$$\frac{1}{2}m\omega^2 - \frac{\hbar^2}{8m\langle x^2\rangle^2} = 0, \tag{4.94}$$

hence

$$\langle x^2\rangle = \frac{\hbar}{2m\omega} = \frac{x_0^2}{2}. \tag{4.95}$$

The second derivative is

$$\frac{\hbar^2}{4m\langle x^2\rangle^3} > 0, \tag{4.96}$$

consequently this extremum is a minimum. Substituting $\frac{\hbar}{2m\omega}$ for $\langle x^2 \rangle$ in the inequality, we have

$$E \geq \frac{1}{2}\hbar\omega \tag{4.97}$$

and

$$E_0 = \frac{1}{2}\hbar\omega. \tag{4.98}$$

The zero-point energy is a manifestation of the wave properties of the microparticles. The zero-point energy was first observed experimentally in the scattering of X-rays by crystals at low temperatures. If there were no lattice vibrations ($E_0 = 0$), at low temperatures (as T tends to the absolute zero) as predicted by classical mechanics, there would be no interaction between the X-rays and the crystal lattice, and no scattering would occur.

If on the other hand, the minimal energy were different from zero ($E_0 \neq 0$), the scattering cross section at low temperatures should approach a finite limit. Experiments have shown that the scattering cross section does not vanish and tends to a finite limit as $T \to 0$ confirming the prediction of the *Schrödinger* wave theory.

Exercise 4.1 *The oscillator is in the ground state u_0. Find $\langle x^2 \rangle, \langle p^2 \rangle$.*

5

Motion in Central Fields

In this chapter, motion in central fields is considered and separation of the radial and angular parts of the eigenstates is given. *Legendre* and associated *Legendre* polynomials are found as a factor of the spherical harmonics presenting the angular part of the state. Parity of spherical harmonic is found. The components of the angular momentum and also the square of angular momentum are obtained in terms of the spherical polar angles. The orbital angular momentum quantum number and the magnetic quantum number are given.

5.1 Separation of Radial and Angular Parts of the Eigenstates

The Hamiltonian operator of a particle in a central field is

$$H = \frac{p^2}{2m} + U(r). \tag{5.1}$$

It does not depend on time, therefore, a particular solution of the time-dependent *Schrödinger* equation is an energy eigenfunction multiplied by a function of time equal to $e^{-\frac{i}{\hbar}Et}$ where E is the corresponding eigenvalue. The general solution is any superposition (linear combination) of these particular solutions.

The central potential does not depend on the angles θ and φ, hence, it is worth to choose the spherical coordinates to work with. In these coordinates the variables are separable. The energy eigenvalue problem in spherical coordinates reads:

$$\left[\frac{1}{r^2}\frac{\partial}{\partial r}r^2\frac{\partial}{\partial r} + \frac{1}{r^2\sin\theta}\frac{\partial}{\partial\theta}\sin\theta\frac{\partial}{\partial\theta} + \frac{1}{r^2\sin^2\theta}\frac{\partial^2}{\partial\varphi^2}\right]u + \frac{2m_0}{\hbar^2}[E - U(r)]u = 0. \tag{5.2}$$

We seek for u a solution of the form

$$u(r,\theta,\varphi) = R(r)Y(\theta,\varphi). \tag{5.3}$$

DOI: 10.1201/9781003677109-5

Substituting for u in Eq. (5.2), dividing by it and multiplying by r^2, we have

$$\{\frac{1}{R}\frac{d}{dr}r^2\frac{dR}{dr}+\frac{2m_0r^2}{\hbar^2}(E-U(r))\}+\{\frac{1}{Y\sin\theta}\frac{\partial}{\partial\theta}\sin\theta\frac{\partial Y}{\partial\theta}+\frac{1}{Y\sin^2\theta}\frac{\partial^2 Y}{\partial\varphi^2}\}=0. \tag{5.4}$$

The first term in the above equation depends only on r and does not depend on the angles θ and φ, the second depends only on θ and φ, and does not depend on r, therefore their sum can not be zero unless each of them is constant. Put

$$\frac{1}{R}\frac{d}{dr}r^2\frac{dR}{dr}+\frac{2m_0r^2}{\hbar^2}(E-U(r))=\lambda, \tag{5.5}$$

i.e.,

$$\frac{1}{r^2}\frac{d}{dr}r^2\frac{dR}{dr}+\frac{2m_0}{\hbar^2}\left\{E-U-\frac{\hbar^2\lambda}{2m_0r^2}\right\}R=0. \tag{5.6}$$

The second term in Eq. (5.4) must equal $-\lambda$, therefore

$$\left\{\frac{1}{\sin\theta}\frac{\partial}{\partial\theta}\sin\theta\frac{\partial}{\partial\theta}+\frac{1}{\sin^2\theta}\frac{\partial^2}{\partial\varphi^2}\right\}Y=-\lambda Y. \tag{5.7}$$

Notice that the angular equation (Eq. (5.7)) does not depend on the (central) potential, consequently, the angular solution is common to the motion in any central field. The radial equation (Eq. (5.6)) can not be further investigated without giving the concrete form of the potential U, which differs from one application to another.

5.2 The Angular Part—Spherical Harmonics

5.2.1 Separation of θ and φ

The angles θ and φ can be further separated in the angular equation (Eq. (5.7)) by putting

$$Y=\Theta(\theta)\Phi(\varphi). \tag{5.8}$$

Substituting for Y in the angular equation, dividing by it and multiplying by $\sin^2\theta$, we have

$$\left(\frac{\sin\theta}{\Theta}\frac{d}{d\theta}\sin\theta\frac{d\Theta}{d\theta}+\lambda\sin^2\theta\right)+\frac{1}{\Phi}\frac{d^2\Phi}{d\varphi^2}=0. \tag{5.9}$$

The first term in the above equation depends only on θ and does not depend on φ, the second depends only on φ, and does not depend on θ, therefore, their sum can not be zero unless each of them is constant. Put

$$\frac{\sin\theta}{\Theta}\frac{d}{d\theta}\sin\theta\frac{d\Theta}{d\theta}+\lambda\sin^2\theta=m^2, \tag{5.10}$$

i.e.

$$\frac{1}{\sin\theta}\frac{d}{d\theta}\sin\theta\frac{d\Theta}{d\theta}+(\lambda-\frac{m^2}{\sin^2\theta})\Theta=0. \tag{5.11}$$

(The constant m^2 up till now is any complex number).

The second term in Eq. (5.9) must equal $-m^2$, i.e.,

$$\frac{d^2\Phi}{d\varphi^2}=-m^2\Phi. \tag{5.12}$$

The normalization condition on u will be

$$\begin{aligned} 1=(u,u) &= \int R^*\Theta^*\Phi^* R\Theta\Phi r^2\sin\theta dr d\theta d\varphi \\ &= \int_0^\infty R^*Rr^2dr\int_0^\pi \Theta^*\Theta\sin\theta d\theta\int_0^{2\pi}\Phi^*\Phi d\varphi, \end{aligned} \tag{5.13}$$

which shows that each function can be normalized separately:

$$\int_0^\infty R^*Rr^2dr=1,\quad \int_0^\pi \Theta^*\Theta\sin\theta d\theta=1,\quad \int_0^{2\pi}\Phi^*\Phi d\varphi=1. \tag{5.14}$$

Eq. (5.12) has the immediate solution:

$$\Phi=Ce^{im\varphi}. \tag{5.15}$$

Since, (for given r and θ) φ and $\varphi+2\pi$ correspond to the same position, Φ must satisfy:

$$\Phi(\varphi+2\pi)=\Phi(\varphi), \tag{5.16}$$

therefore

$$e^{2im\pi}=1, \tag{5.17}$$

and m can be only an integer

$$m=0,\pm1,\pm2,\pm3,\ldots \tag{5.18}$$

The normalization condition gives $C=\frac{1}{\sqrt{2\pi}}$, and

$$\Phi_m=\frac{1}{\sqrt{2\pi}}e^{im\varphi}. \tag{5.19}$$

It is easily shown that the Φ_m's are orthonormal:

$$(\Phi_m,\Phi_{m'})=\frac{1}{2\pi}\int_0^{2\pi}e^{i(m'-m)\varphi}d\varphi=\delta_{mm'}. \tag{5.20}$$

5.2.2 Solution for Θ

Substituting the new variable x for $\cos\theta$ in Eq. (5.11) we obtain

$$\frac{d}{dx}(1-x^2)\frac{d\Theta}{dx}+(\lambda-\frac{m^2}{1-x^2})\Theta=0. \tag{5.21}$$

To eliminate the singularity in the coefficient of Θ in the above equation, substitute in Eq. (5.21) for Θ:

$$\Theta=(1-x^2)^{s/2}w, \tag{5.22}$$

to obtain

$$(1-x^2)w''-2(s+1)xw'+(\lambda-s^2-s+\frac{s^2-m^2}{1-x^2})w=0. \tag{5.23}$$

The singularity in this equation is removed by taking $s=\pm m$.

Since the solutions $\Theta(\pm m)$ satisfy the same equation (Eq. (5.21) for the same m^2), we must have

$$\Theta(-m)=A\Theta(m), \tag{5.24}$$

and it is enough to find $\Theta(m\geq 0)$. Substitute m for s in Eq. (5.23) and get:

$$(1-x^2)w''-2(m+1)xw'+(\lambda-m(m+1))w=0. \tag{5.25}$$

To obtain its solution in series, put

$$w=\sum_{k=0}^{\infty}a_kx^k \tag{5.26}$$

and substitute for w in Eq. (5.25), to obtain the recurrence relation

$$a_{k+2}=\frac{(k+m+1)(k+m)-\lambda}{(k+2)(k+1)}a_k, \tag{5.27}$$

beginning from a certain value of k all the coefficients are positive. Also we can obtain:

$$\frac{a_k}{a_{k-2}}=1+\frac{2(m-1)}{k}-\frac{\lambda-m(m-1)}{k(k-1)}. \tag{5.28}$$

1. For $m\neq 0$ this ratio is asymptotically equal to the corresponding one in the expansion of the function $(1-x^2)^{-m}$, indeed

$$\begin{aligned}(1-x^2)^{-m}=\sum_{j=0}^{\infty}\frac{(m-1+j)!}{(m-1)!j!}x^{2j}&=\sum_{k=0,2,\ldots}^{\infty}\frac{(m-1+\frac{k}{2})!}{(m-1)!(\frac{k}{2})!}x^k\\&=\sum_{k=0,2,\ldots}^{\infty}b_kx^k,\end{aligned} \tag{5.29}$$

hence

$$\frac{b_k}{b_{k-2}} = \frac{\frac{k}{2}+m-1}{\frac{k}{2}} = 1 + \frac{2(m-1)}{k}, \tag{5.30}$$

therefore, in the neighborhood of $x = \pm 1$, w diverges as $(1-x^2)^{-m}$ and Θ diverges as $(1-x^2)^{-m/2}$ (remember that $m > 0$).

2. For $m = 0$,

$$\frac{a_k}{a_{k-2}} = 1 - \frac{2}{k} - \frac{\lambda}{k(k-1)}. \tag{5.31}$$

which is asymptotically equal to the corresponding ratio in the expansion of the function $\ln(1-x^2)$:

$$\ln(1-x^2) = -\sum_{j=1}^{\infty} \frac{x^{2j}}{j} = -\sum_{k=2,4,6,\ldots}^{\infty} \frac{x^k}{\frac{k}{2}} = \sum b_k x^k, \tag{5.32}$$

hence

$$\frac{b_k}{b_{k-2}} = \frac{\frac{k}{2}-1}{\frac{k}{2}} = 1 - \frac{2}{k}, \tag{5.33}$$

therefore, in the neighborhood of $x = \pm 1$, w and also Θ ($\Theta = w$, for $m = 0$), diverge.

That is in all cases Θ tends to ∞ as $x \to \pm 1$ unless the series in Eq. (5.26) terminates at a some maximum power $k_{\max} = q$, that is

$$a_q \neq 0, \quad a_{q+2} = 0. \tag{5.34}$$

Therefore,

$$\lambda = (q+m)(q+m+1), \quad q = 0, 1, 2, \ldots, \tag{5.35}$$

or

$$\lambda = \ell(\ell+1), \quad \ell = 0, 1, 2, \ldots, \tag{5.36}$$

$$m \leq \ell, \tag{5.37}$$

since

$$m = \ell - q, \quad q \geq 0. \tag{5.38}$$

To obtain the closed form for w, consider the function

$$v = (x^2-1)^\ell \tag{5.39}$$

which satisfies the equation

$$(1-x^2)v' + 2\ell x v = 0. \tag{5.40}$$

Differentiating this equation $(\ell + m + 1)$ times, we obtain

$$(1-x^2)w_1'' - 2(m+1)x w_1' + (\ell - m)(\ell + m + 1)w_1 = 0, \tag{5.41}$$

where

$$w_1 = \frac{d^{\ell+m}}{dx^{\ell+m}}(x^2-1)^\ell. \tag{5.42}$$

Eq. (5.41) is the same as Eq. (5.25) (with $\lambda = \ell(\ell+1)$), therefore $w = \text{const } w_1$. We take this constant to be equal to $\frac{1}{\ell!2^\ell}$, hence

$$w = \frac{d^{\ell+m}}{dx^{\ell+m}}\frac{(x^2-1)^\ell}{\ell!2^\ell}, \tag{5.43}$$

with this choice of the constant, w for $m = 0$ will be the *Legendre* polynomial

$$P_\ell = \frac{d^\ell}{dx^\ell}\frac{(x^2-1)^\ell}{\ell!2^\ell}. \tag{5.44}$$

Therefore, for Θ we have

$$\Theta_\ell^m = C_\ell^m P_\ell^m(x), \tag{5.45}$$

where

$$P_\ell^m = (1-x^2)^{m/2}\frac{d^{\ell+m}}{dx^{\ell+m}}\frac{(x^2-1)^\ell}{\ell!2^\ell}. \tag{5.46}$$

(which is the associated *Legendre* polynomial), C_ℓ^m is the normalization constant.

Since P_ℓ^{-m} and P_ℓ^m satisfy the same equation (Eq. (5.21)), they must be proportional to each other and we write

$$P_\ell^{-m} = (1-x^2)^{-m/2}\frac{d^{\ell-m}}{dx^{\ell-m}}\frac{(x^2-1)^\ell}{\ell!2^\ell} = A(1-x^2)^{m/2}\frac{d^{\ell+m}}{dx^{\ell+m}}\frac{(x^2-1)^\ell}{\ell!2^\ell}, \tag{5.47}$$

or

$$\frac{d^{\ell-m}}{dx^{\ell-m}}(x^2-1)^\ell = A(1-x^2)^m\frac{d^{\ell+m}}{dx^{\ell+m}}(x^2-1)^\ell. \tag{5.48}$$

Equating the coefficients of $x^{\ell+m}$ in both sides of the above equation, we obtain

$$\frac{2\ell!}{(\ell+m)!} = (-1)^m\frac{2\ell!}{(\ell-m)!}A, \tag{5.49}$$

or

$$A = (-1)^m\frac{(\ell-m)!}{(\ell+m)!}, \tag{5.50}$$

and

$$P_\ell^{-m} = (-1)^m\frac{(\ell-m)!}{(\ell+m)!}P_\ell^m. \tag{5.51}$$

Remember that m was positive in inequality (5.37), hence $-m \geq -\ell$ and the quantum number m is

$$m = 0, \pm1, \pm2, \ldots, \pm\ell. \tag{5.52}$$

The normalization constant C_ℓ^m is found from the normalization condition:

$$1 = \int_0^\pi \Theta_\ell^m \Theta_\ell^m \sin\theta d\theta = \int_{-1}^1 \Theta_\ell^m(x)\Theta_\ell^m(x)dx, \tag{5.53}$$

that is

$$1 = \frac{(-1)^m}{(\ell!2^\ell)^2}\frac{(\ell+m)!}{(\ell-m)!}|C_\ell^m|^2 \int_{-1}^1 \{\frac{d^{\ell-m}}{dx^{\ell-m}}(x^2-1)^\ell\}\{\frac{d^{\ell+m}}{dx^{\ell+m}}(x^2-1)^\ell\}dx. \tag{5.54}$$

Notice that for $k < \ell$, $\frac{d^k}{dx^k}(x^2-1)^\ell$ equals to $(x^2-1)^{\ell-k}$ times a polynomial, hence it vanishes at the two bounds of integration, and we can transfer the differentiation from the first factor in the integrand to the second one $\ell - m$ times ($m \geq 0$) to have

$$1 = \frac{1}{(\ell!2^\ell)^2}\frac{(\ell+m)!}{(\ell-m)!}|C_\ell^m|^2 \int_{-1}^1 (1-x^2)^\ell \frac{d^{2\ell}}{dx^{2\ell}}(x^2-1)^\ell dx, \tag{5.55}$$

since

$$\frac{d^{2\ell}}{dx^{2\ell}}x^{2\ell} = (2\ell)!, \quad \int_{-1}^1 (1-x^2)^\ell dx = \frac{(\ell!)^2 2^{2\ell+1}}{(2\ell+1)!}, \tag{5.56}$$

we obtain:

$$C_\ell^m = \sqrt{\frac{2\ell+1}{2}\frac{(\ell-m)!}{(\ell+m)!}}, \tag{5.57}$$

and

$$\Theta_\ell^m = \sqrt{\frac{2\ell+1}{2}\frac{(\ell-m)!}{(\ell+m)!}} P_\ell^m(\cos\theta). \tag{5.58}$$

For spherical harmonics (satisfying Eq. (5.7)) we have

$$Y_\ell^m(\theta,\varphi) = \Theta_\ell^m \Phi_m = \sqrt{\frac{2\ell+1}{4\pi}\frac{(\ell-m)!}{(\ell+m)!}} P_\ell^m(\cos\theta)e^{im\varphi}. \tag{5.59}$$

The orthonormality condition for the spherical harmonics is fulfilled:

$$(Y_{\ell'}^{m'}, Y_\ell^m) = \int_0^\pi \int_0^{2\pi} (Y_{\ell'}^{m'})^* Y_\ell^m \sin\theta d\theta d\varphi = \delta_{\ell'\ell}\delta_{m'm}. \tag{5.60}$$

Indeed, first perform the integral $\frac{1}{2\pi}\int_0^{2\pi} e^{i(m-m')\varphi}d\varphi$ to obtain $\delta_{mm'}$, then in integrating $P_{\ell'}^{m'}P_\ell^m$ over θ we can put $m' = m$ and take $\ell' > \ell$. For $\ell' = \ell$, the integral is one as was shown above, for $\ell' > \ell$ the integral will vanish since

$$\frac{d^{\ell'+\ell}}{dx^{\ell'+\ell}}(x^2-1)^\ell = 0, \ \forall \ell' > \ell. \tag{5.61}$$

5.2.3 Parity of the Spherical Harmonics

The space reflection is effected in spherical coordinates by the transformation:

$$r \to r, \quad \theta \to \pi - \theta, \quad \varphi \to \pi + \varphi. \tag{5.62}$$

Now

$$Y_\ell^m(\pi - \theta, \pi + \varphi) = \Theta_\ell^m(-x)\Phi_m(\pi + \varphi). \tag{5.63}$$

But

$$\Phi_m(\pi + \varphi) = \frac{1}{\sqrt{2\pi}} e^{im\pi} e^{im\varphi} = (-1)^m \Phi_m(\varphi), \tag{5.64}$$

and

$$P_\ell^m(-x) = (1 - x^2)^{m/2}(-1)^{\ell+m} \frac{d^{\ell+m}}{dx^{\ell+m}} \frac{(x^2 - 1)^\ell}{\ell! 2^\ell} = (-1)^{\ell+m} P_\ell^m(x), \tag{5.65}$$

therefore

$$Y_\ell^m(\pi - \theta, \pi + \varphi) = (-1)^{\ell+2m} Y_\ell^m(\theta, \varphi) = (-1)^\ell Y_\ell^m(\theta, \varphi), \tag{5.66}$$

i.e., the spherical harmonics Y_ℓ^m under space reflection has a definite parity equals to $(-1)^\ell$.

5.3 Significance of the Separation Constants ℓ and m—Angular Momentum

If we compare Eq. (5.6)

$$\frac{1}{r^2}\frac{d}{dr} r^2 \frac{dR}{dr} + \frac{2m_0}{\hbar^2}\left\{E - U - \frac{\hbar^2 \lambda}{2m_0 r^2}\right\} R = 0$$

with the corresponding classical equation

$$\frac{1}{2} m_0 \dot{r}^2 = E - \left(U + \frac{L^2}{2m_0 r^2}\right), \tag{5.67}$$

where L is the angular momentum of the particle, we would expect that $\hbar^2\lambda = \hbar^2\ell(\ell + 1)$ is an eigenvalue of the square of the angular momentum and in virtue of Eq. (5.7) we expect that the operator representing L^2 (in spherical coordinates) is

$$L^2 = -\hbar^2 \left(\frac{1}{\sin\theta}\frac{\partial}{\partial\theta}\sin\theta\frac{\partial}{\partial\theta} + \frac{1}{\sin^2\theta}\frac{\partial^2}{\partial\varphi^2}\right). \tag{5.68}$$

To clarify this, let us find in spherical coordinates, the components L_x, L_y, L_z of the angular momentum $\vec{L}$:

$$x = r\sin\theta\cos\phi, \quad y = r\sin\theta\sin\phi, \quad z = r\cos\theta, \quad \rho = r\sin\theta, \tag{5.69}$$

therefore

$$\frac{\partial}{\partial\theta} = \frac{\partial x}{\partial\theta}\frac{\partial}{\partial x} + \frac{\partial y}{\partial\theta}\frac{\partial}{\partial y} + \frac{\partial z}{\partial\theta}\frac{\partial}{\partial z} = z\left[\cos\phi\frac{\partial}{\partial x} + \sin\phi\frac{\partial}{\partial y}\right] - \rho\frac{\partial}{\partial z}, \tag{5.70}$$

$$\frac{\partial}{\partial\phi} = \rho\left(-\sin\phi\frac{\partial}{\partial x} + \cos\phi\frac{\partial}{\partial y}\right), \tag{5.71}$$

$$\cot\theta\frac{\partial}{\partial\phi} = z\left(-\sin\phi\frac{\partial}{\partial x} + \cos\phi\frac{\partial}{\partial y}\right). \tag{5.72}$$

Eliminating $\frac{\partial}{\partial x}$ and $\frac{\partial}{\partial y}$ between Eq. (5.70) and Eq. (5.72) we have

$$\sin\phi\frac{\partial}{\partial\theta} + \cot\theta\cos\phi\frac{\partial}{\partial\phi} = -\left(y\frac{\partial}{\partial z} - z\frac{\partial}{\partial y}\right) \tag{5.73}$$

$$\cos\phi\frac{\partial}{\partial\theta} - \cot\theta\sin\phi\frac{\partial}{\partial\phi} = z\frac{\partial}{\partial x} - x\frac{\partial}{\partial z}, \tag{5.74}$$

also from Eq. (5.71)

$$\frac{\partial}{\partial\phi} = x\frac{\partial}{\partial y} - y\frac{\partial}{\partial x}. \tag{5.75}$$

Therefore,

$$L_x = yp_z - zp_y = -\frac{\hbar}{i}\left(\sin\phi\frac{\partial}{\partial\theta} + \cot\theta\cos\phi\frac{\partial}{\partial\phi}\right), \tag{5.76}$$

$$L_y = zp_x - xp_z = \frac{\hbar}{i}\left(\cos\phi\frac{\partial}{\partial\theta} - \cot\theta\sin\phi\frac{\partial}{\partial\phi}\right), \tag{5.77}$$

$$L_z = xp_y - yp_x = \frac{\hbar}{i}\frac{\partial}{\partial\phi}. \tag{5.78}$$

Notice that L_z is the partial derivative (times $\frac{\hbar}{i}$) with respect to the angle of rotation about the z-axis. This rule is general: the component of the angular momentum in any direction is the partial derivative (times $\frac{\hbar}{i}$) with respect to the angle of rotation about this direction.

Exercise 5.1 *Show that* $\frac{\hbar}{i}\frac{\partial}{\partial\theta}$ *is the component of* $\vec{L}$ *in the direction normal to the plane* $(\vec{e}_z, \vec{e}_r)$.

The easiest way to obtain L^2 in spherical coordinates is as follows:

$$\begin{aligned} L^2 &= (\vec{r}\wedge\vec{p})\cdot(\vec{r}\wedge\vec{p}) = \vec{r}\cdot[\vec{p}\wedge(\vec{r}\wedge\vec{p})] = x_\alpha p_\beta x_\alpha p_\beta - x_\alpha p_\beta x_\beta p_\alpha \\ &= r^2p^2 - (\vec{r}\cdot\vec{p})(p_\beta x_\beta) + 2\frac{\hbar}{i}\vec{r}\cdot\vec{p} \\ &= r^2p^2 - (\vec{r}\cdot\vec{p})^2 - \frac{\hbar}{i}\vec{r}\cdot\vec{p}. \end{aligned} \tag{5.79}$$

Since $\vec{r}\cdot\vec{p}=\frac{\hbar}{i}r\frac{\partial}{\partial r}$, therefore,

$$L^2=-\hbar^2(r^2\nabla^2-r\frac{\partial}{\partial r}r\frac{\partial}{\partial r}-r\frac{\partial}{\partial r})=-\hbar^2\left(\frac{1}{\sin\theta}\frac{\partial}{\partial\theta}\sin\theta\frac{\partial}{\partial\theta}+\frac{1}{\sin^2\theta}\frac{\partial^2}{\partial\varphi^2}\right), \tag{5.80}$$

as was expected at the beginning of this section.

Exercise 5.2 *Square the operators L_x, L_y, L_z in Eqs. (5.76-5.78) and obtain Eq. (5.80).*

Now

$$L^2Y_\ell^m=\hbar^2\ell(\ell+1)Y_\ell^m,\quad L_zY_\ell^m=\hbar mY_\ell^m, \tag{5.81}$$

that is Y_ℓ^m is a simultaneous eigenstate of L^2 and L_z, corresponding to the eigenvalues $\hbar^2\ell(\ell+1)$ and $\hbar m$, where

$$\ell=0,1,2,\ldots,\quad m=0,\pm1,\pm2,\ldots,\pm\ell \tag{5.82}$$

i.e.,

$$L^2=\hbar^2\ell(\ell+1)=(L_z)^2_{\max}+\hbar^2\ell. \tag{5.83}$$

Where the additional term $\hbar^2\ell$ came from? It is again the uncertainty principle.

In fact

$$\langle\hat{L}^2\rangle=\langle\hat{L}_z^2\rangle+\langle\hat{L}_x^2\rangle+\langle\hat{L}_y^2\rangle, \tag{5.84}$$

therefore

$$L^2=L_z^2+\langle\hat{L}_x^2\rangle+\langle\hat{L}_y^2\rangle=(L_z^2)_{\max}+\langle\hat{L}_x^2\rangle_{\min}+\langle\hat{L}_y^2\rangle_{\min}. \tag{5.85}$$

Since $\hat{L}_x$ and $\hat{L}_y$ do not commute with $\hat{L}_z$, the corresponding angular momentum components can not have definite values when $L_z=(L_z)_{\max}=\hbar\ell$, and ΔL_x and ΔL_y do not vanish but have certain minimal values satisfying relation (3.45).

It can be proved that in any L_z eigenstate

$$\langle\hat{L}_x\rangle=0,\quad\langle\hat{L}_y\rangle=0, \tag{5.86}$$

hence

$$(\Delta L_x)^2=\langle\hat{L}_x^2\rangle,\quad(\Delta L_y)^2=\langle\hat{L}_y^2\rangle, \tag{5.87}$$

and since

$$[L_x,L_y]=i\hbar L_z, \tag{5.88}$$

therefore, substituting in the uncertainty relation (3.45) we have

$$\langle\hat{L}_x^2\rangle_{\min}\langle\hat{L}_y^2\rangle_{\min}=\frac{\hbar^2(L_z^2)_{\max}}{4}=\frac{\hbar^4\ell^2}{4}. \tag{5.89}$$

Here the situation is symmetric with respect to the x and y axes, thus

$$\langle \hat{L}_x^2 \rangle_{\rm min} = \langle \hat{L}_y^2 \rangle_{\rm min} = \frac{1}{2}\hbar^2 \ell, \tag{5.90}$$

and

$$\langle \hat{L}_x^2 \rangle_{\rm min} + \langle \hat{L}_y^2 \rangle_{\rm min} = \hbar^2 \ell, \tag{5.91}$$

which is just the extra angular momentum square to be added to $(L_z^2)_{\rm max}$ to obtain L^2. The numbers ℓ and m are called the orbital angular momentum quantum number and the magnetic quantum number.

Exercise 5.3

1. *In an L_z eigenstate, show that $\langle L_x \rangle = 0$ and $\langle L_y \rangle = 0$.*
2. *In an L_z eigenstate, show why, while $\Delta L_z = 0$, ΔL_x and ΔL_y still have finite values although $\hat{L}_x$ and $\hat{L}_y$ do not commute with $\hat{L}_z$.*

6

The Rotator

In this chapter, the rotator is considered which is a good approximation to a diatomic molecule when radial vibrations can be neglected. As a result of very simple calculations, the energy eigenvalues and the degree of degeneracy of the energy levels are found. The s-state is considered as an example of a state with constant probability density (independent of the coordinates). In the p-state, the eigenstates are given and discussed to show what the most probable trajectories of the particle are, and a comparison with the classical situation is given.

6.1 The Energy Eigenvalues

A rotator is a particle moving freely on a sphere. Hence, for a rotator, $r = a$, where a is the radius of the sphere. The motion of the rotator can be considered as a motion in a central field. The central potential in this case is

$$U(r) = U(a) = \text{const.}$$

Since the potential can be measured from an (arbitrary) level, this constant can be taken to be zero. Substituting this for U in the radial equation (5.6) and putting $\lambda = \ell(\ell+1)$, we have:

$$\frac{1}{r^2}\frac{d}{dr}r^2\frac{dR}{dr} + \left(\frac{2m_0E}{\hbar^2} - \frac{\ell(\ell+1)}{a^2}\right)R = 0. \tag{6.1}$$

Since $R(r) = R(a) = \text{const.}$, the first term in Eq. (6.1) vanishes and the energy eigenvalues E_ℓ is easily found to be

$$E_\ell = \frac{\hbar^2\ell(\ell+1)}{2m_0a^2} = \frac{L^2}{2I}, \tag{6.2}$$

where L^2 is the eigenvalue of the square of the angular momentum and $I = m_0a^2$ is the moment of inertia of the rotator about the center of the sphere (compare with the corresponding classical formula).

The rotator model is successfully used for describing the motion of diatomic molecules (when the radial vibrations of the molecule can be neglected). In

DOI: 10.1201/9781003677109-6

this case the moment of inertia should be taken to be

$$I = m_1 r_1^2 + m_2 r_2^2,$$

where m_1, m_2 are the masses of the atoms and r_1, r_2 are their distances from their center of mass. Also, the rotator model describes successfully the rotational motion of nuclei.

6.2 Degeneracy

Equation (6.2) shows that the energy eigenvalue E_ℓ depends only on the orbital quantum number ℓ, while the energy eigenstate Y_ℓ^m depends on both the orbital and the magnetic quantum numbers ℓ and m. Therefore, except for $\ell = 0$, there are more than one (namely $2\ell+1$) linearly independent eigenstates corresponding to the same energy level. In this case, the energy levels are said to be degenerate. The degree of degeneracy of the level E_ℓ is $2\ell + 1$.

The states corresponding to $\ell = 0, 1, 2, 3, 4$ are called the s, p, d, f, g states, respectively. All the energy levels of a rotator are degenerate except the s-level which is not degenerate. For example, the s and p levels are considered. In the s-level $\ell = 0, m = 0$; therefore, the eigenvalue $E_0 = 0$ (Eq. (6.2)) and the eigenstate is Y_0^0.

$$Y_0^0 = \text{const.}$$

this constant, for normalization, must be taken to be

$$Y_0^0 = \frac{1}{\sqrt{4\pi}}.$$

For the probability distribution

$$|Y_0^0|^2 = \frac{1}{4\pi} \quad \text{(probability density per unit solid angle)},$$

i.e., the probability density of finding the rotator anywhere on the sphere in the s-state does not depend neither on θ nor on ϕ and since $L^2 = \hbar^2\ell(\ell+1) = 0$ in the s-state, the particle is at rest on the sphere and is equally likely to be anywhere on its surface. The s-state does not have a classical analogue .

In the p-state $\ell = 1, m = 0, \pm 1$, hence there are 3 eigenstates corresponding to the energy level $E = \frac{\hbar^2}{I}$:

$$Y_1^{+1} = \sqrt{\frac{3}{8\pi}} \sin\theta \, e^{i\phi}, \quad Y_1^0 = \sqrt{\frac{3}{4\pi}} \cos\theta, \quad Y_1^{-1} = -\sqrt{\frac{3}{8\pi}} \sin\theta \, e^{-i\phi}, \quad (6.3)$$

(see Eqs. (5.46), (5.59)).

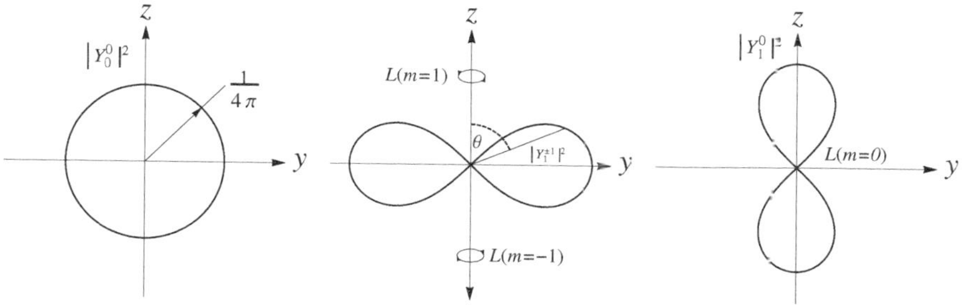

FIGURE 6.1: *Probability density as a function of θ.*

The probability densities (per unit solid angle) in these states are:

$$|Y_1^{\pm}|^2 = \frac{3}{8\pi}\sin^2\theta, \quad |Y_1^0|^2 = \frac{3}{4\pi}\cos^2\theta,$$

hence the most probable of all trajectories of the particle on the sphere in the p-state with $m = \pm 1$ is the trajectory:

$$\theta = \frac{\pi}{2}, \tag{6.4}$$

i.e., the trajectory is in the xy-plane. In the p-state with $m = 0$, it can be shown that[1] the most probable trajectory of the rotator lies in a plane passing through the z-axis, and the angular momentum is perpendicular to it. The classical analog of the states Y_1^1 and Y_1^{-1} are the states of a rotator orbiting on the great circle on the sphere laying in the xy-plane, the two states differ only in the direction of rotation. The first state is a right-handed rotation with the angular momentum parallel to the z-axis, while the second one is a left-handed rotation with the angular momentum antiparallel to the z-axis. It can be shown that[1] the p-state with $m = 0$ is a superposition of the two states of rotation about an axis in the xy-plane with equal probabilities (each equals $\frac{1}{2}$), and the classical analog of this state is a state with the angular momentum perpendicular to the z-axis and the orbit is a great circle in a plane passing through the z-axis (see Fig. 6.1 where the probability density functions are plotted against θ. They are given in the yz-plane. To obtain the complete picture, rotate the graph about the z-axis (since the probability densities do not depend on ϕ).

[1]See Example 11.1.

7

The Hydrogen-Like Atom

In this chapter, the hydrogen-like atom is considered. The radial equation with *Coulomb* potential is solved, and the energy spectrum, the energy eigenstates, and the degree of degeneracy are found. The electron radius expectation value is calculated. The quantum analog of the classical motion in circular orbits is shown. The two-body problem in quantum mechanics is studied to find the effect of the nucleus motion on the energy spectrum of the electron. This effect is the basis of the work of the mass spectrometer which enables to confirm by spectroscopic methods, the existence of the hydrogen isotopes: deuterium and tritium. The effect of the nucleus motion could also confirm the existence of helium in the spectroscopic studies of the solar spectrum.

7.1 The Energy Eigenvalues

The hydrogen-like atom is a single electron (of charge $-e$) bounded to a nucleus (of charge Ze). The (central) potential in this case equals to the electrostatic *Coulomb* potential $-\frac{Ze^2}{r}$. Substituting this for U in the radial equation (5.6) and putting $\lambda = \ell(\ell+1)$, we have:

$$\frac{1}{r^2}\frac{d}{dr}r^2\frac{dR}{dr} + \frac{2m_0}{\hbar^2}\{E + \frac{Ze^2}{r} - \frac{\hbar^2\ell(\ell+1)}{2m_0r^2}\}R = 0. \tag{7.1}$$

Put $R = \frac{\chi}{r}$, then

$$r^2\frac{dR}{dr} = r\frac{d\chi}{dr} - \chi, \qquad \frac{1}{r^2}\frac{d}{dr}r^2\frac{dR}{dr} = \frac{1}{r}\frac{d^2\chi}{dr^2}, \tag{7.2}$$

therefore

$$\frac{d^2\chi}{dr^2} + \frac{2m_0}{\hbar^2}\{E + \frac{Ze^2}{r} - \frac{\hbar^2\ell(\ell+1)}{2m_0r^2}\}\chi = 0. \tag{7.3}$$

As $r \to \infty$, this equation reduces to

$$\frac{d^2\chi_\infty}{dr^2} + \frac{2m_0E}{\hbar^2}\chi_\infty = 0. \tag{7.4}$$

DOI: 10.1201/9781003677109-7

1. For $E > 0$, Eq. (7.4) has oscillatory solutions which correspond to unbound (scattering) states, in which the electron can go to infinity.
2. For $E < 0$, this equation has decreasing exponential solutions which describes bound states.

Therefore, we take $E < 0$ and write:

$$\frac{d^2\chi}{dr^2} + \{-A + \frac{2B}{r} - \frac{\ell(\ell+1)}{r^2}\}\chi = 0, \tag{7.5}$$

where

$$A = -\frac{2m_0 E}{\hbar^2} > 0, \qquad B = \frac{m_0 Ze^2}{\hbar^2} > 0, \tag{7.6}$$

put $\rho = 2\sqrt{A}r$, then

$$\chi'' + \{-\frac{1}{4} + \frac{B}{\sqrt{A}\rho} - \frac{\ell(\ell+1)}{\rho^2}\}\chi = 0, \tag{7.7}$$

where $\chi'' = \frac{d^2\chi}{d\rho^2}$. To find the asymptotic behavior of χ as $\rho \to \infty$, that is when the last two terms in the coefficient of χ are negligible compared with the first one, we have

$$\chi''_\infty - \frac{1}{4}\chi_\infty = 0, \tag{7.8}$$

that is

$$\chi_\infty = c_1 e^{-\rho/2} + c_2 e^{\rho/2}. \tag{7.9}$$

The second term in this expression tends to ∞ as $\rho \to \infty$; therefore, c_2 must be taken equal to zero. Coefficient c_1 can be taken equal to 1.

For the asymptotic solution as $\rho \to 0$, that is when the first two terms in the coefficient of χ (in Eq. (7.7)) are negligible compared with the last one, we have:

$$\chi''_0 - \frac{\ell(\ell+1)}{\rho^2}\chi_0 = 0, \tag{7.10}$$

we try a solution in the form $\chi_0 = \rho^\lambda$, therefore

$$\lambda(\lambda-1) - \ell(\ell+1) = 0, \tag{7.11}$$

hence

$$(\lambda - \ell - 1)(\lambda + \ell) = 0, \tag{7.12}$$

and

$$\lambda = \ell + 1, -\ell. \tag{7.13}$$

The general solution of Eq. (7.10) is

$$\chi_0 = c_1 \rho^{\ell+1} + c_2 \rho^{-\ell}. \tag{7.14}$$

The second term in the above expression tends to ∞ as $\rho \to 0$; therefore, c_2 must be taken equal to zero. Coefficient c_1 can be taken equal to 1.

The solution of Eq. (7.7) can be written in the form

$$\chi = \chi_0 \chi_\infty u = \rho^{\ell+1} e^{-\rho/2} u = vu, \tag{7.15}$$

therefore

$$\begin{aligned} \chi'' &= vu'' + 2v'u' + v''u \\ &= [\rho u'' + \{2(\ell+1) - \rho\}u' + \{\frac{\rho}{4} - (\ell+1) + \frac{\ell(\ell+1)}{\rho}\}u]\rho^\ell e^{-\rho/2}. \end{aligned} \tag{7.16}$$

Substituting this in Eq. (7.7), we have

$$\rho u'' + \{2(\ell+1) - \rho\}u' + (\frac{B}{\sqrt{A}} - \ell - 1)u = 0, \tag{7.17}$$

we solve this equation in series, putting

$$u = \sum_{k=0}^{\infty} a_k \rho^k \tag{7.18}$$

substituting for u in Eq. (7.17), and equating to zero the coefficient of ρ^k we have:

$$a_{k+1} = \frac{k + \ell + 1 - \frac{B}{\sqrt{A}}}{(k + 2\ell + 2)(k+1)} a_k, \tag{7.19}$$

beginning from $k > \dfrac{B}{\sqrt{A}} - \ell - 1$ all the coefficients are positive, and

$$\frac{a_{k+1}}{a_k} \xrightarrow{k \to \infty} \frac{1}{k+1}, \tag{7.20}$$

which is the same as that for the series

$$e^\rho = \sum_{k=0}^{\infty} \frac{1}{k!} \rho^k. \tag{7.21}$$

If the series (7.18) does not terminate, we shall have for $\rho \to \infty$, $u \to e^\rho$ and $R = \rho^\ell e^{-\rho/2} u \to \rho^\ell e^{\rho/2} \to \infty$. Therefore, the power series (7.18) must terminate to be reduced to a polynomial of degree q, say, that is $a_q \neq 0$, $a_{q+1} = 0$. This can be achieved only if

$$\frac{B}{\sqrt{A}} = q + \ell + 1 = n, \tag{7.22}$$

n and q are called the principal and radial quantum numbers. Since

$$q = 0, 1, 2, \ldots, \quad \ell = 0, 1, 2, \ldots \tag{7.23}$$

then

$$n = 1, 2, 3, \ldots \tag{7.24}$$

The energy spectrum for a hydrogen-like atom is found from Eqs. (7.6), (7.22)

$$E_n = -\frac{\hbar^2}{2m_0}A = -\frac{\hbar^2 B^2}{2m_0 n^2} = -\frac{m_0 Z^2 e^4}{2\hbar^2 n^2} = -\frac{Z^2 e^2}{2a_0 n^2} = -\frac{R\hbar Z^2}{n^2}, \tag{7.25}$$

where R is the *Rydberg* constant $R = \frac{m_0 e^4}{2\hbar^3}$, and a_0 is the radius of the *Bohr*'s first orbit $a_0 = \frac{\hbar^2}{m_0 e^2}$. Therefore, the radiation frequency $\omega_{nn'}$ is given by

$$\omega_{nn'} = \frac{E_n - E_{n'}}{\hbar} = Z^2 R\left(\frac{1}{n'^2} - \frac{1}{n^2}\right). \tag{7.26}$$

7.2 The Energy Eigenstates

To obtain the energy eigenstates for the hydrogen-like atom, put $\frac{B}{\sqrt{A}} = q + \ell + 1$ in Eq. (7.19) and obtain

$$a_{k-j} = (-1)^j \frac{k!(s+k)!(q-k)!}{(k-j)!(s+k-j)!(q-k+j)!} a_k, \tag{7.27}$$

where $s = 2\ell + 1$.

Take the coefficient a_q of the highest power to be $(-1)^q$, therefore

$$a_{q-j} = (-1)^{q+j} \frac{q!(s+q)!}{j!(q-j)!(s+q-j)!}, \tag{7.28}$$

and

$$\begin{aligned} u = Q_q^s(\rho) &= \sum_{j=0}^{q} (-1)^{q+j} \frac{q!(s+q)!}{j!(q-j)!(s+q-j)!} \rho^{q-j} \\ &= \sum_{j=0}^{q} (-1)^j \frac{q!(s+q)!}{j!(q-j)!(s+j)!} \rho^j, \end{aligned} \tag{7.29}$$

Q_q^s is an associated *Laguerre* polynomial of degree q and can be represented in a closed form

$$Q_q^s(\rho) = e^{\rho} \rho^{-s} \frac{d^q}{d\rho^q} e^{-\rho} \rho^{q+s}, \tag{7.30}$$

which can easily be proved by just calculating the qth derivative of $e^{-\rho}\rho^{q+s}$.

Thus, for the radial function $R_{n\ell}$, we have

$$R_{n\ell} = C_{n\ell}\rho^{\ell} e^{-\rho/2} Q_{n-\ell-1}^{2\ell+1}(\rho), \tag{7.31}$$

where $\rho = 2\sqrt{A}r$. Substituting $\sqrt{A}$ from Eqs. (7.6) and (7.22), we obtain

$$\rho = \frac{2Zm_0e^2}{n\hbar^2} r = \frac{2Z}{na_0} r. \tag{7.32}$$

To find $C_{n\ell}$, we have:

$$1 = C_{n\ell}^2 \int_0^\infty \rho^{2\ell} e^{-\rho} Q_q^s Q_q^s r^2 dr = C_{n\ell}^2 (\frac{na_0}{2Z})^3 \int_0^\infty \rho^{s+1} e^{-\rho} Q_q^s Q_q^s d\rho. \tag{7.33}$$

Let $I = \int_0^\infty \rho^{s+1} e^{-\rho} Q_q^s Q_q^s d\rho$. Substituting the closed form of the *Laguerre* polynomial, to have

$$I = \int_0^\infty \rho \, Q_q^s \frac{d^q}{d\rho^q} e^{-\rho} \rho^{s+q} d\rho. \tag{7.34}$$

Notice that for $k < q$, the kth derivative of $e^{-\rho}\rho^{s+q}$ is $e^{-\rho}$ times a polynomial, the lowest power of which is greater than s; hence, this (k^{th}) derivative vanishes at both bounds of the above integral, and integrating q times by parts, we have

$$I = (-1)^q \int_0^\infty e^{-\rho} \rho^{s+q} \frac{d^q}{d\rho^q} \rho Q_q^s d\rho, \tag{7.35}$$

using Eq. (7.29), to obtain

$$\begin{aligned} I &= q! \int_0^\infty e^{-\rho}\rho^{s+q}\{(q+1)\rho - q(s+q)\} d\rho \\ &= q!(s+q)!\{(q+1)(s+q+1) - q(s+q)\} \\ &= q!(s+q)!(2q+s+1) \\ &= (n-\ell-1)!(n+\ell)!(2n), \end{aligned} \tag{7.36}$$

therefore

$$C_{n\ell} = \left(\frac{Z}{na_0}\right)^{3/2} \sqrt{\frac{4}{(n-\ell-1)!(n+\ell)!n}}, \tag{7.37}$$

thus $R_{n\ell}$ is found, and when multiplied by Y_ℓ^m, the energy eigenstate $u_{n\ell m}$ of the electron in the hydrogen-like atom is obtained.

7.3 Parity

What is the parity of $u_{n\ell m} = R_{n\ell} Y_\ell^m$?

7.4 Degeneracy

Equation (7.25) shows that the energy eigenvalue depends only on the principal quantum number n (that is only on the sum of the orbital and radial quantum numbers ℓ and q), while the energy eigenstate $u_{n\ell m} = R_{n\ell}Y_\ell^m$ depends on each of the three quantum numbers n, ℓ, and m. Therefore, (except for $(n,\ell,m) = (1,0,0)$) there are more than one linearly independent eigenstates corresponding to the same energy eigenvalue. In this case the energy levels are said to be degenerate. The degree of degeneracy of an enegy level is the number of eigenstates corresponding to this level. For the energy level E_n in hydrogen, since ℓ varies from 0 to $n-1$ in unit steps and m varies from $-\ell$ to ℓ in unit steps also, therefore, the degree of degeneracy is

$$\sum_{\ell=0}^{n-1}\sum_{m=-\ell}^{\ell} 1 = \sum_{\ell=0}^{n-1} 2\ell + 1 = n^2. \tag{7.38}$$

Thus, all the energy levels in hydrogen are degenerate except the ground state $(n = 1)$.

7.5 Electron Radius Expectation Value

Let the electron of the hydrogen-like atom be in the state $u_{n\ell m}$, hence,

$$\langle r\rangle = (u_{n\ell m}, ru_{n\ell m}) = \int_0^\infty rr^2R_{n\ell}^2dr \oint (Y_\ell^m)^*Y_\ell^m d\Omega = \int_0^\infty rr^2R_{n\ell}^2dr. \tag{7.39}$$

Therefore, $r^2R_{n\ell}^2$ is the radial probability density (distribution). Denote it by D, that is

$$D(r) = r^2R_{n\ell}^2. \tag{7.40}$$

Let $\ell = n-1$ (thus $q = 0$). Since $Q_0^{2\ell+1} = 1$, $\forall\ell$ (see Eq. (7.29)) and from Eq. (7.29), we have

$$R_{n,n-1} = \text{const. } \rho^{n-1}e^{-\rho/2}, \tag{7.41}$$

therefore

$$D(r) = r^2R_{n\ell}^2 = \text{const } \rho^{2n}e^{-\rho}, \tag{7.42}$$

(notice that the constant in the above equation is positive). It is easy to show that D has only one extremum which is a maximum; therefore, the states with $q = 0$ (no radial excitations) are the quantum mechanical analogue of the classical motion in circular orbits; see Fig. 7.1. The value r_n of r at which D is maximum, that is the most probable value of r, is given by

$$D'(r) = 0, \tag{7.43}$$

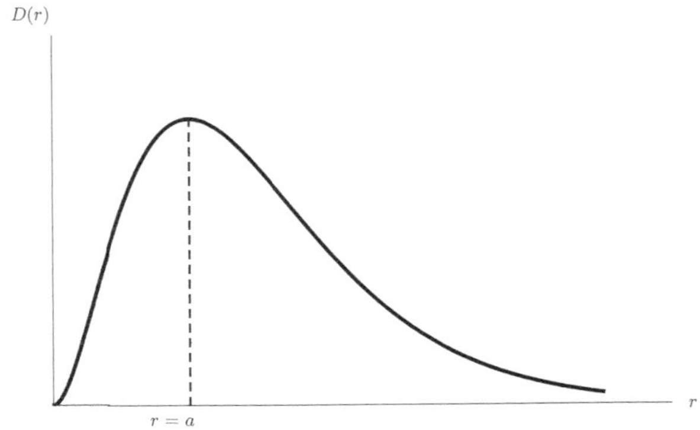

FIGURE 7.1: *Radial probability density (distribution) in the case of circular orbits.*

hence $2n - \rho_n = 0$, $\rho_n = 2n$, and from Eq. (7.32),

$$r_n = \frac{n^2}{Z} a_0, \tag{7.44}$$

when there are radial excitations ($q \neq 0, \ell < n - 1$), D has several maxima and the state will be the quantum mechanical analogue of the classical motion in elliptic orbits.

7.6 Effects of the Nucleus Motion. The Mass Spectrometer

In obtaining Eq. (7.1), we considered the motion in a central potential, i.e., we implicitly assumed that the nucleus does not move. This would be rigorously correct only for a nucleus of infinite mass. In fact, the hydrogen-like atom is a two-body problem. The hamiltonian for it is

$$H = \frac{p_1^2}{2m_0} + \frac{p_2^2}{2M} + U(|\vec{r}_1 - \vec{r}_2|), \tag{7.45}$$

where m_0 and M are the masses of the electron and the nucleus, $\vec{r}_1$ and $\vec{r}_2$ are their position vectors, therefore

$$H = -\frac{\hbar^2}{2m_0}\nabla_1^2 - \frac{\hbar^2}{2M}\nabla_2^2 + U(|\vec{r}_1 - \vec{r}_2|). \tag{7.46}$$

The six coordinates $\vec{r}_1$ and $\vec{r}_2$ are inseparable in *Schrödinger* equation, since they all appear explicitly in the potential U. Introducing $\vec{R}$, the position vector of the center of mass of the atom and $\vec{r}$, the position vector of the electron relative to the nucleus:

$$\vec{R} = \frac{m_0\vec{r}_1 + M\vec{r}_2}{m_0 + M}, \qquad \vec{r} = \vec{r}_1 - \vec{r}_2, \tag{7.47}$$

we have:

$$\frac{\partial}{\partial x_1} = \frac{\partial X}{\partial x_1}\frac{\partial}{\partial X} + \frac{\partial x}{\partial x_1}\frac{\partial}{\partial x} = \frac{m_0}{m_0 + M}\frac{\partial}{\partial X} + \frac{\partial}{\partial x}, \tag{7.48}$$

$$\frac{\partial^2}{\partial x_1^2} = \frac{m_0^2}{(m_0 + M)^2}\frac{\partial^2}{\partial X^2} + \frac{\partial^2}{\partial x^2} + \frac{2m_0}{m_0 + M}\frac{\partial^2}{\partial X \partial x}. \tag{7.49}$$

Similarly,

$$\frac{\partial^2}{\partial x_2^2} = \frac{M^2}{(m_0 + M)^2}\frac{\partial^2}{\partial X^2} + \frac{\partial^2}{\partial x^2} - \frac{2M}{m_0 + M}\frac{\partial^2}{\partial X \partial x}, \tag{7.50}$$

hence,

$$\frac{1}{m_0}\frac{\partial^2}{\partial x_1^2} + \frac{1}{M}\frac{\partial^2}{\partial x_2^2} = \frac{1}{m_0 + M}\frac{\partial^2}{\partial X^2} + \frac{m_0 + M}{m_0 M}\frac{\partial^2}{\partial x^2}, \tag{7.51}$$

therefore

$$H = -\frac{\hbar^2}{2\mathcal{M}}\nabla_{\vec{R}}^2 - \frac{\hbar^2}{2\mu}\nabla_{\vec{r}}^2 + U(r), \tag{7.52}$$

where $\mu = \frac{m_0 M}{m_0 + M}$ is the reduced mass and $\mathcal{M} = m_0 + M$ is the total mass of the atom.

The time-independent *Schrödinger* equation is

$$Hu(\vec{R}, \vec{r}) = \mathcal{E}u(\vec{R}, \vec{r}). \tag{7.53}$$

To separate the variables $\vec{R}$ and $\vec{r}$, try a solution in the form

$$u(\vec{R}, \vec{r}) = W(\vec{R})w(\vec{r}). \tag{7.54}$$

Substituting this for u in Eq. (7.53), we obtain

$$-\frac{\hbar^2}{2\mathcal{M}}\frac{1}{W}\nabla_{\vec{R}}^2 W + \left(-\frac{\hbar^2}{2\mu}\frac{1}{w}\nabla_{\vec{r}}^2 w + U\right) = \mathcal{E}. \tag{7.55}$$

The first term in the L.H.S. of this equation depends only on $\vec{R}$ and does not depend on $\vec{r}$; the second depends only on $\vec{r}$ and does not depend on $\vec{R}$; therefore, their sum can not be a constant ($\mathcal{E}$) unless each of them is a constant. Denote these two constants by E and E', and write

$$-\frac{\hbar^2}{2\mathcal{M}}\nabla_{\vec{R}}^2 W = E'W, \tag{7.56}$$

$$-\frac{\hbar^2}{2\mu}\nabla_{\vec{r}}^2 w + Uw = Ew. \tag{7.57}$$

(Notice that $E + E' = \mathcal{E}$).

Thus, the two-body problem is reduced to two problems: The first is the problem of motion of a free fictitious particle with the total mass $\mathcal{M}$ at the center of mass of the atom, the second is the problem of motion in a central field of another fictitious particle with the reduced mass μ at a distance r from the center of the potential. The solution of these two one-body problems is equivalent to the solution of the original two-body problem.

It is easily shown that

$$W(\vec{R}) = \text{const. } e^{\frac{i}{\hbar}\vec{P}\cdot\vec{R}}, \tag{7.58}$$

where $\vec{P} = \langle\vec{p}_1\rangle + \langle\vec{p}_2\rangle$, is the total momentum of the atom.

Exercise 7.1 *Prove that*

1. $\vec{\nabla}_{\vec{R}} = \vec{\nabla}_{\vec{r}_1} + \vec{\nabla}_{\vec{r}_2}$,
2. $\frac{d}{dt}\langle\vec{p}_1\rangle \neq 0,\ \frac{d}{dt}\langle\vec{p}_2\rangle \neq 0$ *but* $\frac{d}{dt}\langle\vec{p}_1 + \vec{p}_2\rangle = 0$,
3. $E' = \frac{P^2}{2\mathcal{M}}$.

When the atom (its center of mass) is at rest,

$$\vec{P} = 0, \quad W = \text{const.}, \tag{7.59}$$

therefore

$$u(\vec{R}, \vec{r}) = w(\vec{r}). \tag{7.60}$$

Eq. (7.57) is just the same as the corresponding Eq. (5.2) with the mass m_0 of the electron replaced by the reduced mass μ. Separating the radial and angular parts of Eq. (7.57), we obtain for the angular part an equation identical with Eq. (5.7), and for the radial part we obtain Eq. (7.1) with m_0 replaced by μ. Replacing m_0 by the reduced mass

$$\mu = \frac{m_0}{1 + \frac{m_0}{M}} \simeq m_0\left(1 - \frac{m_0}{M}\right), \tag{7.61}$$

the *Rydberg* constant becomes

$$R_M = \frac{\mu e^4}{2\hbar^3} = R_\infty\left(1 - \frac{m_0}{M}\right), \tag{7.62}$$

where $R_\infty = \frac{m_0 e^4}{2\hbar^3}$ is the *Rydberg* constant for a nucleus of an infinite mass ($M \to \infty$) and R_M is the *Rydberg* constant for a nucleus with finite mass M. Therefore, the radiation frequency is given by

$$\omega_{nn'} = Z^2 R_\infty\left(1 - \frac{m_0}{M}\right)\left(\frac{1}{n'^2} - \frac{1}{n^2}\right), \tag{7.63}$$

which differs from $\omega_{nn'}$ in Eq. (7.26) by the factor $(1-\frac{m_0}{M})$. This factor depends on the nucleus mass, therefore it gives rise to slight differences between the spectrum of different isotopes of the same chemical element. The very high resolving power (great precision) achieved in atomic spectroscopy admits the measurement of these small differences allowing to measure atomic masses by spectroscopic methods. This is the physics, the work of the mass spectrometer is based on. Chemical methods give the average atomic weights of a chemical element (relative to carbon-12). The mass spectrometer can measure the atomic weight of each atom, it gave a different value for the atomic mass of hydrogen:

$$\frac{M_{\text{chem}} - M_{\text{spec}}}{M_{\text{chem}}} = 0.015\%. \tag{7.64}$$

This led *Birge* and *Menzel* (1931) to predict the existence of a hydrogen isotope, the deuterium (heavy hydrogen), with an atomic mass twice that of hydrogen ($D =^2_1H$). The presence of deuterium in natural hydrogen explains the greater atomic mass obtained by chemical methods. The existence of deuterium was confirmed by spectroscopic studies which showed that the *Balmer* sequence ($n' = 2$) not only consists of the lines

$$\omega_H = R_\infty \left(1 - \frac{1}{1840}\right)\left(\frac{1}{2^2} - \frac{1}{n^2}\right), \tag{7.65}$$

but that each of these lines is associated with a second close line to its right. This second sequence of lines [5, Fig. 13.5] is described by

$$\omega_D = R_\infty \left(1 - \frac{1}{3680}\right)\left(\frac{1}{2^2} - \frac{1}{n^2}\right), \tag{7.66}$$

which is obtained from Eq. (7.63) by setting M equal to twice the hydrogen nucleus mass and $Z = 1$.

Another isotope of hydrogen, the tritium ($T =^3_1H$) is also known. The position of the spectral lines of tritium are slightly displaced to the right of the deuterium lines, and are given by

$$\omega_T = R_\infty \left(1 - \frac{1}{5520}\right)\left(\frac{1}{2^2} - \frac{1}{n^2}\right), \tag{7.67}$$

Deuterium combines with oxygen to form heavy water D_2O, discovered by *Urey* and *Osbon* (1932). On the account of the large difference between the masses of the deuterium and hydrogen atoms, their physical and chemical properties differ more than for the isotopes of the other elements. Thus although heavy water is the same in appearance as ordinary water, its melting and boiling points at atmospheric pressure are 3.82° C and 101.4° C, its viscosity is greater and it is a poorer solvent for salts. Heavy water is particularly important because it is a good moderator for fast neutrons and can be used as a source of deuterium. Deuterium can be obtained by electrolytic decomposition of water: the rate of evolution of hydrogen at the cathode exceeds the rate of evolution of deuterium, thus the concentration of deuterium in the residual electrolyte increases. It is practically impossible to detect deuterium in natural water because of its low concentration. The ratio of deuterium atoms to hydrogen atoms in natural water is 1/6800, whereas the ratio of tritium atoms to hydrogen atoms is approximately 10^{-18} (tritium also combines with oxygen to form tritium water T_2O). A mixture of deuterium with tritium is a very important substance for thermonuclear reactions (nuclear fusion): reaction between ${}_1^2H$ and ${}_1^3H$ nuclei gives ${}_2^4He$ and one neutron. Each such reaction releases more than 17 MeV of energy.[1] Fusion of deuterium and tritium into helium, however, is possible only if the potential barrier of the *Coulomb* repulsion between the D and T nuclei is overcome. High temperature of order 10^8 degrees[2] is needed to start such reaction. In the case of nuclear fission, slow neutrons at low temperatures, can easily penetrate the nucleus, a mixture of deuterium and tritium must first be heated to a temperature of tens of millions of degrees. Such a temperature may be created in atomic explosion. A controlled thermonuclear reaction (nuclear fusion) can be carried out in confined plasma, the obtaining of which is a chief subject of plasma physics. Tritium is also a beta emitter (with half-lifetime of 12 years) and consequently is used as a radioactive indicator in chemical and biological investigations.

Another important consequence of the nucleus motion, was the discovery of ionized helium, first detected (1868) in the spectroscopic studies of the sun. The solar spectrum was found to contain a sequence of lines, with positions described by the equation

$$\omega_{2n_1} = R\left(\frac{1}{2^2} - \frac{1}{n_1^2}\right), \tag{7.68}$$

where n_1 takes the values

$$n_1 = \frac{5}{2}, 3, \frac{7}{2}, 4, \dots \tag{7.69}$$

This sequence is, in effect, the *Balmer* hydrogen sequence ($n_1 = 3, 4, 5, \dots$) with a number of intermediate lines, which form the *Pickering* sequence, characterized by the half-odd integral quantum numbers ($n_1 = 5/2, 7/2, 9/2, \dots$). At first, the *Pickering* sequence was explained by the hypothesis that hydrogen is in special state in the sun, so that the quantum number n can assume half-odd integral values. The second hypothesis was that the observed spectrum comes from singly ionized helium $({}_2^4He)^+$ whose nuclear mass is $M = 7360\, m_0$, and nuclear charge $Z = 2$. According to Eq. (7.63), its radiation frequencies

[1] This means that a mixture of one mole (4 g) of deuterium and one mole (6 g) of tritium (the mass of the mixture is 10 g) when exhausted in this reaction releases about 328×10^{10} J of energy which is enough to raise to the melting point and melt a mass of iron equal to 3750 Ton. ($N_A = 6.02 \times 10^{23}$, MeV$=1.602 \times 10^{-13}$ J, room temperature $= 30°$C, for iron: melting point $= 1530°$C, specific heat $= 500$ J/kg.degree and latent heat $= 126$ J/kg.)

[2] The temperature of the surface of the sun is 6000 K.

are

$$\omega_{He} = 2^2 R_{He} \left(\frac{1}{n'^2} - \frac{1}{n^2} \right) . \tag{7.70}$$

Setting $n' = 4$, we have

$$\omega_{He} = R_{He} \left(\frac{1}{2^2} - \frac{1}{(\frac{n}{2})^2} \right) , \tag{7.71}$$

where $n = 5, 6, 7, \ldots$

To decide which of the two hypotheses is the correct one, the value of the *Rydberg* constant was to be found experimentally to see whether it is that of hydrogen

$$R_H = R_\infty (1 - \frac{1}{1840}), \tag{7.72}$$

or that of helium

$$R_{He} = R_\infty (1 - \frac{1}{7360}). \tag{7.73}$$

The value of R obtained by experiment agreed with R_{He}, and thus it was shown that the *Pickering* sequence gives the spectrum of helium.

Part II

Matrix Mechanics

8

Some Elements of the Transformation Theory

In this chapter, the representation of states and operators on them in an orthonormal basis of a linear Hermitian operator is given, in addition to the transformation between different representations. The projection operator is explained, and the closure property is expressed in terms of the projection operators on the basis states. The *Dirac* bra and ket notations are established and explained. The coordinate and momentum representations are treated in more detail. The relation between *Fourier* transformation and the transformation from the coordinate representation to the momentum representation is established. The coordinate and momentum operators on functions of the coordinates are found.

8.1 The $\mathcal{A}$-Representation

The set $\{u_m\}$ is the complete orthonormal set of eigenstates of a linear Hermitian operator $\mathcal{A}$. $\{u_m\}$ is complete, hence any state f in a vector space can be linearly expanded in terms of them:

$$f = \sum_m \mathrm{f}'_m u_m, \qquad \mathrm{f}'_m = (u_m, f). \tag{8.1}$$

f is said to be represented in the $\mathcal{A}$-representation by the one column matrix

$$\mathrm{f}' = \begin{pmatrix} \mathrm{f}'_1 \\ \mathrm{f}'_2 \\ \vdots \end{pmatrix} = \begin{pmatrix} (u_1, f) \\ (u_2, f) \\ \vdots \end{pmatrix}. \tag{8.2}$$

Exercise 8.1 *Find u_m in this representation, i.e., find u'_m (one of the basis states) in this basis.*

DOI: 10.1201/9781003677109-8

It is easily seen that αf and $f + \mathrm{g}$ are represented by $\alpha \mathrm{f}'$ and $\mathrm{f}' + g'$, respectively, where α is any number and g is any state represented by g'. Also:

$$(f, \mathrm{g}) \;=\; \sum_m (\mathrm{f}'_m u_m, \mathrm{g}) = \sum_m \mathrm{f}'^*_m (u_m, \mathrm{g}) = \sum_m \mathrm{f}'^*_m g'_m = \mathrm{f}'^\dagger g'. \quad (8.3)$$

Let $\mathcal{K}$ be a linear operator, $\mathcal{K} f = \mathrm{g}$. Then

$$g'_\ell = (u_\ell, \mathcal{K} f) = \sum_m (u_\ell, \mathcal{K} u_m)\, \mathrm{f}'_m = \sum_m \mathrm{k}'_{\ell m} \mathrm{f}'_m, \quad (8.4)$$

where

$$\mathrm{k}'_{jm} = (u_j, \mathcal{K} u_m), \quad (8.5)$$

i.e.,

$$g' = K' \mathrm{f}', \quad (8.6)$$

where

$$K' = [\mathrm{k}'_{jm}]. \quad (8.7)$$

Also,

$$(h, \mathcal{K} f) = (h, \mathrm{g}) = \mathrm{h}'^\dagger g' = \mathrm{h}'^\dagger K' \mathrm{f}' = (K'^\dagger \mathrm{h}')^\dagger \mathrm{f}'. \quad (8.8)$$

Exercise 8.2

1. *Prove that a linear operator is represented in its own representation by a diagonal matrix, the diagonal elements of which are its eigenvalues.*

2. *Prove that* $\mathcal{O}, \mathcal{I}, \alpha \mathcal{K}, \mathcal{K} + \mathcal{L}$ *and* $\mathcal{K}\mathcal{L}$ *are represented by* $O', I', \alpha K', K' + L'$*, and* $K'L'$*, where* $\mathcal{O}$ *and* $\mathcal{I}$ *are the null and the unit operators,* $\mathcal{L}$ *is any operator represented by* L'*, and* O' *and* I' *are the zero and unit matrices in this representation.*

3. *Prove that* $\mathcal{K}^\dagger$ *and* $\mathcal{K}^{-1}$ *(if exists) are represented by* $K'^\dagger$ *and* K'^{-1}.

4. *Prove that: If* $\mathcal{K}$ *is Hermitian or unitary, then* K' *is also so.*

8.2 Transformation to Another Representation

Let $\{v_\mu\}$ be the complete orthonormal set of eigenstates of another linear Hermitian operator $\mathcal{B}$. In this representation, f is represented by f'' as follows

$$f = \sum_\mu \mathrm{f}''_\mu v_\mu, \quad \mathrm{f}''_\mu = (v_\mu, f). \quad (8.9)$$

v_μ is linearly expanded in terms of $\{u_m\}$ as

$$v_\mu = \sum_m \mathrm{v}'_{m\mu}\, u_m, \qquad \mathrm{v}'_{m\mu} = (u_m, v_\mu). \tag{8.10}$$

Therefore,

$$f = \sum_m \mathrm{f}'_m\, u_m = \sum_\mu \mathrm{f}''_\mu\, v_\mu = \sum_{m\,\mu} \mathrm{v}'_{m\mu}\, \mathrm{f}''_\mu\, u_m. \tag{8.11}$$

Thus,

$$\mathrm{f}'_m = \sum_\mu \mathrm{v}'_{m\mu}\, \mathrm{f}''_\mu. \tag{8.12}$$

That is,

$$\mathrm{f}' = V'\mathrm{f}'', \quad \text{where} \quad V' = [\mathrm{v}'_{m\mu}] = [(u_m, v_\mu)]. \tag{8.13}$$

If we consider the old representation to be the $\mathcal{B}$-representation and the $\mathcal{A}$-representation to be the new one, i.e., if we interchange u_m and v_μ, we get

$$\mathrm{f}'' = U''\mathrm{f}', \quad U'' = [\mathrm{u}''_{\mu m}] = [(v_\mu, u_m)] = V'^{\dagger}, \tag{8.14}$$

that is the inverse transformation is performed by $U'' = V'^{\dagger}$. From Eq. (8.13), $\mathrm{f}'' = V'^{-1}\mathrm{f}'$, therefore,

$$U'' = V'^{-1} = V'^{\dagger} \qquad \text{(i.e. } V' \text{ is unitry)}, \tag{8.15}$$

$$\mathrm{f}'' = V'^{\dagger}\mathrm{f}'. \tag{8.16}$$

Exercise 8.3 *Make the analogy with 3-dimensional vector space.*

The scalar product is invariant under unitary transformations (rotations):

$$\mathrm{f}'^{\dagger} g' = (V'\mathrm{f}'')^{\dagger}\, V' g'' = \mathrm{f}''^{\dagger}\, V'^{\dagger}\, V' g'' = \mathrm{f}''^{\dagger} g'' = (f, \mathrm{g}). \tag{8.17}$$

Notice that, $V' = [\mathrm{v}'_1\ \mathrm{v}'_2\ \ldots]$, i.e., the columns of the transformation matrix from the $\mathcal{B}$ to the $\mathcal{A}$-representation, are just the v's represented in the $\mathcal{A}$-representation and those of U'' are the u's represented in the $\mathcal{B}$-representation: $U'' = [\mathrm{u}''_1\ \mathrm{u}''_2\ \ldots]$.

How an operator is transformed? If $\mathcal{K} f = \mathrm{g}$, then $K'\mathrm{f}' = g'$, $K''\mathrm{f}'' = g''$ where K'' is the $\mathcal{B}$-representation of $\mathcal{K}$.
Thus,

$$K'\mathrm{f}' = V' K''\mathrm{f}'' = V' K'' V'^{\dagger}\mathrm{f}', \quad \forall \mathrm{f}'. \tag{8.18}$$

Therefore,

$$K' = V' K'' V'^{\dagger}. \tag{8.19}$$

And the eigenvalue problem of a linear Hermitian operator is equivalent to the diagonalization of a Hermitian matrix by a unitary one. Notice that the columns of the transformation matrix V' from the diagonal form to the given one are just the required eigenstates in the given representation, and the elements of the diagonalized matrix are the required eigenvalues.[1]

[1] See Example 11.1.

8.3 Projection Operator

The projection operator $\mathcal{P}_f$ on a state f is defined as

$$\mathcal{P}_f\, \mathrm{g} = \frac{(f, \mathrm{g})}{(f, f)} f. \tag{8.20}$$

Exercise 8.4 *1. Prove that $\mathcal{P}_f$ is linear, Hermitian and*

$$\mathcal{P}_f f = f, \quad \mathcal{P}_f^2 = \mathcal{P}_f. \tag{8.21}$$

2. Prove that

$$P_f' = \frac{f' f'^{\dagger}}{f'^{\dagger} f'}. \tag{8.22}$$

The matrix elements of the projection operator $\mathcal{P}_m$ on one of the states u_m are

$$(P_m')_{ij} = (\mathrm{u}_m' \, \mathrm{u}_m'^{\dagger})_{ij} = \delta_{im}\, \delta_{mj} \text{ and } (P_m'')_{\mu\nu} = (\mathrm{u}_m'' \, \mathrm{u}_m''^{\dagger})_{\mu\nu} = (v_\mu, u_m)\,(u_m, v_\nu). \tag{8.23}$$

Since

$$\sum_m P_m' = \sum_m \mathrm{u}_m' \, \mathrm{u}_m'^{\dagger} = I', \tag{8.24}$$

then

$$\sum_m \mathcal{P}_m = \mathcal{I}, \tag{8.25}$$

this is a mathematical expression of the closure property of the complete orthonormal set of eigenfunctions of linear Hermitian operators.

Formally speaking, all what preceded is kept, without requiring that m or μ to be finite or even discrete. It may be in part discrete and in part continuous. In the continuous part of the spectrum, $\sum_m$ is replaced by $\int dm$, f_m' by $\mathrm{f}'(m)$ and δ_{jm} by $\delta(j - m)$. Moreover, j can mean ordered sets of numbers like (n, l, m) of the energy-spectrum of the hydrogen-like atoms, and here n, l, m are integers, $-l \leq m \leq l$, $0 \leq l \leq n-1$, $n \geq 1$ as is well known. The spectrum of the free particle momentum is the ordered triplet (p_x, p_y, p_z) each number of which is a real number ranging from $-\infty$ to ∞.

Exercise 8.5 *Show by direct calculations that* $\sum_m P_m'' = I''$.

8.4 *Dirac* Bra and Ket Notations

The scalar product (f, g) is denoted by $\langle f|\mathrm{g}\rangle$ which can be looked at as the product of two vectors from two different spaces, the ket space $\{|f\rangle\}$ and the dual to it space $\{\langle f|\}$ which is called the bra space. The state to the right is taken from the ket space. $\langle f|g\rangle$ is a bracket, the names of the two kinds of vectors came from this word.

In the ket space we define addition and multiplication by a number as

$$\begin{aligned} |f\rangle + |g\rangle &= |f + g\rangle, \\ \alpha|f\rangle &= |\alpha f\rangle. \end{aligned} \tag{8.26}$$

i.e., the sum of two kets is the ket of the sum of the two corresponding states, and the ket multiplied by a number is the ket of the corresponding state multiplied by this number. This definition of addition and multiplication by a number satisfies the conditions for the ket space to be a linear one. Also, the definition of operators in the ket space is

$$\mathcal{K}|f\rangle = |\mathcal{K}f\rangle, \tag{8.27}$$

i.e., the ket after been operated on by an operator is just the ket of the corresponding state after been operated on by the same operator.

Remember that, in a certain representation

$$\langle f|g\rangle = \mathrm{f}^\dagger g, \tag{8.28}$$

that is, in any representation, the ket $|f\rangle$ is represented by the one-column matrix f and the bra $\langle f|$ is represented by its adjoint one-row matrix $\mathrm{f}^\dagger$.

To preserve the scalar product properties, addition and multiplication by a number in the bra space must be defined as

$$\langle f| + \langle g| = \langle f + g|, \tag{8.29}$$

$$\alpha\langle f| = \langle \alpha^* f|. \tag{8.30}$$

This definition satisfies the condition fro the bra space to be antilinear.

The definition of operators in the bra space is:

$$\langle f|\mathcal{K} = \langle \mathcal{K}^\dagger f|. \tag{8.31}$$

(Notice that operators in the bra space act on the bras from the right is just as the square matrices multiply row matrices only from the right.)

By the definition, the Hermitian adjoint operator $\mathcal{K}^\dagger$ of an operator $\mathcal{K}$ is $(\mathcal{K}^\dagger f, h) = (f, \mathcal{K}h)\ \forall f, h$.

In bra and ket notations this reads:

$$\langle \mathcal{K}^\dagger f|h\rangle = \langle f|\mathcal{K}h\rangle, \tag{8.32}$$

that is (by the definition of operators in the bra and ket spaces), we have:

$$(\langle f|\mathcal{K})|h\rangle = \langle f|(\mathcal{K}|h\rangle), \tag{8.33}$$

and there is no need to write the brackets, so we write

$$(f, \mathcal{K}h) = \langle f|\mathcal{K}|h\rangle, \tag{8.34}$$

and we can, equally well, consider $\mathcal{K}$ to act either from the left on $|h\rangle$ or from the right on $\langle f|$, so we can always take the more convenient choice for the calculations.

Notice that

$$\langle f|\mathcal{K}|\mathrm{g}\rangle^* = \langle \mathrm{g}|\mathcal{K}^\dagger|f\rangle. \tag{8.35}$$

If we define $|f\rangle^\dagger$, $\langle f|^\dagger$ such that $(\langle f||\mathrm{g}\rangle)^\dagger = |\mathrm{g}\rangle^\dagger\langle f|^\dagger$, then (remember that $\langle f|\mathrm{g}\rangle$ is a complex number):

$$\langle f|\mathrm{g}\rangle^* = \langle f|\mathrm{g}\rangle^\dagger = |\mathrm{g}\rangle^\dagger\langle f|^\dagger = \langle \mathrm{g}|f\rangle, \quad \text{hence}$$

$$|\mathrm{g}\rangle^\dagger = \langle \mathrm{g}| \quad \text{and} \quad \langle f|^\dagger = |f\rangle,$$

that is, the kets are the adjoints of the corresponding bras and vice versa, and e.g.,

$$\langle f|\mathcal{K}|\mathrm{g}\rangle^* = \langle f|\mathcal{K}|\mathrm{g}\rangle^\dagger = \langle \mathrm{g}|\mathcal{K}^\dagger|f\rangle. \tag{8.36}$$

Example 8.1

1. Eq. (8.20) in the bra and ket notations is

$$|\mathcal{P}_f\, g\rangle = \mathcal{P}_f|g\rangle = \frac{\langle f|g\rangle}{\langle f|f\rangle}|f\rangle = \frac{|f\rangle\langle f|}{\langle f|f\rangle}|g\rangle \quad \forall g, \tag{8.37}$$

therefore, $\mathcal{P}_f = \frac{|f\rangle\langle f|}{\langle f|f\rangle}$ *(compare with Eq. (8.22)).*

2. $\mathcal{P}_m = |m\rangle\langle m|$ *and Eq. (8.25) will have the form*

$$\sum_m |m\rangle\langle m| = \mathcal{I}. \tag{8.38}$$

3. The solution of Exercise 8.4.1 in the bra and ket notations is extremely simple.

Exercise 8.6 *Rewrite the equations from Eq. (8.1) to Eq. (8.25) in the bra and ket notation. (Most of these equations in this notation become obvious identities.)*

8.5 Coordinate and Momentum-Representations

8.5.1 The Coordinate-Representation

Let $\psi_{\vec{r}}$ be the orthonormal eigenstates of the position operator $\vec{\mathcal{R}}$, i.e.,

$$\vec{\mathcal{R}}\,\psi_{\vec{r}} = \vec{r}\,\psi_{\vec{r}}, \qquad (\psi_{\vec{r}}, \psi_{\vec{r}\,'}) = \langle \vec{r}|\vec{r}\,'\rangle = \delta(\vec{r} - \vec{r}\,'), \tag{8.39}$$

(the ket $|\psi_{\vec{r}}\rangle$ is denoted by $|\vec{r}\rangle$ for simplicity). Therefore, in the coordinate representation:

$$f = \int \mathrm{f}(\vec{r})|\vec{r}\rangle\, d^3r, \quad \mathrm{f}(\vec{r}) = \langle\vec{r}\,|f\rangle, \tag{8.40}$$

$\mathrm{f}(\vec{r})$ (the coordinate representation of f) can be looked at as the matrix element of the one column matrix f in the row labeled by $\vec{r} = (x, y, z)$.

Also,

$$\langle f|\mathrm{g}\rangle = \langle f| \int |\vec{r}\rangle\langle\vec{r}\,|\mathrm{g}\rangle d^3r = \int \langle f|\vec{r}\rangle\langle\vec{r}\,|\mathrm{g}\rangle d^3r = \int \mathrm{f}^*(\vec{r})\, g(\vec{r})\, d^3r. \tag{8.41}$$

If $\mathcal{K} f = \mathrm{g}$ and $\hat{\mathrm{k}}(\vec{r})\mathrm{f}(\vec{r}) = g(\vec{r})$, then

$$\begin{aligned} g(\vec{r}) &= \int K_{\vec{r}\vec{r}\,'}\, \mathrm{f}(\vec{r}\,')\, d^3r' = \hat{\mathrm{k}}(\vec{r})\, \mathrm{f}(\vec{r}) = \hat{\mathrm{k}}(\vec{r}) \int \mathrm{f}(\vec{r}\,')\, \delta(\vec{r} - \vec{r}\,')\, d^3r' \\ &= \int \mathrm{f}(\vec{r}\,')\hat{\mathrm{k}}(\vec{r})\, \delta(\vec{r} - \vec{r}\,')\, d^3r'. \end{aligned} \tag{8.42}$$

Thus,

$$K_{\vec{r}\vec{r}\,'} = \hat{\mathrm{k}}(\vec{r})\, \delta(\vec{r} - \vec{r}\,'). \tag{8.43}$$

Example 8.2 *In the coordinate representation Eq. (8.38) reads:*

$$I_{\vec{r}\vec{r}\,'} = \langle\vec{r}\,|\mathcal{I}|\vec{r}\,'\rangle = \sum_m \langle\vec{r}\,|m\rangle\langle m|\vec{r}\,'\rangle = \sum_m u_m(\vec{r}\,)u_m^*(\vec{r}\,') \tag{8.44}$$

$$= \langle\vec{r}\,|\vec{r}\,'\rangle = \delta(\vec{r} - \vec{r}\,'), \tag{8.45}$$

which is the same as Eq. (3.18).

Exercise 8.7 *Show that* $(h, \mathcal{K} f) = \int h^*(\vec{r})\, \hat{k}(\vec{r})\, f(\vec{r})\, d^3r$.

8.5.2 The Momentum-Representation

Let $\varphi_{\vec{p}}$ be the orthonormal eigenstates of the momentum operator $\vec{\mathcal{P}}$, i.e.,

$$\vec{\mathcal{P}}\,\varphi_{\vec{p}} = \vec{p}\,\varphi_{\vec{p}}, \quad (\varphi_{\vec{p}}, \varphi_{\vec{p}\,'}) = \langle\vec{p}\;\vec{p}\,'\rangle = \delta(\vec{p} - \vec{p}\,'), \tag{8.46}$$

(the ket $|\varphi_{\vec{p}}\rangle$ is denoted by $|\vec{p}\rangle$ for simplicity). If $\tilde{\mathrm{f}}$ is the momentum representation of f, then

$$f = \int \tilde{\mathrm{f}}(\vec{p})\,|\vec{p}\rangle\, d^3p, \quad \tilde{\mathrm{f}}(\vec{p}) = \langle\vec{p}\,|f\rangle. \tag{8.47}$$

Since

$$(\psi_{\vec{r}}, \varphi_{\vec{p}}) = \langle\vec{r}\,|\vec{p}\rangle = \varphi_{\vec{p}}(\vec{r}), \tag{8.48}$$

and $\phi_{\vec{p}}$ is the coordinate representation of the momentum eigenstate $\varphi_{\vec{p}}$, i.e.,

$$\phi_{\vec{p}}(\vec{r}) = \frac{1}{(2\pi\hbar)^{3/2}} e^{\frac{i}{\hbar}\vec{p}\cdot\vec{r}}, \tag{8.49}$$

(see Eq. (3.85)) and the comment after Eq. (3.86). Therefore,

$$\mathrm{f}(\vec{r}) = \langle \vec{r}\,|f\rangle = \int \tilde{\mathrm{f}}(\vec{p})\,\langle \vec{r}\,|\vec{p}\rangle\, d^3p, \tag{8.50}$$

i.e.,

$$\mathrm{f}(\vec{r}) = \int \tilde{\mathrm{f}}(\vec{p})\,\phi_{\vec{p}}(\vec{r})\,d^3p = \frac{1}{(2\pi\hbar)^{3/2}} \int \tilde{\mathrm{f}}(\vec{p})\,e^{\frac{i}{\hbar}\vec{p}\cdot\vec{r}}\,d^3p. \tag{8.51}$$

That is, $\tilde{\mathrm{f}}(\vec{p})$ is the *Fourier* transform of $\mathrm{f}(\vec{r})$. Similarly,

$$\begin{aligned} \tilde{\mathrm{f}}(\vec{p}) &= \langle \vec{p}\,|f\rangle = \int \mathrm{f}(\vec{r})\,\langle \vec{p}\,|\vec{r}\rangle\,d^3r = \int \mathrm{f}(\vec{r})\,\phi^*_{\vec{p}}(\vec{r})\,d^3r \\ &= \frac{1}{(2\pi\hbar)^{3/2}} \int \mathrm{f}(\vec{r})\,e^{-\frac{i}{\hbar}\vec{p}\cdot\vec{r}}\,d^3r, \end{aligned} \tag{8.52}$$

which is the inverse *Fourier* transformation.
Eqs. (8.51) and (8.52) can formally be written as

$$\mathrm{f} = \Phi\,\tilde{\mathrm{f}}, \quad \Phi_{\vec{r}\vec{p}} = \langle \vec{r}\,|\vec{p}\rangle = \phi_{\vec{p}}(\vec{r}), \tag{8.53}$$

$$\tilde{\mathrm{f}} = \tilde{\Psi}\,\mathrm{f}, \quad \tilde{\Psi}_{\vec{p}\vec{r}} = \langle \vec{p}\,|\vec{r}\rangle = \tilde{\psi}_{\vec{r}}(\vec{p}) = \phi^*_{\vec{p}}(\vec{r}). \tag{8.54}$$

That is,

$$\tilde{\Psi} = \Phi^\dagger = \Phi^{-1}, \tag{8.55}$$

(compare these equations with Eqs. (8.13)–(8.16)).
Also,

$$K = \Phi\,\tilde{K}\,\Phi^\dagger. \tag{8.56}$$

That is,

$$K_{\vec{r}\vec{r}\,'} = \int \phi_{\vec{p}}(\vec{r})\tilde{K}_{\vec{p}\vec{p}\,'}\,\phi^*_{\vec{p}\,'}(\vec{r}\,')\,d^3p'\,d^3p = \frac{1}{(2\pi\hbar)^3} \int \tilde{K}_{\vec{p}\vec{p}\,'}\,e^{\frac{i}{\hbar}(\vec{p}\cdot\vec{r}-\vec{p}\,'\cdot\vec{r}\,')}\,d^3p'\,d^3p. \tag{8.57}$$

8.5.3 The Position Operator on Functions of the Coordinates

It is immediate to prove that the position operator $\hat{\vec{r}}$ acting on functions of the coordinates is just the multiplication of these functions by the position vector. Let

$$\vec{\mathcal{R}}f = \vec{g} \quad \text{and} \quad \hat{\vec{r}}\,\mathrm{f}(\vec{r}) = \vec{g}(\vec{r}), \quad \text{hence}$$

$$\hat{\vec{r}}\,\mathrm{f}(\vec{r}) = \vec{g}(\vec{r}) = \langle \vec{r}\,|(\vec{\mathcal{R}}|f\rangle) = (\langle \vec{r}\,|\vec{\mathcal{R}})|f\rangle = \vec{r}\langle \vec{r}\,|f\rangle = \vec{r}\,\mathrm{f}(\vec{r}) \quad \forall\,\mathrm{f},$$

i.e. $\hat{\vec{r}}$ is the multiplication operator by $\vec{r}$, and

$$\vec{R}_{\vec{r}\vec{r}\,'} = \hat{\vec{r}}\,\delta(\vec{r}-\vec{r}\,') = \vec{r}\,\delta(\vec{r}-\vec{r}\,'),$$

that is, $\vec{R}$ in its own representation is diagonal as expected.

8.5.4 The Momentum Operator on Functions of the Coordinates

A particle in the momentum eigenstate $\varphi_{\vec{p}}$, is a particle with definite momentum $\vec{p}$ and energy E. Therefore, according to *De Broglie* hypothesis, its motion is described by a wave with definite frequency $\frac{1}{\hbar}E$ and definite wave vector $\frac{1}{\hbar}\vec{p}$, which is the plane monochromatic wave $e^{-\frac{i}{\hbar}(Et-\vec{p}\cdot\vec{r})}$. Therefore, the position factor of this function is the coordinate representation of $\varphi_{\vec{p}}$, i.e.,

$$\phi_{\vec{p}}(\vec{r}) = \langle\vec{r}\,|\vec{p}\rangle = \frac{1}{(2\pi\hbar)^{3/2}}\,e^{\frac{i}{\hbar}\vec{p}\cdot\vec{r}} \qquad (\frac{1}{(2\pi\hbar)^{3/2}} \text{ is a normalization factor}).$$

By this formula for $\langle\vec{r}\,|\vec{p}\rangle$, which is just a mathematical expression of *De Broglie* hypothesis, one can prove that the momentum operator $\hat{\vec{p}}$ on functions of the coordinates is $\hat{\vec{p}} = \frac{\hbar}{i}\frac{\partial}{\partial\vec{r}}$.

Let $\vec{\mathcal{P}}f = \vec{g}$, $\hat{\vec{p}}\mathrm{f}(\vec{r}) = \vec{g}(\vec{r})$, hence $\hat{\vec{p}}\mathrm{f}(\vec{r}) = \vec{g}(\vec{r}) = \langle\vec{r}\,|\vec{\mathcal{P}}|f\rangle$.

Insert the unit operator $\int|\vec{p}\rangle\langle\vec{p}|d^3p\int|\vec{r}\,'\rangle\langle\vec{r}\,'|d^3r'$ between $\vec{\mathcal{P}}$ and f to have

$$\begin{aligned}\hat{\vec{p}}\mathrm{f}(\vec{r}) &= \int\langle\vec{r}\,|\vec{\mathcal{P}}|\vec{p}\rangle\langle\vec{p}\,|\vec{r}\,'\rangle\langle\vec{r}\,'|f\rangle d^3p\,d^3r' = \int\vec{p}\,\langle\vec{r}\,|\vec{p}\rangle\langle\vec{p}\,|\vec{r}\,'\rangle\langle\vec{r}\,'|f\rangle d^3p\,d^3r' \\ &= \frac{1}{(2\pi\hbar)^3}\int\vec{p}\,e^{\frac{i}{\hbar}\vec{p}\cdot(\vec{r}-\vec{r}\,')}\,\mathrm{f}(\vec{r}\,')\,d^3p\,d^3r' = \frac{\hbar}{i}\frac{\partial}{\partial\vec{r}}\frac{1}{(2\pi\hbar)^3}\int\mathrm{f}(\vec{r}\,')\,d^3r'\int e^{\frac{i}{\hbar}\vec{p}\cdot(\vec{r}-\vec{r}\,')}\,d^3p \\ &= \frac{\hbar}{i}\frac{\partial}{\partial\vec{r}}\int\mathrm{f}(\vec{r}\,')\,\delta(\vec{r}-\vec{r}\,')\,d^3r' = \frac{\hbar}{i}\vec{\nabla}\mathrm{f}(\vec{r}) \quad \forall\mathrm{f}, \text{ therefore } \quad \hat{\vec{p}} = \frac{\hbar}{i}\vec{\nabla}.\end{aligned}$$

Exercise 8.8 *Prove that the momentum and the position operators acting on functions of the momentum are the multiplication operator by* $\vec{p}$ *and the operator* $i\hbar\frac{\partial}{\partial\vec{p}}$.

8.5.5 Transformation from the $\mathcal{A}$ or $\mathcal{B}$-Representation to the Coordinate-Representation

Since

$$f = \sum_m \mathrm{f}_m^{'}\,u_m$$

hence

$$\langle\vec{r}\,|f\rangle = \sum_m \mathrm{f}_m^{'}\,\langle\vec{r}\,|m\rangle,$$

i.e.,

$$\mathrm{f}(\vec{r}) = \sum_m \mathrm{f}_m^{'}\,\mathrm{u}_m\,(\vec{r}). \tag{8.58}$$

That is,

$$\mathrm{f} = U\mathrm{f}^{'}, \quad U_{\vec{r}m} = \langle\vec{r}\,|m\rangle = \mathrm{u}_m(\vec{r}). \tag{8.59}$$

Similarly,

$$\mathrm{f} = V\,\mathrm{f}^{''}, \quad V_{\vec{r}\mu} = \langle \vec{r}\,|\mu\rangle = \mathrm{v}_{\mu}(\vec{r}). \tag{8.60}$$

Therefore,

$$\mathrm{f}^{'} = U^{\dagger}\,V\,\mathrm{f}^{''}, \;\; \mathrm{f}^{''} = V^{\dagger}\,U\,\mathrm{f}^{'}, \tag{8.61}$$

i.e.,

$$V' = U^{\dagger}\,V, \;\; U'' = V^{\dagger}\,U. \tag{8.62}$$

Exercise 8.9

1. Prove by direct calculations that $V' = U^{\dagger}\,V, \quad U'' = V^{\dagger}\,U$.

2. Find the transformation from the $\mathcal{A}$ *or* $\mathcal{B}$ *representation to the momentum representation and make the analog of Eqs. (8.61), (8.62).*

9

Equation of Motion

In this chapter, *Schrödinger* and *Heisenberg* pictures are explained, and it is shown that they are connected by a unitary transformation. The *Heisenberg* equation of motion is deduced. The matrix elements of operators not explicitly depending on time are found in terms of their initial values and time. It is shown that the scalar products, expectation values, matrix elements, and the commutation relations are picture independent owing to the close resemblance between *Heisenberg* and *Poisson* equations of motion, *Dirac* proved that the quantum mechanical analogue of the Poisson bracket of two dynamical variables is the commutator of the two operators representing them divided by $i\hbar$. The *Lorentz* force is deduced in quantum mechanics using commutator algebra.

9.1 *Poisson* Equation

In classical mechanics, a dynamical variable A, being a function of the mechanical state, depends on time even if it does not depend on it explicitly, indeed

$$\frac{dA}{dt} = \frac{\partial A}{\partial t} + \sum_{i=1}^{s} \frac{\partial A}{\partial q_i}\dot{q}_i + \frac{\partial A}{\partial p_i}\dot{p}_i, \tag{9.1}$$

where s is the number of the degrees of freedom of the mechanical system. It is important to remember that $p_i = \dfrac{\partial L}{\partial \dot{q}_i}$ is the canonical momentum conjugate to the generalized coordinate q_i. From *Hamilton*'s equations:

$$\dot{q}_i = \frac{\partial H}{\partial p_i}, \quad \dot{p}_i = -\frac{\partial H}{\partial q_i}, \tag{9.2}$$

(H is the Hamiltonian of the system),

$$\begin{aligned}\frac{dA}{dt} &= \frac{\partial A}{\partial t} + \sum_{i=1}^{s} \frac{\partial A}{\partial q_i}\frac{\partial H}{\partial p_i} - \frac{\partial A}{\partial p_i}\frac{\partial H}{\partial q_i} \\ &= \frac{\partial A}{\partial t} + \{A, H\},\end{aligned} \tag{9.3}$$

DOI: 10.1201/9781003677109-9

where $\{A,H\}$ is an abbreviation of

$$\sum_{i=1}^{s} \frac{\partial A}{\partial q_i}\frac{\partial H}{\partial p_i} - \frac{\partial A}{\partial p_i}\frac{\partial H}{\partial q_i}, \tag{9.4}$$

and is known as the *Poisson* bracket of A,H. For any two dynamical variables

$$\{A,B\} = \sum_{i=1}^{s} \frac{\partial A}{\partial q_i}\frac{\partial B}{\partial p_i} - \frac{\partial A}{\partial p_i}\frac{\partial B}{\partial q_i} \tag{9.5}$$

is the *Poisson* bracket of A,B. It is clear that $\{q_i, q_j\} = 0$, $\{p_i, p_j\} = 0$, $\{q_i, p_j\} = \delta_{ij}$. Eq. (9.3) is the equation of motion of the dynamical variable A in *Poisson* brackets. One of the privileges of this equation is that it gives an immediate proof of the following theorem: A dynamical variable is a constant of motion (integral of motion, conserved) if it does not depend explicitly on time and its *Poisson* bracket with the Hamiltonian is zero.

9.2 *Schrödinger* and *Heisenberg* Pictures

In quantum mechanics, dynamical variables are represented mathematically by operators and physical states by ket and bra vectors. *Schrödinger* equation reads:

$$i\hbar\frac{d}{dt}|\alpha\rangle = H|\alpha\rangle. \tag{9.6}$$

If the Hamiltonian operator H does not depend explicitly on time, this equation has the formal solution

$$|\alpha(t)\rangle = e^{-\frac{it}{\hbar}H}|\alpha(0)\rangle, \tag{9.7}$$

where the exponential operator e^A is defined as

$$e^A = \sum_{k=0}^{\infty}\frac{1}{k!}A^k. \tag{9.8}$$

It is easily seen that this solution satisfies *Schrödinger* equation by differentiating the power series, term by term.

It is easy to notice that in general:

$$e^A e^B \neq e^{A+B}. \tag{9.9}$$

But if $[A,B] = 0$, then

$$e^A e^B = e^{A+B}. \tag{9.10}$$

Exercise 9.1 *Prove that if A is Hermitian, then e^A is Hermitian and e^{iA} is unitary.*

Now consider the matrix element of an operator A between two states α and β, i.e., consider the scalar product $\langle\alpha|A|\beta\rangle$ at any time t: From (9.7)

$$\begin{aligned}\langle\alpha(t)|A|\beta(t)\rangle &= (\langle\alpha(0)|e^{\frac{it}{\hbar}H})A(e^{-\frac{it}{\hbar}H}|\beta(0)\rangle) \\ &= \langle\alpha(0)|e^{\frac{it}{\hbar}H}Ae^{-\frac{it}{\hbar}H}|\beta(0)\rangle. \qquad (9.11)\end{aligned}$$

Here, we have two equivalent points of view, either to bear the dependence on time of the matrix element on the states $e^{-\frac{it}{\hbar}H}|\alpha(0)\rangle$, $e^{-\frac{it}{\hbar}H}|\beta(0)\rangle$, and the operator A then does not depend on time except for a possible explicit dependence (first line of Eq. (9.11)), or to bear the dependence on time of the matrix element on the operator $e^{\frac{it}{\hbar}H}Ae^{-\frac{it}{\hbar}H}$ and the states $|\alpha(0)\rangle$, $|\beta(0)\rangle$ then do not depend on time (second line of Eq. (9.11)). The first point of view is the *Schrödinger* picture, and the second one is the *Heisenberg* picture. The states and operators are distinguished by subscript S (*Schrödinger*) in the first picture and subscript H (*Heisenberg*) in the second one. It is clear that the two pictures are connected by the unitary transformation $e^{\frac{it}{\hbar}H}$ (since H is Hermitian, then $e^{\frac{it}{\hbar}H}$ is unitary):

$$|\alpha_{\rm H}(t)\rangle = e^{\frac{it}{\hbar}H}|\alpha_{\rm S}(t)\rangle = \alpha_{\rm S}(0)\rangle, \qquad (9.12)$$

$$A_{\rm H} = e^{\frac{it}{\hbar}H}A_{\rm S}e^{-\frac{it}{\hbar}H}. \qquad (9.13)$$

Therefore:

$$\frac{dA_{\rm H}}{dt} = e^{\frac{it}{\hbar}H}\frac{\partial A_{\rm S}}{\partial t}e^{-\frac{it}{\hbar}H} + \frac{i}{\hbar}\left[e^{\frac{it}{\hbar}H}HA_{\rm S}e^{-\frac{it}{\hbar}H} - e^{\frac{it}{\hbar}H}A_{\rm S}He^{-\frac{it}{\hbar}H}\right], \qquad (9.14)$$

i.e.,

$$\frac{dA_{\rm H}}{dt} = \frac{\partial A_{\rm H}}{\partial t} + \frac{1}{i\hbar}\left[A_{\rm H}, H\right], \qquad (9.15)$$

which is the *Heisenberg* equation of motion.

Notice that

$$\frac{\partial A_{\rm H}}{\partial t} = e^{\frac{it}{\hbar}H}\frac{\partial A_{\rm S}}{\partial t}e^{-\frac{it}{\hbar}H},$$

from the definition, $e^{\pm\frac{it}{\hbar}H}$ commutes with H, therefore $H_{\rm H} = H_{\rm S} = H$.

From *Heisenberg* equation of motion, a sufficient condition for a dynamical variable to be conserved is that it commutes with the Hamiltonian and does not depend explicitly on time.

In the energy representation, for operators not explicitly depending on time:

$$\begin{aligned}(A_{\rm H}(t))_{mn} &= \langle m(0)|e^{\frac{it}{\hbar}H}A_{\rm S}e^{-\frac{it}{\hbar}H}|n(0)\rangle \\ &= \langle m(0)|A_{\rm S}|n(0)\rangle e^{i\frac{E_m-E_n}{\hbar}t} \\ &= (A_{\rm H}(0))_{mn}e^{i\frac{E_m-E_n}{\hbar}t} \qquad (9.16)\end{aligned}$$

(for $t = 0$, states and operators in the two pictures coincide (Eqs. (9.12), (9.13))).

Since the *Heisenberg* and *Schrödinger* pictures are transformed one to the other by a unitary transformation, the scalar products, expectation values, matrix elements, and the commutation relations are picture independent.

9.3 Quantization Procedure

The very close resemblance between *Heisenberg* and *Poisson* equations of motion shows that *Bohr*'s correspondence principle between quantum and classical mechanics is simply performed by the substitution: $\{A, B\} \to \frac{1}{i\hbar}[A, B]$.

Since the fundamental *Poisson* brackets are

$$\{q_i, q_j\} = 0, \quad \{p_i, p_j\} = 0, \quad \{q_i, p_j\} = \delta_{ij}, \tag{9.17}$$

then the fundamental commutators would be

$$[q_i, q_j] = 0, \quad [p_i, p_j] = 0, \quad [q_i, p_j] = i\hbar\delta_{ij} \tag{9.18}$$

which are called also the quantum conditions.

It is important to know that the canonical variables must be expressed in Cartesian coordinates and, after being considered as operators in agreement with the quantum conditions, can be, if desired, transformed accurately to any other coordinates ($p^2 = -\hbar^2\nabla^2$ as an example) (for details see e.g., [12, Ch. 15, § 4]). *Dirac* proved that the quantum mechanical analogue of the *Poisson* bracket of two dynamical variables is the commutator of the two operators representing them divided by $i\hbar$ and assumed that at least the most simple of the quantum *Poisson* brackets have the same values as the corresponding classical ones. Thus, he was led to the quantum conditions Eq. (9.18) [13].

Heisenberg (1925) without introducing the notion of an operator, associated with each physical observable a square array of numbers and deduced on experimental grounds (*Ritz* combination principle) the rule of calculating the array corresponding to the product of two observables from the two arrays corresponding to these two observables. *Born* and *Jordan* noticed that this rule is just the rule of matrix multiplication, and thus the *Heisenberg* arrays were considered as infinite matrices. This form of the new theory was known as Matrix Mechanics.

The first to consider the wave function $\psi(\vec{r}, t)$ was *De Broglie* (1924–1925) who introduced the idea of the associated wave for the electron and thus founded Wave Mechanics. *Schrödinger* (1926) made the formulation of the theory by introducing his famous equation and proved in the same year that both matrix and wave mechanics are two different formalisms of one and the same theory "Quantum Mechanics" (see § 8).

Exercise 9.2

1. Prove that q_j is a multiplication operator and $p_j = \frac{\hbar}{i}\frac{\partial}{\partial q_j}$ or $q_j = i\hbar\frac{\partial}{\partial p_j}$ and p_j is a multiplication operator, are solutions of Eq. (9.18).

2. Find the classical Lagrangian of a closed system of two interacting particles of masses m_1, m_2 in terms of the center of mass and the relative position vectors and the corresponding velocities. Find the Hamiltonian as a function of these position vectors and the canonically conjugate momenta. Hence use the quantum conditions Eq. (9.18) to find the Hamiltonian operator of the system. Compare the result with Eq. (7.52).

9.4 Example: Motion of a Particle in an Electromagnetic Field

m_0, e are the mass and charge of the particle and $\varphi, \vec{A}$ are the scalar and vector potentials of the field. The nonrelativistic Lagrange function of the particle is

$$L = \frac{1}{2}m_0 v^2 + \frac{e}{c}\vec{A}\cdot\vec{v} - e\varphi. \tag{9.19}$$

The canonical momentum $\vec{p}$ of the particle is

$$\vec{p} = \frac{\partial L}{\partial \vec{v}} = m_0\vec{v} + \frac{e}{c}\vec{A} = \vec{p}_{\text{kin}} + \frac{e}{c}\vec{A} \tag{9.20}$$

where $\vec{p}_{\text{kin}} = m_0\vec{v}$ is the kinetic momentum of the particle. The classical Hamiltonian of the particle is:

$$H = \vec{v}\cdot\vec{p} - L = m_0 v^2 + \frac{e}{c}\vec{A}\cdot\vec{v} - L = \frac{1}{2}m_0 v^2 + e\varphi = \frac{p_{\text{kin}}^2}{2m_0} + e\varphi. \tag{9.21}$$

The operator

$$\frac{p_{\text{kin}}^2}{2m_0} + e\varphi$$

is not the Hamiltonian, although it is equal to it, because the Hamiltonian must be expressed not in terms of the kinetic momentum but it must be expressed in terms of the canonical momentum $\vec{p}$ the components of which will satisfy the quantum conditions in the quantization procedure. That is,

$$H = \frac{\left(\vec{p} - \frac{e}{c}\vec{A}\right)^2}{2m_0} + e\varphi. \tag{9.22}$$

We shall work in the *Heisenberg* picture and omit the subscript H,

$$\begin{aligned}\frac{dx_\alpha}{dt} &= \frac{1}{i\hbar}[x_\alpha, H] = \frac{1}{2im_0\hbar}\left[x_\alpha, \sum_{\nu=1}^{3}(p_\nu - \frac{e}{c}A_\nu)^2\right] \\ &= \frac{1}{2im_0\hbar}\sum_{\nu=1}^{3}\left[x_\alpha, (p_\nu - \frac{e}{c}A_\nu)\right](p_\nu - \frac{e}{c}A_\nu) + (p_\nu - \frac{e}{c}A_\nu)\left[x_\alpha, (p_\nu - \frac{e}{c}A_\nu)\right] \\ &= \frac{1}{m_0}\sum_{\nu=1}^{3}(p_\nu - \frac{e}{c}A_\nu)\delta_{\nu\alpha} = \frac{p_\alpha - \frac{e}{c}A_\alpha}{m_0},\end{aligned} \tag{9.23}$$

i.e.,

$$\frac{d\vec{r}}{dt} = \frac{\vec{p} - \frac{e}{c}\vec{A}}{m_0}, \tag{9.24}$$

which agrees with the classical relation. Since the states in the *Heisenberg* picture are time-independent, then

$$\frac{d}{dt}\langle\vec{r}\,\rangle = \langle\frac{d\vec{r}}{dt}\rangle = \frac{\langle\vec{p} - \frac{e}{c}\vec{A}\,\rangle}{m_0}, \tag{9.25}$$

which is the first part of *Ehrenfest*'s theorem (putting $\vec{A} = 0$).

Exercise 9.3 *Find the Lorentz force* $(m_0\frac{d^2\vec{r}}{dt^2})$.

10

Linear Harmonic Oscillator in the Energy Representation

In this chapter, the linear harmonic oscillator is studied in the energy representation, and it is shown that the energy spectrum is bounded from below. The energy raising and lowering operators are found, and the energy spectrum is obtained. The energy eigenstates are found in terms of the ground state and powers of the energy raising operator (*Fock* representation). The energy eigenstates in this representation are transformed to the coordinate representation, and the *Hermite* polynomials are reobtained.

10.1 The Energy Lowering and Raising Operators

For a linear harmonic oscillator, H is given by

$$\begin{aligned} H &= \frac{m^2\omega^2x^2 + p^2}{2m} = \frac{(m\omega x - ip)(m\omega x + ip) + m\hbar\omega}{2m} \\ &= \left[\frac{(m\omega x - ip)(m\omega x + ip)}{2m\hbar\omega} - \frac{1}{2}\right]\hbar\omega = \left(a^\dagger a + \frac{1}{2}\right)\hbar\omega, \end{aligned} \tag{10.1}$$

where

$$a = \frac{m\omega x + ip}{\sqrt{2m\hbar\omega}}. \tag{10.2}$$

The expectation value (and as a special case the eigenvalues) of any operator in the form $B^\dagger B$ is non-negative:

$$(u, B^\dagger B\, u) = (B\, u, B\, u) = \|B\, u\|^2 \geq 0 \quad \forall u.$$

Therefore, any eigenvalue E_k of H satisfies

$$E_k \geq \frac{1}{2}\hbar\omega, \tag{10.3}$$

i.e., the energy spectrum is bounded from below.
It is easily shown that

$$[a, a^\dagger] = 1, \quad [H, a] = -\hbar\omega a, \quad [H, a^\dagger] = \hbar\omega a^\dagger. \tag{10.4}$$

DOI: 10.1201/9781003677109-10

If $H|k\rangle = E_k|k\rangle$, then

$$Ha|k\rangle = (aH - a\hbar\omega)|k\rangle = (E_k - \hbar\omega)a|k\rangle, \tag{10.5}$$

thus $a|k\rangle$ is also an energy eigenket, but it corresponds to a lower eigenvalue $E_k - \hbar\omega$, a is called the energy lowering operator. Similarly, it is easily shown that $a^\dagger$ is the raising operator by the energy quantum $\hbar\omega$.

Exercise 10.1

1. *Prove Eq. (10.4).*
2. *Prove (by induction) that* $[H, a^n] = -n\hbar\omega a^n$ *and* $[H, (a^\dagger)^n] = n\hbar\omega(a^\dagger)^n$.

10.2 The Energy Spectrum

$$Ha^n|k\rangle = (a^n H - na^n\hbar\omega)|k\rangle = (E_k - n\hbar\omega)a^n|k\rangle, \tag{10.6}$$

thus, $a^n|k\rangle$ is an energy eigenket corresponding to the energy eigenvalue $E_k - n\hbar\omega$, and however large E_k is, there is a non-negative integer n_0 such that

$$E_k - n\hbar\omega < \frac{1}{2}\hbar\omega \qquad \forall n > n_0, \tag{10.7}$$

which contradicts inequality (10.3). Therefore, there must exist a ground state $|0\rangle$ such that $|0\rangle \neq 0$, $a|0\rangle = 0$: The corresponding energy eigenvalue to this state (the least energy eigenvalue) is $\frac{1}{2}\hbar\omega$. Indeed:

$$H|0\rangle = \hbar\omega(a^\dagger a + \frac{1}{2})|0\rangle = \frac{1}{2}\hbar\omega|0\rangle. \tag{10.8}$$

Operating on $|0\rangle$ by $(a^\dagger)^n$, we obtain the eigenket (unnormalized) $(a^\dagger)^n|0\rangle$ corresponding to the eigenvalue $(n + \frac{1}{2})\hbar\omega$:

$$H(a^\dagger)^n|0\rangle = (n + \frac{1}{2})\hbar\omega(a^\dagger)^n|0\rangle, \tag{10.9}$$

that is the energy spectrum is

$$E_n = (n + \frac{1}{2})\hbar\omega, \tag{10.10}$$

where $n = 0, 1, 2, \ldots$ (a non-negative integer).

10.3 The Energy Eigenstates

Let $|n\rangle$ be the (normalized) eigenket corresponding to the eigenvalue E_n:

$$H|n\rangle = (n + \frac{1}{2})\hbar\omega|n\rangle. \tag{10.11}$$

Therefore,

$$a^\dagger a|n\rangle = n|n\rangle. \tag{10.12}$$

Since $a|n\rangle$ corresponds to the eigenvalue E_{n-1}, then

$$a|n\rangle = \lambda_n|n-1\rangle. \tag{10.13}$$

Therefore,

$$|\lambda_n|^2 = \parallel au_n \parallel^2 = (au_n, au_n) = (u_n, a^\dagger au_n) = n. \tag{10.14}$$

Thus, we can take λ_n to be $\sqrt{n}$, that is

$$a|n\rangle = \sqrt{n}\,|n-1\rangle, \tag{10.15}$$

operating on this equation by $a^\dagger$, we have

$$a^\dagger|n-1\rangle = \sqrt{n}\,|n\rangle$$

that is

$$|n\rangle = \frac{1}{\sqrt{n}}a^\dagger|n-1\rangle = \sqrt{\frac{(n-2)!}{n!}}\,(a^\dagger)^2|n-2\rangle,$$

and by induction

$$|n\rangle = \frac{1}{\sqrt{n!}}(a^\dagger)^n|0\rangle. \tag{10.16}$$

This is called by *Dirac* the *Fock* representation, which leads by slight generalization to the *Fock* space in quantum field theory, where this space is spanned from vacuum by operating with different powers of the creation operator on the vacuum.

Exercise 10.2 *In the energy representation find $a, a^\dagger, x,$ and p at any time t.*

10.4 The Energy Eigenstates in the Coordinate Representation

We have $a|0\rangle = 0$, therefore

$$\begin{aligned} 0 &= \langle x|a|0\rangle = \int_{-\infty}^{\infty} \langle x|a|x'\rangle\langle x'|0\rangle\, dx' = \int_{-\infty}^{\infty} a_{xx'}\langle x'|0\rangle\, dx' \\ &= \hat{a}(x)\int_{-\infty}^{\infty} \langle x'|0\rangle\delta(x-x')\, dx' = \hat{a}\langle x|0\rangle = \hat{a}\mathrm{u}_0(x). \end{aligned}$$

$$\hat{a} = \frac{1}{\sqrt{2}}\left(\sqrt{\frac{m\omega}{\hbar}}\,x + \sqrt{\frac{\hbar}{m\omega}}\frac{d}{dx}\right) = \frac{1}{\sqrt{2}}\left(\xi + \frac{d}{d\xi}\right),$$

where

$$\xi = \frac{x}{x_0}, \quad x_0 = \sqrt{\frac{\hbar}{m\omega}}.$$

Therefore,

$$\frac{d\mathrm{u}_0}{d\xi} + \xi\mathrm{u}_0 = 0,$$

with the immediate solution

$$\mathrm{u}_0 = C_0\, e^{-\xi^2/2}.$$

Now

$$\begin{aligned} 1 &= \langle 0|0\rangle = \int_{-\infty}^{\infty} \langle 0|x\rangle\, dx\langle x|0\rangle = \int_{-\infty}^{\infty} \mathrm{u}_0^*(x)\mathrm{u}_0(x)\, dx \\ &= |C_0|^2\, x_0 \int_{-\infty}^{\infty} e^{-\xi^2}\, d\xi = x_0\sqrt{\pi}\,|C_0|^2. \end{aligned}$$

Therefore, C_0 can be taken to be $\frac{1}{\sqrt{x_0\sqrt{\pi}}}$,

$$\mathrm{u}_0(x) = \frac{1}{\sqrt{x_0\sqrt{\pi}}}\, e^{-\xi^2/2}. \tag{10.17}$$

Exercise 10.3

1. *Prove that* $(AB)_{op} = \hat{A}\hat{B}$.
2. *Prove (by induction) that* $(a^\dagger)^n_{op} = (\hat{a}^\dagger)^n$.

Since $|n\rangle = \frac{1}{\sqrt{n!}}(a^\dagger)^n|0\rangle$, then

$$\mathrm{u}_n(x) = \langle x|n\rangle = \frac{1}{\sqrt{n!}}(a^\dagger)^n_{\mathrm{op}}\mathrm{u}_0(x) = \frac{1}{\sqrt{n!}}(\hat{a}^\dagger)^n\mathrm{u}_0(x), \qquad \hat{a}^\dagger = \frac{-1}{\sqrt{2}}\left(\frac{d}{d\xi} - \xi\right).$$

Therefore,

$$\hat{a}^\dagger \, \mathrm{f}(\xi) = \frac{-1}{\sqrt{2}} \, e^{\xi^2/2} \, \frac{d}{d\xi} \, e^{-\xi^2/2} \, \mathrm{f}(\xi), \;\; \forall \mathrm{f}(\xi),$$

that is

$$\hat{a}^\dagger = \frac{-1}{\sqrt{2}} \, e^{\xi^2/2} \, \frac{d}{d\xi} \, e^{-\xi^2/2}, \tag{10.18}$$

$$(\hat{a}^\dagger)^n = \frac{(-1)^n}{\sqrt{2^n}} \, e^{\xi^2/2} \, \frac{d^n}{d\xi^n} \, e^{-\xi^2/2}. \tag{10.19}$$

Therefore,

$$\mathrm{u}_n(x) = \frac{(-1)^n}{\sqrt{n! \, 2^n}} \, e^{\xi^2/2} \, \frac{d^n}{d\xi^n} \, e^{-\xi^2/2} \, \mathrm{u}_0(x) = \frac{1}{\sqrt{n! \, 2^n \, x_0 \sqrt{\pi}}} \, e^{-\xi^2/2} \, H_n(\xi), \tag{10.20}$$

where H_n is the *Hermite* polynomial of degree n.

11

Angular Momentum

A theorem about simultaneous diagonalization of two operators is provided. The eigenvalues and eigenstates of the square angular momentum and of the z-component of the angular momentum are found. For a rotator in the p-state with zero magnetic quantum number, it is proved that the most probable orbit is any orbit lying in a plane passing through the z-axis and the angular momentum is perpendicular to the z-axis. It is shown how to add angular momenta, and the *Clebsch-Gordan* coefficients, which are used to perform the explicit direct sum decomposition of the tensor product of two irreducible representations into irreducible ones are given. The *Bloch* sphere is defined. An experimental background to the discovery of spin is given, and the *Pauli* matrices are found. Also, the spin states and how the *Pauli* matrices operate on them are shown. This chapter is ended by giving the *Pauli* equation.

11.1 A Theorem

Theorem 11.1 *Two Hermitian operators A, B can have a common complete set of eigenstates iff they mutually commute.*

Proof

1. **Necessity** Let $|ab\rangle$ be the complete set of simultaneous eigenstates of A, B, i.e.,

$$\begin{aligned} A|ab\rangle &= a|ab\rangle, \\ B|ab\rangle &= b|ab\rangle, \ \forall a, b \text{ in the spectrum of } A, B, \end{aligned}$$

therefore:

$$\begin{aligned} AB|ab\rangle &= ab|ab\rangle, \\ BA|ab\rangle &= ba|ab\rangle, \end{aligned}$$

therefore:

$$[A, B]|ab\rangle = 0, \ \forall a, b.$$

Since any state $|f\rangle$ is a linear combination of the basis states $\{|ab\rangle\}$, then $[A, B]|f\rangle = 0, \ \forall |f\rangle$, i.e., $[A, B] = 0$.

DOI: 10.1201/9781003677109-11

2. **Sufficiency** Let $[A, B] = 0$, $A|a\rangle = a|a\rangle$. If a is nondegenerate level, i.e., if there is only one eigenstate corresponding to this eigenvalue, then $AB|a\rangle = BA|a\rangle = aB|a\rangle$, i.e., $B|a\rangle$ is also an eigenstate of A corresponding to the same eigenvalue a, and since there is only one such a state, then $B|a\rangle$ is proportional to $|a\rangle$, i.e., $B|a\rangle = b|a\rangle$, that is, $|a\rangle$ is also an eigenstate of B corresponding to the eigenvalue b, it is therefore better to denote it by $|a, b\rangle$.

   If a is n-fold degenerate level in the spectrum of A, then the subspace V_a corresponding to this level is n-dimensional. Let $\{|a\rangle_i, \forall i = 1, \ldots, n\}$ be an orthonormal basis in it. In any basis in V_a, A is represented by aI, where I is the $n \times n$ unit matrix. B does not take any state belonging to V_a out of it, indeed

$$AB|a\rangle_i = BA|a\rangle_i = aB|a\rangle_i, \ \forall i = 1, \ldots, n$$

   i.e., $B|a\rangle_i \in V_a$, but $|a\rangle_i$ need not to be eigenstates of B, i.e., B is not necessarily diagonal in this basis. Diagonalize B in V_a and denote by $\{|a, b_i\rangle, \forall i = 1, \ldots, n\}$, the basis of V_a where B is diagonal:

$$B|a, b_i\rangle = b_i|a, b_i\rangle,$$

   therefore, $\{|a, b_i\rangle, \forall a, b_i \text{ in the spectrum of } A, B\}$ is the required common set of eigenstates of A, B.

11.2 Eigenvalues and Eigenstates of Angular Momentum

Consider a vector (in spatial 3-space) Hermitian operator $\vec{J}$, the components of which satisfy the commutator algebra: $[J_\alpha, J_\beta] = i\hbar J_\gamma$ with (α, β, γ) being any even permutation of (x, y, z). It is easily proved that $[J_\alpha, J^2] = 0$, $\alpha = x, y, z$. Therefore, a representation can be found in which J^2 and only one of the components J_α are diagonal. More than one of the components can not be diagonalized simultaneously with J^2 because the components do not commute with each other. Take J_α to be J_z and define

$$J_\pm = J_x \pm iJ_y. \tag{11.1}$$

Notice that $J_\pm^\dagger = J_\mp$.

It is easily proved that

$$[J_z, J_\pm] = \pm\hbar J_\pm, \quad [J_-, J_-] = 2\hbar J_z. \tag{11.2}$$

Denote the eigenvalues of J^2 by $\hbar^2\lambda$ and those of J_z by $\hbar m$, and the simultaneous eigenstates by $|\lambda m\rangle$, i.e.,

$$\begin{aligned} J^2|\lambda\, m\rangle &= \hbar^2\lambda|\lambda\, m\rangle, \\ J_z|\lambda\, m\rangle &= \hbar m|\lambda\, m\rangle. \end{aligned} \tag{11.3}$$

Since

$$J^2 = J_+J_- + J_z^2 - \hbar J_z = J_-J_+ + J_z^2 + \hbar J_z = \frac{J_+J_- + J_-J_+}{2} + J_z^2. \quad (11.4)$$

Therefore,

$$\hbar^2\lambda = \langle\lambda\, m|\frac{J_+J_- + J_-J_+}{2}|\lambda\, m\rangle + \hbar^2 m^2 \geq \hbar^2 m^2 \text{ (why?)}, \quad (11.5)$$

i.e.,

$$m^2 \leq \lambda. \quad (11.6)$$

Thus, for a given λ, the spectrum of J_z has both upper and lower bounds.

$$J_zJ_\pm|\lambda\, m\rangle = (J_\pm J_z \pm \hbar J_\pm)|\lambda\, m\rangle = \hbar(m \pm 1)J_\pm|\lambda\, m\rangle, \quad (11.7)$$

i.e.,

$$J_\pm|\lambda\, m\rangle = C_\pm\,(\lambda, m)|\lambda, m \pm 1\rangle. \quad (11.8)$$

Let j, j' be the greatest and lowest values of m, respectively.
Therefore:

$$|\lambda\, j\rangle \neq 0, \quad |\lambda\, j'\rangle \neq 0 \quad \text{and} \quad J_+|\lambda\, j\rangle = 0, \; J_-|\lambda\, j'\rangle = 0.$$

From Eq. (11.4):

$$0 = J_-J_+|\lambda\, j\rangle = (J^2 - J_z^2 - \hbar J_z)|\lambda\, j\rangle = \hbar^2(\lambda - j^2 - j)|\lambda\, j\rangle.$$

Therefore,

$$\lambda = j(j+1). \quad (11.9)$$

Similarly, it is proved that $\lambda = j'(j'-1)$. Therefore,

$$\begin{aligned} j'(j'-1) - j(j+1) &= 0, \\ (j'+j)(j'-j-1) &= 0, \\ j' = -j \quad \text{or} \quad j' &= j+1. \end{aligned}$$

The second solution ($j' = j+1$) is ruled out by the assumption that j is the greatest value of m, therefore $|\lambda\, j'\rangle = |\lambda, -j\rangle$. Since $|\lambda\, j\rangle$ is obtained from $|\lambda, -j\rangle$ by applying J_+, $2j$ times. Therefore, $2j$ must be a non-negative integer, i.e.,

$$j = 0, \frac{1}{2}, 1, \frac{3}{2}, \ldots \quad (11.10)$$

For a given j: $m = -j, -j+1, \ldots, j$, i.e., m ranges from $-j$ to j in unit steps, and it has $2j+1$ different values for each value of j. The ket $|\lambda\, m\rangle$ is better denoted by $|j\, m\rangle$.
From Eq. (11.8),

$$\begin{aligned} |C_+(j,m)|^2 &= \langle j\, m|J_-J_+|j\, m\rangle \\ &= \hbar^2\left[j(j+1) - m(m+1)\right] \\ &= \hbar^2(j-m)(j+m+1). \end{aligned}$$

Therefore,

$$J_+|j\,m\rangle = \hbar\sqrt{(j-m)(j+m+1)}|j,m+1\rangle, \tag{11.11}$$

operating on this equation by J_-, we obtain

$$J_-|j,m+1\rangle = \hbar\sqrt{(j-m)(j+m+1)}\,|j\,m\rangle, \tag{11.12}$$

hence

$$J_-|j,m\rangle = \hbar\sqrt{(j+m)(j-m+1)}\,|j,m-1\rangle. \tag{11.13}$$

Exercise 11.1

1. *Prove Eq. (11.2).*
2. *Prove Eqs. (11.4).*
3. *Prove that* $\lambda = j'(j'-1)$.
4. *Find* J_x, J_y, J_z, *and* J^2 *in the above representation for* $j = 0, \frac{1}{2}, 1$. ***Hint.*** *Find first* $J_+, J_- \equiv J_+^\dagger$, *then obtain* J_x, J_y.

Notice that, for $\vec{J} = \vec{L} = \vec{r}\wedge\vec{p}$ *(orbital angular momentum)* $j = l$ *can not have half-integer values and can have only the values* $0, 1, 2, \ldots$.

5. *Find the eigenvalues and eigenstates of* J_x, J_y *in the simultaneous representation of* J^2, J_z *for* $j = 1$. ***Hint.*** *Diagonalize* J_x, J_y.

Example 11.1 *A rotator in the p-state with zero magnetic quantum number. Prove that the most probable orbit of the rotator is any orbit lying in a plane passing through the z-axis, and in this case, the angular momentum is perpendicular to the z-axis.*

Being a particle in a central field, the rotator must move in a plane passing through the center of the field, which in this case is the center of the sphere on which the rotator moves. Therefore, the possible orbits of the rotator are the great circles of this sphere. Let the normal to one of these circles be the z'*-axis with spherical angles* α, γ*; see Fig. 11.1. From the spherical symmetry of the problem, the eigenvalues of* $L_{z'}$ *are the same as those for* L_z*. Therefore, in the p-state they are* $0, \pm 1$*. A particle rotating right-handedly on the great circle normal to the* z'*-axis is in the eigenstate* $|1,1\rangle_{z'}$*, that is, with angular momentum parallel to the* z'*-axis. To calculate the probability for this state, assuming that the rotator is in the* $|1,0\rangle_z$ *state, find the eigenstates of* $L_{z'}$*:*

$$L_{z'} = L_x \sin\alpha\cos\gamma + L_y \sin\alpha\sin\gamma + L_z\cos\alpha.$$

In the p-state $\ell = 1$*, therefore, in the* L_z *representation:*

$$L_{z'} = \hbar\begin{pmatrix} \cos\alpha & \frac{\sin\alpha}{\sqrt{2}}e^{-i\gamma} & 0 \\ \frac{\sin\alpha}{\sqrt{2}}e^{i\gamma} & 0 & \frac{\sin\alpha}{\sqrt{2}}e^{-i\gamma} \\ 0 & \frac{\sin\alpha}{\sqrt{2}}e^{i\gamma} & -\cos\alpha \end{pmatrix}.$$

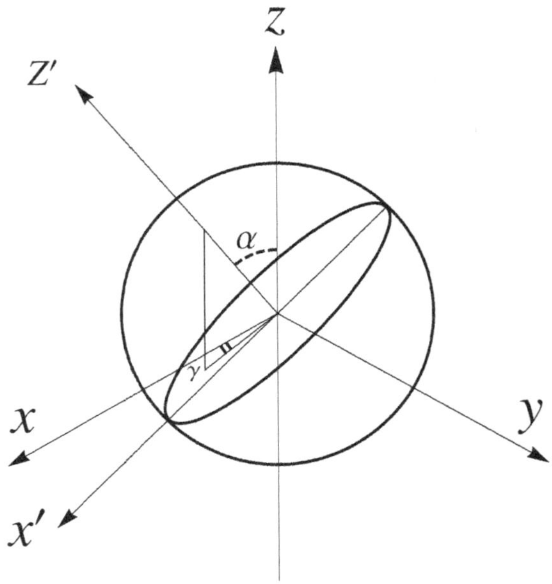

FIGURE 11.1: Orbits of the rotator which are the great circles.

The eigenvalue equation of $L_{z'}$ is:

$$(L_{z'} - \mu I)\begin{pmatrix} a_{1\mu} \\ a_{0\mu} \\ a_{-1,\mu} \end{pmatrix} = 0.$$

Its characteristic equation is

$$|L_{z'} - \mu I| = \mu(\mu^2 - 1) = 0, \tag{11.14}$$

therefore $\mu = 0, \pm 1$ as expected. Substitute these values for μ in turn in the eigenvalue equation to obtain:

$$|1,1\rangle_{z'} = \begin{pmatrix} \cos^2\frac{\alpha}{2} \\ \frac{\sin\alpha}{\sqrt{2}} e^{i\gamma} \\ \sin^2\frac{\alpha}{2} e^{2i\gamma} \end{pmatrix}, \quad |1,0\rangle_{z'} = \begin{pmatrix} \frac{\sin\alpha}{\sqrt{2}} \\ -\cos\alpha e^{i\gamma} \\ -\frac{\sin\alpha}{\sqrt{2}} e^{2i\gamma} \end{pmatrix}, \quad |1,-1\rangle_{z'} = \begin{pmatrix} \sin^2\frac{\alpha}{2} \\ -\frac{\sin\alpha}{\sqrt{2}} e^{i\gamma} \\ \cos^2\frac{\alpha}{2} e^{2i\gamma} \end{pmatrix},$$

notice that these 3 states are normalized and mutually orthogonal. Therefore,[1]

$$|1,0\rangle_z = e^{-i\gamma}\left[\frac{\sin\alpha}{\sqrt{2}}|1,1\rangle_{z'} - \cos\alpha|1,0\rangle_{z'} - \frac{\sin\alpha}{\sqrt{2}}|1,-1\rangle_{z'}\right] = \begin{pmatrix} 0 \\ 1 \\ 0 \end{pmatrix} \quad \text{as must be.}$$

[1] See the comment after Eq. (8.19).

Therefore, for a rotator in the p-state with zero magnetic quantum number, the probabilities for the states $|1,1\rangle_{z'}$, $|1,0\rangle_{z'}$ and $|1,-1\rangle_{z'}$ are $\frac{1}{2}\sin^2\alpha$, $\cos^2\alpha$, and $\frac{1}{2}\sin^2\alpha$, respectively. The maximum value for the probability of the state $|1,1\rangle_{z'}$ or the state $|1,-1\rangle_{z'}$ is attained at $\alpha=\frac{\pi}{2}$ and equals $\frac{1}{2}$. In this case the probability for $|1,0\rangle_{z'}$ is zero. That is, the most probable orbit is one passing through the z-axis ($\alpha=\frac{\pi}{2}$) with probability $\frac{1}{2}$ for rotating in each direction. The angular momentum in this case is in the direction of the z'-axis, which is perpendicular to the z-axis.

11.3 Addition of Angular Momenta

$\vec{J}_1$ and $\vec{J}_2$ are two vector operators, any components of $\vec{J}_1$ commutes with all components of $\vec{J}_2$ and the components of each operator separately satisfy the commutator algebra of angular momentum components. $|j_1m_1\rangle$ are the orthonormal eigenstates of J_1^2, J_{1z} and $\vec{J}_2$ has no effect on them. Also, $|j_2m_2\rangle$ are the orthonormal eigenstates of J_2^2, J_{2z}, and $\vec{J}_1$ has no effect on them. The orthonormal simultaneous eigenstates of $J_1^2, J_2^2, J_{1z}, J_{2z}$ are denoted by $|j_1j_2m_1m_2\rangle$; that is, $|j_1j_2m_1m_2\rangle = |j_1m_1\rangle|j_2m_2\rangle$, and it is a basis in the product space.
Notice that

$$\begin{aligned}\langle j_1j_2m_1m_2|j_1'j_2'm_1'm_2'\rangle &= \langle j_1m_1|j_1'm_1'\rangle\langle j_2m_2|j_2'm_2'\rangle \\ &= \delta_{j_1j_1'}\delta_{j_2j_2'}\delta_{m_1m_1'}\delta_{m_2m_2'}.\end{aligned}$$

Put $\vec{J}=\vec{J}_1+\vec{J}_2$.

Exercise 11.2 *1. Prove that the components of $\vec{J}$ satisfy the commutator algebra of angular momentum components.*

2. Prove that $[J^2,J_1^2]$, $[J^2,J_2^2]$, $[J_z,J_1^2]$ and $[J_z,J_2^2]$ all vanish.

3. Find $[J^2,J_{1z}]$, $[J^2,J_{2z}]$.

The eigenstates of J^2 and J_z can be required to be simultaneous eigenstates of J_1^2, J_2^2 also, but they can not be eigenstates of neither J_{1z} nor J_{2z} (why?). Denote them by $|j_1j_2jm\rangle$, they constitute another basis in the product space.

In what follows, we consider only a subspace corresponding to definite values of j_1 and j_2; its dimensionality is $(2j_1+1)(2j_2+1)$, and we may simplify the notation:

$$\begin{aligned}|j_1j_2m_1m_2\rangle &= |m_1m_2\rangle = |m_1\rangle|m_2\rangle, \\ |j_1j_2jm\rangle &= |jm\rangle.\end{aligned}$$

Since

$$|jm\rangle = |m_1m_2\rangle\langle m_1m_2|jm\rangle \quad \text{(why?)}$$

(summation on m_1, m_2 is implied), then $[\langle m_1m_2|jm\rangle]$ is the transformation matrix from the J^2, J_z representation to the J_{1z}, J_{2z} representation. The transformation matrix elements $\langle m_1m_2|jm\rangle$ are called *Clebsch-Gordan, Wigner*, or vector-coupling coefficients. They are the expansion coefficients of total angular momentum eigenstates in an uncoupled direct product basis. They are used to perform the explicit direct sum decomposition of the tensor product of two irreducible representations into irreducible ones.

Consider

$$\langle m_1m_2|J_z|jm\rangle = \hbar m\langle m_1m_2|jm\rangle,$$

if we instead operate with J_z which is $J_{1z} + J_{2z}$ on the bra on its left, we obtain

$$\langle m_1m_2|J_z|jm\rangle = \hbar(m_1 + m_2)\langle m_1m_2|jm\rangle,$$

subtracting, we obtain

$$(m_1 + m_2 - m)\langle m_1m_2|jm\rangle = 0,$$

$\langle m_1m_2|jm\rangle$ vanishes if $m \neq m_1 + m_2$. $\langle m_1m_2|jm\rangle$ is also denoted by $C(j_1j_2j; m_1m_2)$. Notice that the maximum value of m is $j_1 + j_2$ and it occurs once, when $m_1 = j_1$ and $m_2 = j_2$, and it is associated with $j = j_1 + j_2$. The second largest value of m is j_1+j_2-1 and it occurs twice ($m_1 = j_1$, $m_2 = j_2-1$ or $m_1 = j_1 - 1$, $m_2 = j_2$). One of the two appropriate linear combinations of these two states is associated with $j = j_1 + j_2$, the other with $j = j_1 + j_2 - 1$. The third largest value of m is $j_1 + j_2 - 2$, and it occurs three times, two of the three appropriate linear combinations of these three states are associated with $j = j_1 + j_2, j_1 + j_2 - 1$, and the third one is associated with a new value $j = j_1 + j_2 - 2$, and so on, up to $m = |j_1 - j_2|$. Choose j_1 to be $\geq j_2$, hence the value $m = |j_1 - j_2| = j_1 - j_2$ will occur $2j_2+1$ times corresponding to all possible values of m_2. $m_1 = j_1 - j_2 - m_2$ will be in the range $(-j_1, j_1)$ for all values of m_2. $2j_2$ linear combinations of the states corresponding to this value of m are associated with $j = j_1 + j_2, \ldots, j_1 - j_2 + 1$ and the $(2j_2+1)^{\text{th}}$ one is associated with a new $j = j_1 - j_2$. The number of occurrence of $m = j_1 - j_2, j_1 - j_2 - 1$ up to $-j_1 + j_2$ is also $2j_2 + 1$ (all the possible values of m_2). m_1 will be in its range for all these values of m and m_2. Hence, there will be no new values for j. For $m = -j_1 + j_2 - 1$, m_2 can not acquire the value j_2 otherwise m_1 should be equal to $-j_1 - 1$ which is out of its range. Hence m_2 will begin with $j_2 - 1$ and decrease down with unit steps to $-j_2$ without m_1 being out of its range. Therefore, the number of states corresponding to $m = -j_1 + j_2 - 1$ is $2j_2$ only and the $2j_2$ linear combinations of these states will be associated with the first $2j_2$ values of j, and so on up to $m = -j_1 - j_2$ which occurs only once and is associated with $j = j_1 + j_2$. Therefore, j has $2j_2 + 1$ ($j_2 \leq j_1$) different values ranging from $|j_1 - j_2|$ to $j_1 + j_2$ in unit steps, and each value occurs only once. For each value of j, m has $2j + 1$ values ranging from $-j$ to j in unit steps; each value of m occurs only once for each value of j.

TABLE 11.1: In this table, the same value of m in each row occurs in different columns corresponding to the different values of j.

j_1+j_2				
j_1+j_2-1	j_1+j_2-1			
$\dots$	$\dots$			
j_1-j_2+1	j_1-j_2+1	$\dots$	j_1-j_2+1	
j_1-j_2	j_1-j_2	$\dots$	j_1-j_2	j_1-j_2
j_1-j_2-1	j_1-j_2-1	$\dots$	j_1-j_2-1	j_1-j_2-1
$\dots$	$\dots$	$\dots$	$\dots$	$\dots$
$-j_1+j_2+1$	$-j_1+j_2+1$	$\dots$	$-j_1+j_2+1$	$-j_1+j_2+1$
$-j_1+j_2$	$-j_1+j_2$	$\dots$	$-j_1+j_2$	$-j_1+j_2$
$-j_1+j_2-1$	$-j_1+j_2-1$	$\dots$	$-j_1+j_2-1$	
$\dots$	$\dots$			
$-j_1-j_2+1$	$-j_1-j_2+1$			
$-j_1-j_2$				

Thus, the number of $|jm\rangle$ for all values of j and m is

$$\sum_{|j_1-j_2|}^{j_2+j_1} 2j+1 = (2j_1+1)(2j_2+1) \tag{11.15}$$

The transformation matrix $[\langle m_1 m_2|jm\rangle]$ is unitary. Indeed:

$$\langle m_1 m_2|jm\rangle\langle jm|m_1' m_2'\rangle = \langle m_1 m_2|m_1' m_2'\rangle = \langle m_1|m_1'\rangle\langle m_2|m_2'\rangle = \delta_{m_1 m_1'}\delta_{m_2 m_2'},$$

$$\langle jm|m_1 m_2\rangle\langle m_1 m_2|j'm'\rangle = \langle jm|j'm'\rangle = \delta_{jj'}\delta_{mm'}.$$

11.4 Spin

11.4.1 The Gyromagnetic Ratio[2]

In classical mechanics and electrodynamics, it is known that a circulating nonrelativistic point charge with (orbital) angular momentum $\vec{L}$ acquires a

[2]See [1, 5, 14, 12].

magnetic dipole moment $\vec{\mu}$:

$$\vec{\mu} = \frac{e}{2m_0 c}\vec{L}, \tag{11.16}$$

where m_0, e are the mass and the charge of the particle, and $\dfrac{e}{2m_0 c}$ is called the gyromagnetic ratio.

The same formula for the gyromagnetic ratio is obtained by *Schrödinger* theory for a particle in an electromagnetic field. In § 9.4, it was shown that the Hamiltonian of the particle, in this case, is

$$H = \frac{\left(\vec{p} - \frac{e}{c}\vec{A}\right)^2}{2m_0} + e\varphi, \tag{11.17}$$

where φ and $\vec{A}$ are the scalar and vector potentials of the field. Therefore,

$$H = \frac{p^2}{2m_0} - \frac{e}{m_0 c}\vec{A}\cdot\vec{p} - \frac{e}{2m_0 c}\frac{\hbar}{i}\text{div}\vec{A} + \frac{e^2}{2m_0 c^2}A^2 + e\varphi. \tag{11.18}$$

If the magnetic field $\vec{\mathcal{H}}$ is constant and homogeneous, we take

$$\vec{A} = \frac{1}{2}\vec{\mathcal{H}} \wedge \vec{r}, \tag{11.19}$$

with this choices $\text{div}\vec{A} = 0$, and if $\vec{\mathcal{H}}$ is small enough to neglect the term proportional to its square, then[3]

$$H = \frac{p^2}{2m_0} + e\varphi - \frac{e}{2m_0 c}(\vec{\mathcal{H}} \wedge \vec{r})\cdot\vec{p} = \frac{p^2}{2m_0} + e\varphi + U^{\text{mag}}, \tag{11.20}$$

where

$$U^{\text{mag}} = -\frac{e}{2m_0 c}\vec{\mathcal{H}}\cdot(\vec{r} \wedge \vec{p}) = -\frac{e}{2m_0 c}\vec{\mathcal{H}}\cdot\vec{L} = -\vec{\mathcal{H}}\cdot\vec{\mu}, \tag{11.21}$$

$$\vec{\mu} = \frac{e}{2m_0 c}\vec{L}, \tag{11.22}$$

that is, the particle acquires beside its charge a dipole magnetic moment proportional to its (orbital) angular momentum with the same classical gyromagnetic ratio of proportionality.

Notice that

$$\mu_z = \frac{e}{2m_0 c}L_z = \frac{e\hbar}{2m_0 c}m, \tag{11.23}$$

therefore, μ_z can have only $2l + 1$ different values corresponding to all the possible values of the magnetic quantum number m for a given value l of the orbital quantum number. That is, the (orbital) magnetic dipole moment $\vec{\mu}$ in *Schrödinger* theory can have only $2l+1$ different orientations in space relative to the quantization axis z. This number of possible orientations is definitely odd, since l is a non-negative integer.

[3]This term is also proportional to $\frac{e^2}{c^2}$.

11.4.2 An Experimental Background

The first experiments that led to the discovery of spin were *Einstein-de Haas* and *Stern-Gerlach* experiments. The *Einstein-de Haas* experiments were based on a remark by *Einstein* that on magnetizing a ferromagnetic rod, it acquires, besides the magnetic moment, an angular momentum to compensate for the change of the angular momentum of its atoms and thus to save the law of conservation of angular momentum. Therefore, when the magnetization of the rod is changed, it must rotate. Hence, a coaxially suspended soft-iron rod inside a coil will undergo torsional oscillations by passing an alternating current through the coil, since this current will at every turn remagnetize the rod. A resonance, which increases the amplitude of the torsional oscillations, will take place if the frequency of the current is taken equal to the intrinsic frequency of the torsional oscillations of the rod. *Einstein* proposed this experiment and then performed it jointly with *de Haas* to verify Eq. (11.16). The measurements of such experiments showed that the sign of the gyromagnetic ratio is negative, which means that the magnetization of the ferromagnetic rod is due to the angular momentum of electrons (negative charges), but it gave for the value of this ratio double the classical value.

Landé expressed the gyromagnetic ratio in the form $g\dfrac{e}{2m_0c}$, where g is known as *Landé* g-factor. That is *Einstein-de Haas* experiments showed that $g = 2$ instead of 1 as being predicted by the classical and by *Schrödinger* theories.

Stern[4] and *Gerlach* (1921) passed a beam of monovalent atoms (hydrogen, lithium, silver) to cross perpendicularly a very inhomogeneous magnetic field with constant direction. The beam of atoms, being moving magnetic dipoles, is expected to split into an odd number of components in this field (see the notice after Eq. (11.23)). If the atoms are in an s-state ($l = 0$), no splitting is expected, if they are in a p-state ($l = 1$), a splitting in a triplet must occur, and so on.

Nevertheless, *Stern* and *Gerlach* observed a splitting of a beam of atoms in the s-state into a doublet, that is the atom in the s-state has a dipole magnetic moment, the projection of which on the direction of the magnetic field has two values. They measured these values to be $\pm\dfrac{e\hbar}{2m_0c}$.

On the basis of these results, *Uhlenbeck* and *Goudsmit* suggested that the electron has an intrinsic angular momentum which they called *spin*. The projection of the electron spin $\vec{s}$ on the quantization axis must have two and only two different values, that is the spin quantum number must be $\frac{1}{2}$. Therefore:

$$s_z = \hbar m = \pm\frac{1}{2}\hbar, \quad s^2 = \hbar^2 s(s+1) = \frac{3}{4}\hbar^2. \tag{11.24}$$

To account for the measured values of μ_z in the *Stern-Gerlach* experiment,

---

[4] *O. Stern* (1888–1969) was awarded the *Nobel* prize in Physics in 1943 for his contribution to the development of the molecular ray method and his discovery of the magnetic moment of the proton.

they took for the *Landé* factor $g = 2$ as measured in the *Einstein-de Haas* experiments, and therefore, for the electron intrinsic magnetic dipole moment $\vec{\mu}$:

$$\vec{\mu} = \frac{e}{m_0 c}\vec{s}, \quad \mu_z = \frac{e}{m_0 c}s_z = \pm\frac{e\hbar}{2m_0 c}. \tag{11.25}$$

Notice that, unlike the orbital quantum numbers, l and m which among other quantum numbers determine the physical state of a particle, the spin quantum number s is a property of the particle, and it among other things identifies the particle itself not its state, i.e., if a particle does not have a spin $\frac{1}{2}$, it definitely cannot be an electron.

The spin $\frac{1}{2}$ for the electron and $g = 2$ for intrinsic magnetic moments were predicted only by the *Dirac* theory which predicts also a value $g = 1$ for orbital magnetic moments.

11.4.3 s^2, s_z Representation

For electrons, $s = \frac{1}{2}$, therefore, s^2 has the unique eigenvalue $\frac{3}{4}\hbar^2$ and s_z has the two eigenvalues $\pm\frac{\hbar}{2}$, hence (in the simultaneous s^2 and s_z representation):

$$s^2 = \frac{3}{4}\hbar^2 I, \quad s_z = \frac{\hbar}{2}\begin{pmatrix} 1 & 0 \\ 0 & -1 \end{pmatrix}, \tag{11.26}$$

where I is the 2×2 unit matrix.
Also, $s_+ \left|\frac{1}{2}\frac{1}{2}\right\rangle = 0$ (why?) and from Eq. (11.11) $s_+ \left|\frac{1}{2}, -\frac{1}{2}\right\rangle = \hbar \left|\frac{1}{2}\frac{1}{2}\right\rangle$, thus

$$s_+ = \hbar\begin{pmatrix} 0 & 1 \\ 0 & 0 \end{pmatrix}, \quad s_- = \hbar\begin{pmatrix} 0 & 0 \\ 1 & 0 \end{pmatrix} \text{ (why?)}, \tag{11.27}$$

therefore

$$s_x = \frac{\hbar}{2}\begin{pmatrix} 0 & 1 \\ 1 & 0 \end{pmatrix}, \quad s_y = \frac{\hbar}{2}\begin{pmatrix} 0 & -i \\ i & 0 \end{pmatrix}, \tag{11.28}$$

that is,

$$\vec{s} = \frac{\hbar}{2}\vec{\sigma}, \tag{11.29}$$

where

$$\sigma_x = \begin{pmatrix} 0 & 1 \\ 1 & 0 \end{pmatrix}, \quad \sigma_y = \begin{pmatrix} 0 & -i \\ i & 0 \end{pmatrix}, \quad \sigma_z = \begin{pmatrix} 1 & 0 \\ 0 & -1 \end{pmatrix} \tag{11.30}$$

are the *Pauli* spin matrices. (The rows and columns of the matrices are labeled by $m = \pm\frac{1}{2}$ in a descending order). Also

$$s_\pm = \frac{\hbar}{2}\sigma_\pm = \frac{\hbar}{2}(\sigma_x \pm i\sigma_y), \tag{11.31}$$

$$\sigma_+ = 2\begin{pmatrix} 0 & 1 \\ 0 & 0 \end{pmatrix}, \quad \sigma_- = 2\begin{pmatrix} 0 & 0 \\ 1 & 0 \end{pmatrix}. \tag{11.32}$$

We denote the spin-state by χ, e.g.,

$$s_z \chi_{z_\pm} = \pm \frac{\hbar}{2} \chi_{z_\pm}. \tag{11.33}$$

In this representation:

$$\chi_{z_+} = \left| \frac{1}{2}, \frac{1}{2} \right\rangle = \begin{pmatrix} 1 \\ 0 \end{pmatrix}, \quad \chi_{z_-} = \left| \frac{1}{2}, -\frac{1}{2} \right\rangle = \begin{pmatrix} 0 \\ 1 \end{pmatrix}, \tag{11.34}$$

which are called spin-up and spin-down states, and are denoted also by $\uparrow, \downarrow$, respectively.
It is clear that

$$\chi_{z_+}^\dagger \chi_{z_+} = \chi_{z_-}^\dagger \chi_{z_-} = 1, \quad \chi_{z_+}^\dagger \chi_{z_-} = \chi_{z_-}^\dagger \chi_{z_+} = 0, \tag{11.35}$$

$\chi_{z_\pm}$ constitute an orthonormal basis and any spin state χ is represented in this representation by a linear combination of $\chi_{z_\pm}$

$$\chi = c_1 \chi_{z_+} + c_2 \chi_{z_-} = \begin{pmatrix} c_1 \\ c_2 \end{pmatrix}, \tag{11.36}$$

where c_1 is the probability amplitude for spin-up state and c_2 is the same for spin-down state (with respect to the quantization axis z). The normalization condition for χ is

$$\chi^\dagger \chi = |c_1|^2 + |c_2|^2 = 1. \tag{11.37}$$

It is clear that special values of c_1 and c_2 will give $\chi_{x_\pm}, \chi_{y_\pm}$.

Exercise 11.3 *Find these values.*

It is easily shown that:

$$\sigma_\alpha^2 = 1, \quad \text{tr } \sigma_\alpha = 0, \ \forall \alpha, \ \alpha = x, y, z, \tag{11.38}$$

$$\{\sigma_\alpha, \sigma_\beta\} = \sigma_\alpha \sigma_\beta + \sigma_\beta \sigma_\alpha = 2\delta_{\alpha\beta}. \tag{11.39}$$

The bracket $\{A, B\}$ is known as the anticommutator of A and B, and if it vanishes, we say that A, B anticommute one with the other.

$$\sigma_\alpha \sigma_\beta = i\sigma_\gamma, \quad (\alpha, \beta, \gamma) \text{ is any even permutation of } (x, y, z). \tag{11.40}$$

Also

$$\sigma_x \uparrow = \downarrow, \quad \sigma_y \uparrow = i \downarrow, \quad \sigma_z \uparrow = \uparrow, \tag{11.41}$$

$$\sigma_x \downarrow = \uparrow, \quad \sigma_y \downarrow = -i \uparrow, \quad \sigma_z \downarrow = - \downarrow \tag{11.42}$$

Exercise 11.4

1. Show that the linear combinations

$$s'_\alpha = \sum_\beta \vec{e}\,'_\alpha \cdot \vec{e}_\beta s_\beta, \quad \alpha, \beta = x, y, z, \tag{11.43}$$

where $\{\vec{e}_x, \vec{e}_y, \vec{e}_z\}$ and $\{\vec{e}\,'_x, \vec{e}\,'_y, \vec{e}\,'_z\}$ are two orthonormal basis of the 3-dimensional space, satisfy the commutator algebra of the angular momentum components.

2. Find the eigenvalues and eigenstates of the spin component along the direction whose spherical angles are θ, ϕ. Investigate the particular cases in which this direction is the x, the y or the z-axis.

3. The electron spin is parallel to the direction with spherical angles θ, ϕ. Find the expectation values of $\vec{S}$, S^2. Is $\langle S^2\rangle = \langle \vec{S}\rangle^2$? Why?

4. The electron spin is parallel to the z-axis. Find the probability that the component of the spin

(a) in a direction making an angle θ with the z-axis, and

(b) in a direction parallel to the x-axis,

will have the values $\frac{1}{2}\hbar$ and $-\frac{1}{2}\hbar$.

In quantum mechanics and quantum computing, the *Bloch* sphere is the unit sphere on which a pure normalized spin state (a two-level quantum state or a 'qubit (quantum bit)') $\chi = c_+\chi_{z_+} + c_-\chi_{z_-}$ is represented by an arrow to its surface and the spin-up and spin-down states $\chi_{z_\pm}$ represent the north and south poles, respectively. In this case, a qubit is written in terms of spherical coordinates as (see Exercise 11.4. See Fig. 11.2)

$$|\nearrow\rangle = \cos(\frac{\theta}{2})|\uparrow\rangle + e^{i\phi}\sin(\frac{\theta}{2})|\downarrow\rangle.$$

11.4.4 *Pauli* Equation

In order to describe spin, *Pauli* introduced two wave functions $\psi_1(\vec{r}, t)$, $\psi_2(\vec{r}, t)$ instead of the single wave function $\psi(\vec{r}, t)$ of *Schrödinger*'s theory. One of the two functions describes a state with one spin orientation and the other describes a state with the opposite spin orientation. Thus, if in Eq. (11.36), instead of the constant coefficients c_1, c_2 in the linear combination of $|\frac{1}{2}, \frac{1}{2}\rangle, |\frac{1}{2}, -\frac{1}{2}\rangle$ for χ, we take the two functions ψ_1 and ψ_2, we then have

$$\psi = \psi_1(\vec{r}, t)\chi_{z_+} + \psi_2(\vec{r}, t)\chi_{z_-} = \begin{pmatrix} \psi_1(\vec{r}, t) \\ \psi_2(\vec{r}, t) \end{pmatrix}, \tag{11.44}$$

the component $\psi_1(\vec{r}, t)$ is the probability amplitude for the electron to be in the position $\vec{r}$ at the moment t with spin up (relative to the quantization

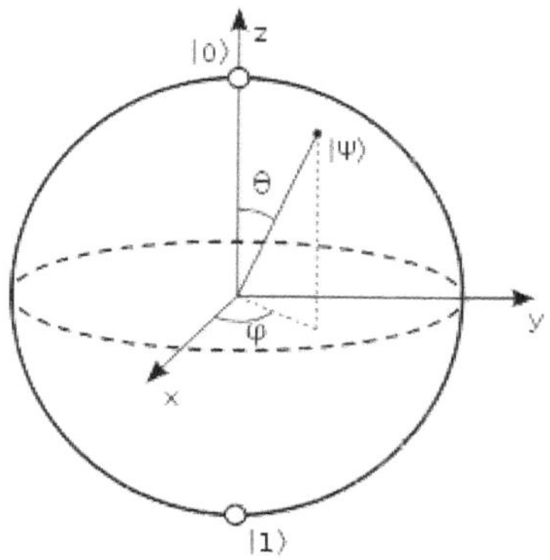

FIGURE 11.2: Bloch Sphere.

axis z) and $\psi_2(\vec{r}, t)$ is the same thing but with spin down (relative to the same axis) and the total probability density for the electron to be in the position $\vec{r}$ at the moment t (irrespective to its spin orientation) is

$$\psi^\dagger \psi = |\psi_1(\vec{r}, t)|^2 + |\psi_2(\vec{r}, t)|^2. \tag{11.45}$$

Pauli added to the Hamiltonian of the *Schrödinger* equation a term representing the interaction of the intrinsic magnetic moment of the electron $\vec{\mu}_s$ and the external magnetic field

$$U^P = -\vec{\mu}_s \cdot \vec{\mathcal{H}}, \quad \vec{\mu}_s = \frac{e\hbar}{2m_0 c}\vec{\sigma}, \tag{11.46}$$

and the time-dependent *Pauli* equation will be

$$\{i\hbar\frac{\partial}{\partial t} - \frac{1}{2m_0}\left(\vec{p} - \frac{e}{c}\vec{A}\right)^2 - e\varphi + \vec{\mu}_s \cdot \vec{\mathcal{H}}\}\psi = 0. \tag{11.47}$$

The operators and quantities in the first three terms of the above equation are understood to be multiplied by the 2×2 unit matrix.

12

Perturbation Theory

Approximation methods are considered in this chapter. Time-independent perturbation theory up to the second approximation is expounded in the nondegenerate and the degenerate cases. *Zeeman* effect and *Stark* effect are explained. The time-dependent perturbation theory is discussed. Harmonic perturbations are considered to illustrate the first-order time-dependent perturbation and to obtain the golden rule. As an example, the ionization probability of hydrogen by a linearly polarized electromagnetic wave is calculated in detail. In addition, the case of unpolarized electromagnetic wave is calculated.

12.1 Time-Independent Perturbation Theory

The equations arising in most of the applications of the quantum theory, as in the classical one, are not accessible to exact solutions. Therefore, approximation methods must be used. One of the widely used approximation methods is the perturbation theory, which was first developed to handle problems in classical celestial mechanics, e.g., Neptune was discovered astronomically after it was predicted by *Leverrier*, comparing the actual observed orbit of Jupiter with the theoretically calculated one. *Leverrier* showed that the deviation of the observed orbit of Jupiter from an ellipse is due to the perturbation caused by Neptune on it.

This theory in quantum mechanics was developed firstly by *Schrödinger* to find successive corrections to a discrete energy eigenvalue and to the corresponding eigenstate. It is assumed that the time-independent Hamiltonian H of the physical system, can be written as

$$H = H_0 + H'$$

where H_0 is such that the solution of its eigenvalue equation is known and H' is small enough compared with H_0, so that it can be considered as perturbation on H_0. The perturbed eigenstate and eigenvalue are expanded as power series in H'. To do this, replace H' by $\lambda H'$ and write the perturbed eigenstate and eigenvalue ψ, W as power series of λ, such that the absolute terms of these two series are the unperturbed eigenstate and eigenvalue, and the coefficients

DOI: 10.1201/9781003677109-12

of the first, second, etc., powers of λ are the first, second, etc., corrections to these unperturbed eigenstate and eigenvalue. Let $\{|k\rangle\}$ be the eigenkets of H_0 corresponding to the eigenvalues E_k, that is:

$$H = H_0 + \lambda H', \qquad H_0|k\rangle = E_k|k\rangle, \tag{12.1}$$

$$H\psi = W\psi, \tag{12.2}$$

$$\psi = \sum_{s=0}^{\infty} \lambda^s \psi_s, \qquad W = \sum_{s=0}^{\infty} \lambda^s W_s. \tag{12.3}$$

Substitute for ψ, W from (12.3) in (12.2):

$$\sum_{s=0}^{\infty} \lambda^s (H_0 + \lambda H')\psi_s = \sum_{i,j=0}^{\infty} \lambda^{i+j} W_j \psi_i. \tag{12.4}$$

Therefore,

$$H_0\psi_0 + \sum_{s=0}^{\infty} \lambda^{s+1}(H_0\psi_{s+1} + H'\psi_s) = W_0\psi_0 + \sum_{s=0}^{\infty} \lambda^{s+1} \sum_{j=0}^{s+1} W_j \psi_{s+1-j}. \tag{12.5}$$

Equating the coefficients of the same power of λ in both sides of this equation, we obtain

$$\begin{aligned}
(H_0 - W_0)\psi_0 &= 0, \\
(H_0 - W_0)\psi_1 &= (W_1 - H')\psi_0, \\
(H_0 - W_0)\psi_s &= (W_1 - H')\psi_{s-1} + \sum_{j=2}^{s} W_j \psi_{s-j}, \ \forall s \geq 2.
\end{aligned} \tag{12.6}$$

Notice that any multiple of ψ_0 can be added to ψ_s $(\forall s)$ without violating Eqs. (12.6), since ψ_s, in the chosen equation, appears only on the left side, which is not affected by this addition. We take this multiple such that

$$(\psi_0, \psi_s) = 0, \ \forall s \geq 1,$$
$$(\psi_0, \psi_0) = 1.$$

12.1.1 Nondegenerate Case

Let

$$W_0 = E_m, \quad \psi_0 = |m\rangle,$$

where E_m is a nondegenerate level (other levels may be continuous and degenerate). Multiply the third of Eqs. (12.6) by $\langle m|$, and notice that $(H_0 - E_m)$ when operates on $\langle m|$ from the right gives zero, and $\langle m|\psi_s\rangle = 0 \ \forall s > 0$, therefore:

$$W_s = \langle m|H'|\psi_{s-1}\rangle, \ \forall s > 0. \tag{12.7}$$

First-Order Correction
From Eq. (12.7)

$$W_1 = \langle m|H'|\psi_0\rangle = \langle m|H'|m\rangle = H'_{mm}, \tag{12.8}$$

which is the expectation value of the perturbation operator in the unperturbed state $|m\rangle$. Working in the unperturbed energy representation (the basis is $\{|k\rangle\}$), hence

$$\psi_1 = \sum a_k^{(1)}|k\rangle. \tag{12.9}$$

Substitution in the second equation of Eqs. (12.6) gives

$$\sum_j (E_j - E_m)a_j^{(1)}|j\rangle = (H'_{mm} - H')|m\rangle,$$

multiplying by $\langle k|$, we obtain

$$(E_k - E_m)a_k^{(1)} = -H'_{km}, \; k \neq m,$$

that is

$$a_k^{(1)} = \frac{H'_{km}}{E_m - E_k}, \; k \neq m.$$

Notice that $a_m^{(1)} = 0$, in order that $\langle m|\psi_1\rangle = 0$.

Second-Order Correction
Again, from Eq. (12.7)

$$W_2 = \langle m|H'|\psi_1\rangle = \langle m|H'|\sum_k a_k^{(1)}|k\rangle = \sum_k H'_{mk}a_k^{(1)}, \tag{12.10}$$

therefore

$$W_2 = \sum_{k\neq m} \frac{H'_{mk}H'_{km}}{E_m - E_k} = \sum_{k\neq m} \frac{|H'_{mk}|^2}{E_m - E_k}.$$

Also, in the basis $\{|k\rangle\}$,

$$\psi_2 = \sum_k a_k^{(2)}|k\rangle.$$

Substituting in the third equation of Eqs. (12.6) (with $s = 2$) gives

$$\sum_j (E_j - E_m)a_j^{(2)}|j\rangle = (H'_{mm} - H')\sum_j a_j^{(1)}|j\rangle + W_2|m\rangle,$$

multiplying by $\langle k|$, we obtain

$$(E_k - E_m)a_k^{(2)} = H'_{mm}a_k^{(1)} - \sum_j H'_{kj}a_j^{(1)}, \; k \neq m,$$

that is

$$a_k^{(2)} = -\frac{H'_{km}H'_{mm}}{(E_m - E_k)^2} + \sum_{j\neq m} \frac{H'_{kj}H'_{jm}}{(E_m - E_k)(E_m - E_j)}, \quad k \neq m,$$

$$a_m^{(2)} = 0.$$

Notice that ψ obtained by these successive approximations is not normalized.

Exercise 12.1 *To the second order in H', find (ψ, ψ).*
Hint. *Prove first that, to this order*

$$(\psi, \psi) = 1 + (\psi_1, \psi_1),$$

then substitute for ψ_1.

Example 12.1 *As an instructive example of the time-independent perturbation applied to a nondegenerate state, in [14] is considered, the perturbation of the mth energy level of a linear harmonic oscillator by an additional energy $H' = \frac{1}{2}bx^2$. The unperturbed Hamiltonian is*

$$H_0 = \frac{p^2}{2\mu} + \frac{1}{2}kx^2 \quad (\mu \textit{ is the mass}). \tag{12.11}$$

It is clear that H' can be considered as perturbation on H_0 if $\frac{b}{k} \ll 1$. The energy W is calculated by successive approximations and the results is compared with the expansion of $\sqrt{k+b}$ in powers of the small quantity $\frac{b}{k}$. (It is simpler for calculating x^2_{km} to begin with x_{mj} then use $x^2_{km} = \sum_j x_{kj}x_{jm}$ to obtain $x^2_{km} = \frac{x_0^2}{2}\left(\sqrt{m(m-1)}\delta_{k,m-2} + (2m+1)\delta_{km} + \sqrt{(m+2)(m+1)}\delta_{k,m+2}\right)$.)

12.1.2 Degenerate Case

If the unperturbed energy level E_m is ν-fold degenerate then, to it corresponds not only one eigenstate but a whole ν-dimensional subspace of eigenkets. With which eigenket of this subspace would we begin the perturbation series (Eq. (12.3))? If we begin by a wrong one, and if the degeneracy is removed (even partially) in some order, the perturbation series would have a discontinuity at $\lambda = 0$, and this method of successive approximations would not give correct results. The correct eigenkets to begin the perturbation series with will depend on the perturbation operator H'. Let $\{|m_1\rangle, |m_2\rangle, \ldots, |m_\nu\rangle\}$ be an orthonormal basis in the eigen subspace corresponding to E_m, then a correct ψ_0 is

$$\psi_0 = \sum_{\beta=1}^{\nu} a_{m_\beta}^{(0)} |m_\beta\rangle,$$

substitute this for ψ_0 and substitute E_m for W_0 in the second of Eqs. (12.6) and multiply by $\langle m_\alpha|$ $(1 \leq \alpha \leq \nu)$ to have:

$$\sum_{\beta=1}^{\nu} a_{m_\beta}^{(0)} \langle m_\alpha | W_1 - H' | m_\beta \rangle = 0, \tag{12.12}$$

therefore,

$$\sum_{\beta=1}^{\nu} (W_1 \delta_{\alpha\beta} - H'_{m_\alpha m_\beta}) a_{m_\beta}^{(0)} = 0, \tag{12.13}$$

where $H'_{m_\alpha m_\beta} = \langle m_\alpha | H' | m_\beta \rangle$. These are ν linear homogeneous algebraic equations for the ν unknowns $(a_{m_1}^{(0)}, a_{m_2}^{(0)}, \ldots, a_{m_\nu}^{(0)})$. This system of equations has a nontrivial solution only if its determinant vanishes, and thus we have the secular equation:

$$|W_1 \delta_{\alpha\beta} - H'_{m_\alpha m_\beta}| = 0. \tag{12.14}$$

This is an algebraic equation of νth degree in the unknown W_1 (the first-order correction to the energy). This equation has ν roots. If all the roots are different, then the degeneracy is completely removed in the first-order, if some of them are repeated, then the degeneracy is only reduced (partially removed) in the first-order. If the secular equation has only one (ν times repeated) root W_1, the degeneracy will not be removed at all in the first-order, in this case, the perturbation operator H' is represented in the eigen subspace corresponding to the energy level E_m by $W_1 I$, where I is the $\nu \times \nu$ unit matrix.

For each unrepeated root of the secular equation, a completely determined set of the coefficients $a_{m_\beta}^{(0)}$ can be found from Eq. (12.13) by substituting in them for W_1, the value of this root, and using the normalization condition on ψ_0. Therefore, the different corrections to the energy lead to different zero-order eigenkets. In a word, diagonalize the perturbation operator H' in the eigen subspace corresponding to the unperturbed energy level E_m. In this subspace, the eigenvalues of H' are the first-order corrections to E_m, and its eigenkets will be the required zero-order kets to begin the perturbation series with.

Notice that in the degenerate case, the condition:

$$(\psi_0, \psi_s) = 0, \ \forall s \geq 1,$$

will mean that

$$a_{m_\beta}^{(s)} = 0, \ \forall s \geq 1, \ \beta = 1, 2, \ldots, \nu.$$

Now multiply the second of Eqs. (12.6) by $\langle k|$ to obtain

$$(E_k - E_m) a_k^{(1)} = -\sum_{\beta=1}^{\nu} H'_{k m_\beta} a_{m_\beta}^{(0)},$$

that is

$$a_k^{(1)} = \frac{1}{E_m - E_k} \sum_{\beta=1}^{\nu} H'_{km_\beta} a_{m_\beta}^{(0)}, \ k \neq m_\beta, \ \beta = 1, 2, \ldots, \nu,$$

$$a_{m_\beta}^{(1)} = 0, \ \beta = 1, 2, \ldots, \nu. \tag{12.15}$$

The calculations can be extended to higher orders as in the nondegenerate case.

If degeneracy is not removed in the first-order, multiply the last of Eqs. (12.6), with $s = 2$, by $\langle m_\alpha|$ $(\alpha = 1, 2, \ldots, \nu)$ to obtain:

$$-\sum_k a_k^{(1)} H'_{m_\alpha k} + \sum_\beta W_2 a_{m_\beta}^{(0)} \langle m_\alpha | m_\beta \rangle = 0. \tag{12.16}$$

Substituting for $a_k^{(1)}$ from Eqs. (12.15), we obtain

$$\sum_{\beta=1}^{\nu} \left(W_2 \delta_{\alpha\beta} - \sum_k \frac{H'_{m_\alpha k} H'_{km_\beta}}{E_m - E_k} \right) a_{m_\beta}^{(0)} = 0, \tag{12.17}$$

the secular equation associated with these equations gives the second-order corrections W_2, and the calculations can be continued as in the first-order.

Example 12.2 (*Zeeman* effect without electron spin) *The Zeeman*[1] *effect is the change of the energy levels of an atom in the presence of a constant, homogeneous magnetic field $\vec{\mathcal{H}}$. We calculate this change in the first-order of the field. Therefore, the additional Hamiltonian H' is*

$$H' = -\vec{\mathcal{H}} \cdot \vec{\mu},$$

where $\vec{\mu}$ is the orbital magnetic moment of the electron (the electron spin, although it gives rise to terms of the same order as H', it is neglected for simplicity, just to illustrate the perturbation calculations in the degenerate case).

Since

$$\vec{\mu} = \frac{e}{2m_0 c} \vec{L} \quad (e, m_0 \text{ are the charge and mass of the electron}),$$

then

$$H' = -\frac{e}{2m_0 c} \vec{\mathcal{H}} \cdot \vec{L},$$

The unperturbed Hamiltonian H_0 of the hydrogen atom is

$$H_0 = \frac{p^2}{2m_0} - \frac{e^2}{r},$$

with the eigenkets $|nlm\rangle$.

---

[1] *P. Zeeman* (1865–1943), jointly with *H. Lorentz* (1853–1928), was awarded the *Nobel* prize in Physics in 1902 in recognition of the extraordinary service they rendered by their researches into the influence of magnetism upon radiation phenomena.

Since $\vec{\mathcal{H}}$ is constant in time and spatial coordinates, the z-axis can be taken parallel to it, therefore

$$H' = -\frac{e\mathcal{H}}{2m_0c}L_z.$$

Taking into account that $|nlm\rangle$ are also eigenkets of L_z, then they will be also eigenkets of H', and H' will be already diagonal in the eigen subspace corresponding to any energy level E_n, therefore the different first-order correction to the energy are $W_1(m)$, where

$$W_1(m) = H'_{mm} = -\frac{e\mathcal{H}}{2m_0c}\langle nlm|L_z|nlm\rangle = -\frac{e\mathcal{H}\hbar}{2m_0c}m,$$
$$m = -l, -l+1, \ldots, l.$$

and the degeneracy of the $2l+1$ states of given n and l is removed in the first-order. Would we choose the z-axis in any other direction, the perturbation H' would be proportional to another component of $\vec{L}$ and the unperturbed energy eigenkets would not be eigenkets of H', and we would go all the way of diagonalizing H' first to obtain the same corrections to E_n.

Example 12.3 (*Stark* effect) *The Stark*[2] *effect is the splitting of the energy levels of an atom when placed in a constant, homogeneous electric field $\vec{\mathscr{E}}$. If this field is much weaker than the field of the nucleus acting on the atomic electron, then the additional potential H' of the electron in the external electric field can be treated as a perturbation on the Hamiltonian H_0 of the hydrogen atom (in the absence of any external field). Following the previous example, take the z-axis in the (constant) direction of the external electric field, therefore:*

$$H' = -e\vec{\mathscr{E}}\cdot\vec{r} = -e\mathscr{E}z = -e\mathscr{E}r\cos\theta,$$

which is odd under space inversion. The eigenkets $|nlm\rangle$ of the unperturbed Hamiltonian H_0 of the hydrogen atom all have the parity of the orbital quantum number l. Therefore, the expectation value of H' is zero for any unperturbed state $|nlm\rangle$. Therefore,

$$\langle 100|H'|100\rangle = 0,$$

and since this ground state $|100\rangle$ is nondegenerate, it has no first-order Stark effect. The first excited level is 4-fold degenerate with the eigenstates

$$|200\rangle, \quad |210\rangle, \quad |211\rangle, \quad |2,1,-1\rangle.$$

To diagonalize H' in this basis, it is useful to remember, from what was mentioned above, that:

1. *All the diagonal elements vanish.*

[2] *J. Stark* (1874–1957) was awarded the *Nobel* prize in Physics in 1919 for his discovery of the Doppler effect in canal rays and the splitting of spectral lines in electric fields.

2. H' connects only states with opposite parities, that is, in this case $|200\rangle$ only, with any of the other three states.

3. Since $[z, L_z] = 0$, therefore, $zL_z = L_z z$, hence

$$\begin{aligned} \langle n'l'm'|zL_z|nlm\rangle &= \langle n'l'm'|L_z z|nlm\rangle, \\ \hbar m\langle n'l'm'|z|nlm\rangle &= \hbar m'\langle n'l'm'|z|nlm\rangle, \end{aligned}$$

therefore

$$(m - m')\langle n'l'm'|z|nlm\rangle = 0,$$

that is, $\langle n'l'm'|z|nlm\rangle$ is proportional to $\delta_{m'm}$, and H' connects only states with the same quantum number m.

To sum up, out of the sixteen matrix elements in this example, only two, namely $\langle 200|H'|210\rangle$ and its conjugate $\langle 210|H'|200\rangle$ do not vanish:

$$\langle 200|H'|210\rangle = -e\mathscr{E}\int u_{200}^* u_{210} r \cos\theta d^3r = 3ea_0\mathscr{E},$$

where a_0 is the radius of the first Bohr's orbit.
The secular equation then will be

$$\begin{vmatrix} W_1 & -3ea_0\mathscr{E} & 0 & 0 \\ -3ea_0\mathscr{E} & W_1 & 0 & 0 \\ 0 & 0 & W_1 & 0 \\ 0 & 0 & 0 & W_1 \end{vmatrix} = 0,$$

therefore,

$$\left(W_1^2 - 9e^2a_0^2\mathscr{E}^2\right) W_1^2 = 0,$$

hence the first corrections of the energy are

$$W_1 = \pm 3ea_0\mathscr{E},\ 0,\ 0. \tag{12.18}$$

It is easy to show that the first two roots correspond to the unperturbed eigenkets

$$\frac{|200\rangle \pm |210\rangle}{\sqrt{2}}, \tag{12.19}$$

and the repeated one corresponds to any two independent linear combinations of $|211\rangle$, $|2, 1, -1\rangle$.

Exercise 12.2

1. Prove that $[z, L_z] = 0$.

2. Calculate $\langle 200|H'|210\rangle$.

3. Find the unperturbed eigenkets of H' in the second energy level $n = 2$.

12.2 Time-Dependent Perturbation Theory

When the Hamiltonian of the physical system depends explicitly on time, there are no stationary solutions of *Schrödinger* equation. If this explicitly time-dependent Hamiltonian can be written as the sum of two parts

$$H = H_0 + H'(t), \tag{12.20}$$

where H_0 does not depend on time and H' is small (over all the time) compared with H_0, the time-dependent perturbation theory as proposed by *Dirac* can be used as follows: The complete *Schrödinger* equation is

$$i\hbar\frac{d}{dt}|\psi(t)\rangle = \{H_0 + H'\}|\psi(t)\rangle. \tag{12.21}$$

The zero-order approximation ($H'(t) = 0$) is

$$|\psi(t)\rangle = \sum_k a_k e^{-i\frac{E_k}{\hbar}t}|k\rangle, \tag{12.22}$$

where a_k are constant amplitudes and the kets $|k\rangle$ are the orthonormal eigenkets of the unperturbed Hamiltonian

$$H_0|k\rangle = E_k|k\rangle, \quad \langle j|k\rangle = \delta_{jk} \text{ or } \delta(j-k). \tag{12.23}$$

Dirac looked for a solution of Eq. (12.21) in the form of Eq. (12.22) with coefficients a_k depending on time, this is why this method is known as the method of variation of constants.

Substituting Eq. (12.22) in Eq. (12.21) gives

$$\sum_k (i\hbar\dot{a}_k + E_k a_k)e^{-i\frac{E_k}{\hbar}t}|k\rangle = \sum_k a_k(E_k + H')e^{-i\frac{E_k}{\hbar}t}|k\rangle. \tag{12.24}$$

That is

$$\sum_k i\hbar\dot{a}_k e^{-i\frac{E_k}{\hbar}t}|k\rangle = \sum_k a_k e^{-i\frac{E_k}{\hbar}t}H'|k\rangle. \tag{12.25}$$

Multiplying both sides by $\langle j|$ gives

$$i\hbar\dot{a}_j = \sum_k a_k e^{i\omega_{jk}t}\langle j|H'|k\rangle, \tag{12.26}$$

where $\omega_{jk} = \dfrac{E_j - E_k}{\hbar}$ is the *Bohr*'s frequency.

Up till now, no approximations were used, and the *Dirac*'s system of Eqs. (12.26) for all j is exactly equivalent to Eq. (12.21).

To perform the algorithm of the perturbation theory replace H' in Eq. (12.26) by $\lambda H'$ and expand the a's as power series of λ

$$a_k = \sum_s a_k^{(s)} \lambda^s, \tag{12.27}$$

assuming that the series is continuous analytic function of λ, $\forall \lambda \in [0, 1]$. Substitute this series for a in Eq. (12.26) to obtain

$$\sum_{s=0}^{\infty} \dot{a}_j^{(s)} \lambda^s = \frac{1}{i\hbar} \sum_{s=0}^{\infty} \lambda^{s+1} \sum_k a_k^{(s)} e^{i\omega_{jk}t} H'_{jk}, \tag{12.28}$$

equating the coefficients of equal powers of λ gives

$$\begin{aligned} \dot{a}_j^{(0)} &= 0, \\ \dot{a}_j^{(s+1)} &= \frac{1}{i\hbar} \sum_k a_k^{(s)} e^{i\omega_{jk}t} H'_{jk}. \end{aligned} \tag{12.29}$$

The first equation of Eqs. (12.29) shows that the zero-order coefficients are constants as they must be, their values are given by the initial state of the system before the perturbation is applied. If the system is in the state $|i\rangle$ at $t = t_0$ (the moment at which the perturbation began to act), then

$$a_j^{(0)} = \langle j|i\rangle = \delta_{ji} \text{ or } \delta(j - i), \tag{12.30}$$

according as the initial energy E_i lies in the discrete or in the continuous sector of the energy spectrum.

Substituting this initial condition in the second of Eqs. (12.29) with $s = 0$, and integrating with respect to time gives

$$a_j^{(1)}(t) = \frac{1}{i\hbar} \int_{t_0}^{t} e^{i\omega_{ji}t'} H'_{ji} dt'. \tag{12.31}$$

The transition probability amplitude from the initial state $|i\rangle$ to a certain final state $|f\rangle$, in the first-order approximation is:

$$a_f^{(1)}(\infty) = \frac{\sqrt{2\pi}}{i\hbar} \frac{1}{\sqrt{2\pi}} \int_{-\infty}^{\infty} H'_{fi}(t') e^{i\omega_{fi}t'} dt' = \frac{\sqrt{2\pi}}{i\hbar} \tilde{H}'_{fi}(\omega_{fi}), \tag{12.32}$$

where $\tilde{H}'_{fi}(\omega_{fi})$ is the *Fourier* transform of the matrix element of the perturbation between these final and initial states $|f\rangle$ and $|i\rangle$.

The total transition probability w per unit time (in the first-order approximation) is

$$w = \frac{d}{dt} \sum_f |a_f(t)|^2. \tag{12.33}$$

To obtain the second-order correction $a_j^{(2)}$ for the transition probability amplitudes, substitute for $a_j^{(1)}$ from Eq. (12.31) in Eqs. (12.29) (with $s = 1$) and integrate with respect to time. Thus, by successive integration of Eqs. (12.29), the correction for the transition probability amplitudes to any order is obtained.

It is clear that the time-dependent perturbation theory can also be applied when H' does not depend on time, and in this case we obtain the same results as in the time-independent perturbation theory 12.1.

There are many applications of the first-order time-dependent perturbation theory based on Eqs. (12.31), (12.33) [5, 14, 12, 15].

12.2.1 Harmonic Perturbation

As an illustration of the first-order time-dependent perturbation, we shall show how to calculate the ionization probability of a hydrogen atom by a transient electromagnetic wave. To do this, we first consider, more generally, simple harmonic (in time) perturbations, which act on the system in a finite interval of time T:

$$H'(t) = 2H' \sin \omega t, \tag{12.34}$$

where H' does not depend explicitly on time (it depends on the position), and ω is positive, therefore

$$H'_{fi}(t) = 2\langle f|H'|i\rangle \sin \omega t. \tag{12.35}$$

Substituting Eq. (12.35) into Eq. (12.31) to obtain

$$a_f^{(1)}(t \geq T) = -\frac{1}{i\hbar} H'_{fi} \left(\frac{e^{i(\omega_{fi}+\omega)T} - 1}{\omega_{fi} + \omega} - \frac{e^{i(\omega_{fi}-\omega)T} - 1}{\omega_{fi} - \omega} \right), \tag{12.36}$$

here we took t_0 to be zero. This amplitude is significantly different from zero only for $\omega_{fi} \to \pm\omega$, i.e., when $E_f \to E_i \pm \hbar\omega$, which means that this perturbation gives to or takes from the system an energy quantum $\hbar\omega$. Notice that, when calculating the probability (by taking the modulus square of this amplitude), the resulting two interference terms are always zero, since each one of them is a product of two factors, when one of them is appreciably different from zero the other is almost zero, therefore each of the two terms of the amplitude Eq. (12.36) can be treated independently. We consider only the second term since we aim finally to calculate an ionization probability where the atom absorbs energy. In this case the transition probability is:

$$|a_f^{(1)}(t \geq T)|^2 = \frac{|H'_{fi}|^2}{\hbar^2} \frac{4 \sin^2 \frac{1}{2}(\omega_{fi} - \omega)T}{(\omega_{fi} - \omega)^2}. \tag{12.37}$$

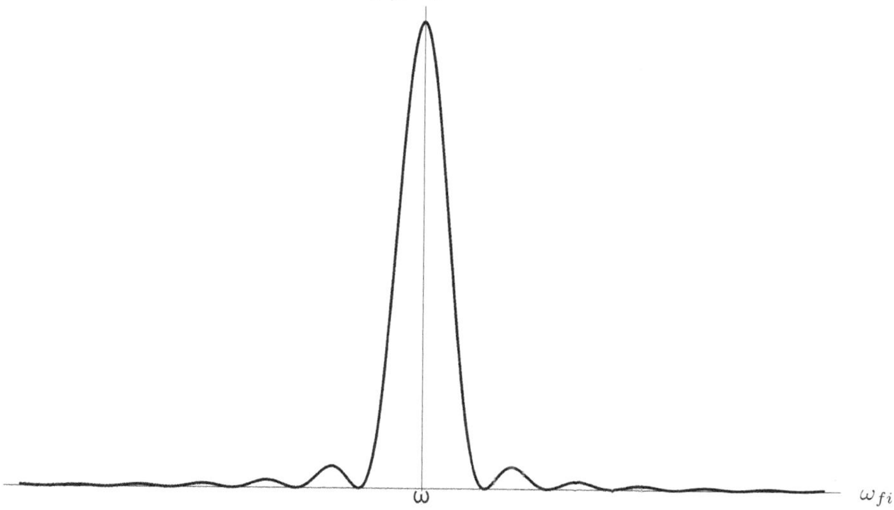

FIGURE 12.1: *The main peak of the curve is at $\omega_{fi} = \omega$ and its height is T^2.*

The factor $\frac{4}{(\omega_{fi}-\omega)^2}\sin^2\frac{1}{2}(\omega_{fi}-\omega)T$ is plotted in Fig. 12.1 as a function[3] of $\omega_{fi}-\omega$. The main peak of the curve is at $\omega_{fi} = \omega$ and its height is T^2. The secondary peaks are very much smaller, the height of the biggest of them (at $\omega_{fi}-\omega \simeq \pm\frac{3\pi}{T}$) is less than 5% of that of the main one. Therefore, transitions from initial state $|i\rangle$ take place with significantly different from zero probabilities only to states $|f\rangle$ for which $\omega_{fi}-\omega$ is under the main peak, that is

$$|\omega_{fi}-\omega| \leq \frac{2\pi}{T}, \tag{12.38}$$

i.e.,

$$|\hbar\omega_{fi}-\hbar\omega| \leq \frac{h}{T},$$

$$-\frac{h}{T} \leq E_f-(E_i+\hbar\omega) \leq \frac{h}{T},$$

$$E_i+\hbar\omega-\frac{h}{T} \leq E_f \leq E_i+\hbar\omega+\frac{h}{T},$$

therefore

$$\Delta E_f \sim \frac{h}{T}, \tag{12.39}$$

[3]Multiply this function by $\frac{1}{\pi T}$ to obtain another example of a family of functions tending to δ-function as $T\to\infty$ (see Example 1.2).

or

$$T\Delta E_f \sim h, \tag{12.40}$$

which is the uncertainty relation for the final energy of the system after the perturbation had passed. The uncertainty of energy is characteristic for any wave process, as in optics, the finite duration of emission causes a widening of the spectral lines.[4] The conservation of energy modified by the uncertainty in energy reads:

$$|E_f - (E_i + \hbar\omega)| \leq \Delta E \sim \frac{h}{T}, \tag{12.41}$$

which came as a result of the calculations. For enough large values of T, the uncertainty in energy tends to zero and (12.41) gives the familiar law of conservation of energy:

$$E_f = E_i + \hbar\omega. \tag{12.42}$$

The total transition probability to all final states $|f\rangle$ is $\sum_f |a_f^{(1)}|^2$, since the contribution to this sum is negligible for all states with energies E_f not satisfying (12.41), then, for large enough T, ΔE is small enough to consider $|H'_{fi}|^2$ independent of f and to take it out of the sum. Also, if the energy levels E_f within ΔE are close enough, the summation is changed to integration[5]

$$\sum_f \to \rho(E_f) \int dE_f, \tag{12.43}$$

where $\rho(E_f)dE_f$ is the number of states with energies between E_f and $E_f + dE_f$. The change of the density of states $\rho(E_f)$ is considered also negligible within the small interval ΔE and $\rho(E_f)$ is taken out of the integral. Therefore,

$$\sum_f |a_f^{(1)}(t \geq T)|^2 = \frac{1}{\hbar}\rho(E_f)|H'_{fi}|^2 \int \frac{4\sin^2 \frac{1}{2}(\omega_{fi} - \omega)T}{(\omega_{fi} - \omega)^2} d\omega_{fi}, \tag{12.44}$$

since the integrand outside the interval of integration $\left[\frac{-2\pi}{T}, \frac{2\pi}{T}\right]$ is almost zero, we can extend this interval to $(-\infty, \infty)$, therefore

$$\begin{aligned}\sum_f |a_f^{(1)}(t \geq T)|^2 &= \frac{2T}{\hbar}\rho(E_f)|H'_{fi}|^2 \int_{-\infty}^{\infty} \frac{\sin^2 x}{x^2} dx \\ &= \frac{2\pi}{\hbar} T\rho(E_f)|H'_{fi}|^2, \end{aligned} \tag{12.45}$$

where $x = (\omega_{fi} - \omega)\frac{T}{2}$. The total transition probability per unit time w is

$$w = \frac{2\pi}{\hbar}\rho(E_f)|H'_{fi}|^2. \tag{12.46}$$

[4] See Eq. (1.72).

[5] Do not forget that the summation (integration) is taken not only on the energy eigenvalues by it is taken over the energy eigenstates.

Fermi called this formula the "golden rule", because very many applications are based on it.

Exercise 12.3 *Find the golden rule for constant perturbations ($\omega = 0$).*

12.2.2 The Ionization Probability of a Hydrogen Atom by a Linearly Polarized Electromagnetic Wave

Neglect the magnetic force, and since the electromagnetic wave is linearly polarized, its electric field is of constant direction, the perturbation $H'(t)$ is

$$H'(t) = -2e\,\vec{r}\cdot\vec{\mathscr{E}}_0 \sin\omega t, \tag{12.47}$$

($2\vec{\mathscr{E}}_0 \sin\omega t$ is the electric field of the incident electromagnetic wave and $e < 0$ is the electron charge.)

Let the initial state of the atom be the ground state:

$$u_i(\vec{r}) = u_{100}(\vec{r}) = \frac{1}{\sqrt{\pi a_0^3}} e^{-r/a_0}, \tag{12.48}$$

(a_0 is the radius of *Bohr*'s first orbit), the final state is the state of a free electron (the ejected electron) with momentum $\hbar\vec{k}$:

$$u_f = \frac{1}{L^{3/2}} e^{i\vec{k}\cdot\vec{r}} \quad \text{(normalization in a box is adopted)}, \tag{12.49}$$

therefore

$$k_\alpha = \frac{2\pi}{L} n_\alpha, \tag{12.50}$$

$$E_f = \frac{\hbar^2 k^2}{2\mu} \quad (\mu \text{ is the electron mass}). \tag{12.51}$$

For a given E_f (given k) there are different states corresponding to the different directions of $\vec{k}$, their number in an element of volume (of the phase space) $dk_x dk_y dk_z$ is $\left(\frac{2\pi}{L}\right)^{-3} dk_x dk_y dk_z = \left(\frac{2\pi}{L}\right)^{-3} k^2 dk \sin\theta d\theta d\phi$, where θ, ϕ are the spherical angles of $\vec{k}$ with respect to some constant direction which we take for convenience to be that of the direction of the momentum of the incident electromagnetic wave[6]

$$\rho(E_f) dE_f = \frac{L^3}{8\pi^3} k^2 dk \sin\theta d\theta d\phi, \tag{12.52}$$

---

[6]This result is obtained also if we take into account that as $L \to \infty$; the Reiman sum over the equipartition $\Delta k_x \Delta k_y \Delta k_z = (\frac{2\pi}{L})^3$ tends to an integration over d^3k, i.e., $\sum_f g(\vec{k}_f)(\frac{2\pi}{L})^3 \to \int g(\vec{k}_f) d^3k_f$.

from Eq. (12.51),

$$dE_f = \frac{\hbar^2}{\mu} k dk, \tag{12.53}$$

therefore

$$\rho = \frac{\mu L^3}{8\pi^3 \hbar^2} k \sin\theta d\theta d\phi. \tag{12.54}$$

H'_{fi} is calculated in the coordinate representation:

$$H'_{fi} = -\frac{e}{\sqrt{\pi a_0^3 L^3}} \int \vec{\mathscr{E}}_0 \cdot \vec{r}\, e^{-(\frac{r}{a_0} + i\vec{k}\cdot\vec{r})} d^3r, \tag{12.55}$$

in calculating this integral, since $\vec{k}$ does not depend on the element of integration, it can be taken as the polar axis. By this choice we avoid the appearance of ϕ' in the exponent, where θ', ϕ' are the spherical angles of $\vec{r}$. Let α, λ be those of $\vec{\mathscr{E}}_0$ and θ'' be the angle between $\vec{\mathscr{E}}_0$ and $\vec{r}$, therefore

$$H'_{fi} = -\frac{e\mathscr{E}_0}{\sqrt{\pi a_0^3 L^3}} \int r^3 e^{-(\frac{r}{a_0} + ikr\cos\theta')} \cos\theta'' \sin\theta' dr d\theta' d\phi', \tag{12.56}$$

$$\cos\theta'' = \cos\alpha\cos\theta' + \sin\alpha\sin\theta'\cos(\phi' - \lambda). \tag{12.57}$$

Perform the integration to obtain

$$H'_{fi} = \frac{32\pi i e\mathscr{E}_0 k a_0^5}{(1 + a_0^2 k^2)^3 \sqrt{\pi a_0^3 L^3}} \cos\alpha, \tag{12.58}$$

and

$$w = A\cos^2\alpha \sin\theta d\theta d\phi, \tag{12.59}$$

with

$$A = \frac{256\mu e^2 \mathscr{E}_0^2 a_0^7 k^3}{\pi\hbar^3 (1 + a_0^2 k^2)^6}. \tag{12.60}$$

w per unit solid angle is proportional to $e^2\mathscr{E}_0^2\cos^2\alpha$, that is the probability amplitude is proportional to $e\mathscr{E}_0\cos\alpha$, the component of the driving force in the direction of ejection. Notice that α depends on θ and ϕ.

The wave number $k = \sqrt{2\mu E_f}/\hbar$ is fixed by the energy conservation $E_f = E_i + \hbar\omega$, since the energy uncertainty ΔE tends to zero as $T \to \infty$ (the atom is exposed to the wave for a time T).

12.2.3 Ionization of a Hydrogen Atom by an Unpolarized Electromagnetic Radiation

The ionization probability in this case is obtained by averaging Eq. (12.59) over the polarization of the incident waves (the waves are incident in one and the same direction). To do this, take the polar axis to be in the direction of the momentum $\hbar\vec{q}$ of the incident ionizing waves. Since an electromagnetic wave is transverse, its electric field is perpendicular to $\vec{q}$ and it is of constant direction (for linearly polarized wave). Take the x-axis in this direction, therefore $\cos\alpha$ in Eq. (12.59) will be $\cos\alpha = \sin\theta\cos\phi$, and w will be

$$w = A\sin^2\theta\cos^2\phi\sin\theta d\vartheta d\phi. \tag{12.61}$$

There are only two independent polarization states for a plane monochromatic electromagnetic wave, and the two states of linear polarization in two different directions can be taken as a basis. For this basis to be orthogonal, the two directions of linear polarization must be mutually orthogonal, and these two directions are perpendicular to $\vec{q}$ as have been mentioned above. Take the x, y axes in these two directions. Denote by w_1, w_2 the ionization probabilities of the hydrogen atom by these two linearly polarized waves, therefore

$$w_1 = A\sin^2\theta\cos^2\phi\sin\theta d\theta d\phi, \tag{12.62}$$
$$w_2 = A\sin^2\theta\sin^2\phi\sin\theta d\theta d\phi, \tag{12.63}$$

hence

$$w_{\text{aver}} = \frac{1}{2}(w_1 + w_2) = \frac{1}{2}A\sin^2\theta\sin\theta d\theta d\phi \tag{12.64}$$

which is the ionization probability (of a hydrogen atom) by unpolarized electromagnetic radiation. $w_{1,2}$ and w_{aver} are the ionization probabilities with the ejected electron emitted within an element $d\Omega = \sin\theta d\theta d\phi$ of a solid angle about $\hbar\vec{k}$. If we are interested in the ionization probability regardless of the direction the electron is emitted in, we integrate over the solid angle. The result is the same for integration of $w_{1,2}$ and w_{aver}. It will be w_{tot}, where:

$$w_{\text{tot}} = \frac{4\pi}{3}A. \tag{12.65}$$

13

Identical Particles

In this chapter, symmetric and antisymmetric wave functions as well as the identical particles are defined. Also, the energy exchange degeneracy is shown how it results. The distinguishability of identical particles and how it affects the results of certain calculations is discussed. For a system of identical particles, when in a zero-order approximation, the interaction can be neglected, and the energy eigenfunction of the system is simply the product of one-particle eigenfunctions. The *Pauli* exclusion principle is obtained. The relation between types of symmetry and statistics is explained. For illustration, the helium atom ground state and the s-state of its first level are studied in a crude approximation.

13.1 Symmetric and Antisymmetric Wave Functions

Definition. A function $\psi(1, 2, \ldots, N; t)$ is symmetric if it is invariant under the interchange of any pair of its arguments $1, 2, \ldots, N$, and it is antisymmetric if such an interchange results only in a change of the sign of ψ.

Any permutation P of N objects can always be decomposed into successive interchanges of pairs of these objects, if the number of these interchanges is even (odd), P is said to be an even (odd) permutation. Therefore, if ψ is symmetric, then

$$\psi(P(1, 2, \ldots, N); t) = \psi(1, 2, \ldots, N; t), \ \forall P, \tag{13.1}$$

and for an antisymmetric function ψ,

$$\psi(P(1, 2, \ldots, N); t) = \begin{cases} \psi(1, 2, \ldots, N; t), & \forall \text{ even } P, \\ -\psi(1, 2, \ldots, N; t), & \forall \text{ odd } P. \end{cases} \tag{13.2}$$

13.2 Identity of Particles

Identical particles can not be distinguished one from the others, by any inherent property, otherwise, they would not be identical. The physical situation

DOI: 10.1201/9781003677109-13

is absolutely unaltered by substituting identical particles for each other.[1] The mathematical expression of this, is that any dynamical variable A of a system of identical particles is always given by a symmetric function $A(1,2,\ldots,N;t)$ of its arguments $1,2,\ldots,N$. Each of these numbers denotes collectively the position and spin of one of the particles. This way, A remains unaltered with the interchange of any pair of the identical particles, or more generally, A is invariant under any permutation of its arguments. That is

$$A(P(1,2,\ldots,N);t) = A(1,2,\ldots,N;t), \ \forall P. \tag{13.3}$$

The *Schrödinger* equation for N particles is

$$i\hbar\frac{\partial}{\partial t}\psi(1,2,\ldots,N;t) = H(1,2,\ldots,N;t)\psi(1,2,\ldots,N;t), \tag{13.4}$$

if the particles are identical, H, as any other dynamical variable, is symmetric in the arguments $1,2,\ldots,N$, so if at any time t the function ψ is symmetric (antisymmetric), the right side of Eq. (13.4) is symmetric (antisymmetric). Therefore, $\dfrac{\partial\psi}{\partial t}$ is also symmetric (antisymmetric). By successive differentiation of Eq. (13.4) with respect to time, we see that all higher derivatives are again symmetric (antisymmetric). Thus, the symmetry property character of ψ (to be symmetric or antisymmetric) is preserved in time. Although there are, theoretically, symmetries other than these two, they are not realized in nature. That is each kind of identical particles in nature is described either by a symmetric or by an antisymmetric wave function (see below).

We show that if $\psi(1,2,\ldots,N;t)$ is a solution of Eq. (13.4), then $\psi(P(1,2,\ldots,N);t)$ is also a solution of the same equation. Since the same permutation applied everywhere in Eq. (13.4) means just a relabelling of the particles, the equation will be still valid, i.e.,

$$i\hbar\frac{\partial}{\partial t}\psi(P(1,2,\ldots,N);t) = H(P(1,2,\ldots,N);t)\psi(P(1,2,\ldots,N);t), \tag{13.5}$$

and since H is symmetric, then

$$i\hbar\frac{\partial}{\partial t}\psi(P(1,2,\ldots,N);t) = H(1,2,\ldots,N;t)\psi(P(1,2,\ldots,N);t), \tag{13.6}$$

[1]A comment of philosophical nature:
The identity of micro-objects: elementary particles, atoms, molecules, to be exact and exhaustive, is a postulate of the theory, that is two individuals of the same kind among these objects have absolutely all their inherent properties to be exactly the same. Can we say, on this basis, that up to this simple structure of matter, the two worlds actual and ideal (Platonic) are the same, and, e.g., the Platonic (ideal) electron is just the actual one, and only when the structure of matter is sophisticated enough, two individuals of the same kind are never exactly identical (two cats, two ducks, two oranges,. . . etc) and always there are differences between individuals of the same kind. May be, for this, *Plato* decided that the actual world is just an approximate (distorted) realisation of the ideal one existing only in his mind, where all the individuals of the same kind are exactly the same (identical).

therefore $\psi(P(1,2,\ldots,N);t)$ is a solution of Eq. (13.4), $\forall P$, and $N!$ solutions are obtained from any one solution. Any linear combination of these solutions is again a solution of the same equation, and thus we can construct the symmetric and antisymmetric solutions as follows:

$$\psi_{S,A}(1,2,\ldots,N;t) = \sum_{P\ \text{even}} \psi(P(1,2,\ldots,N);t) \pm \sum_{P\ \text{odd}} \psi(P(1,2,\ldots,N);t). \tag{13.7}$$

Indeed, the interchange of any pair of the arguments $1,2,\ldots,N$ will change any even permutation in an odd one and vice versa. Therefore, if after this interchange the wave function is denoted by $\psi'_{S,A}$:

$$\psi'_{S,A} = \sum_{P\ \text{odd}} \psi(P(1,2,\ldots,N);t) \pm \sum_{P\ \text{even}} \psi(P(1,2,\ldots,N);t) = \pm\psi_{S,A} \tag{13.8}$$

as it must be. Notice that any antisymmetric function vanishes always whenever it is unaltered under the interchange of any two of its arguments α, β, since it must on the other hand (under this interchange) change sign, being antisymmetric, that is:

$$\begin{aligned}\psi_A(1,\ldots,\alpha,\ldots,\beta,\ldots,N;t) &= \psi_A(1,\ldots,\beta,\ldots,\alpha,\ldots,N;t)\\ &= -\psi_A(1,\ldots,\beta,\ldots,\alpha,\ldots,N;t) = 0.\end{aligned}$$

13.3 The Energy Exchange Degeneracy

If H does not depend explicitly on time, stationary solutions can be found:

$$\psi(1,2,\ldots,N;t) = e^{-i\frac{E}{\hbar}t}u(1,2,\ldots,N), \tag{13.9}$$

where

$$Hu(1,2,\ldots,N) = Eu(1,2,\ldots,N), \tag{13.10}$$

and the solutions derived from any u by permutations of its arguments $1,2,\ldots,N$ are degenerate with the original u. This is called exchange degeneracy. Therefore, if E is originally (without counting the exchange degeneracy) n-fold degenerate then the degree of degeneracy of this level including the exchange degeneracy will be $N!\,n$.

13.4 Distinguishability of Identical Particles

In classical physics, identical particles, though they are identical, are in principle distinguishable by sharply definable trajectories for the individual

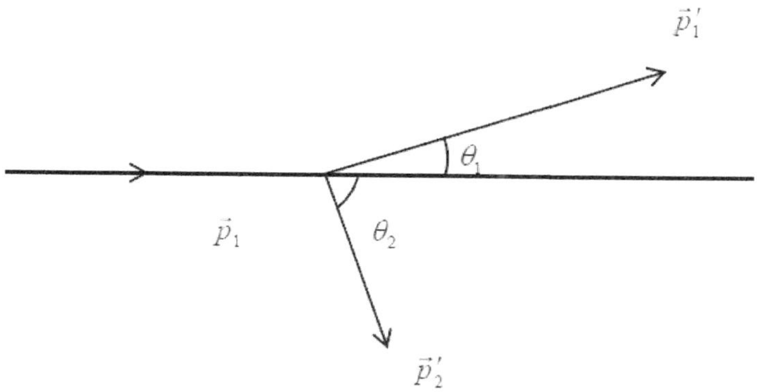

FIGURE 13.1: *Collision of two particles. Before collision one of them is at rest and the other travels towards it with momentum* $\vec{p}_1$. *After collision the two particles possess momenta* $\vec{p}_1{}', \vec{p}_2{}'$.

particles. As an example, consider the elastic collision of two classical particles, one of them (the target) is at rest before collision, after collision the incident particle scatters by an angle θ_1 and the struck one flies at an angle θ_2 with respect to the direction of incidence $\vec{p}_1$, see Fig. 13.1. Two counters, each of which registers one kind of the two particles will count in a unit time, dN_1, dN_2 where

$$dN_1 = nd\sigma_1(\theta_1), \quad dN_2 = nd\sigma_2(\theta_2), \tag{13.11}$$

(n is the intensity of the incident beam). If the two particles are identical, the only difference is that the counter located to register the particles flying at an angle θ (with respect to $\vec{p}_1$) will count (in unit time) the sum dN

$$dN = nd\sigma_1(\theta) + nd\sigma_2(\theta). \tag{13.12}$$

In the quantum mechanical treatment of this example, the connection between the results in these two cases is, in general, not that simple because the identical particles before collision cannot be distinguished into an incident one and one at rest.

Nevertheless, identical particles, in some circumstances, in quantum theory can be distinguished by their positions. For example, if the wave function $u(1,2)$ of two identical particles is such that

$$u(1,2) = 0, \ \forall 1 \notin A \text{ or } 2 \notin B \text{ and } A \cap B = \phi. \tag{13.13}$$

This is the case, for instance, for two electrons in two well separated atoms one in region A and the other in region B, and A, B do not overlap. In the state $u(1,2)$, the probability density is $|u(1,2)|^2$, while in the symmetric or antisymmetric states it is

$$\begin{aligned}|u(1,2) \pm u(2,1)|^2 &= |u(1,2)|^2 + |u(2,1)|^2 \pm 2\text{Re } u^*(1,2)u(2,1) \\ &= |u(1,2)|^2 + |u(2,1)|^2 \end{aligned} \tag{13.14}$$

because if one of the two factors of $u^*(1,2), u(2,1)$ is not zero, the other necessarily is zero owing to Eq. (13.13) and this term is everywhere zero. Therefore, the probability density is independent on whether the wave function is symmetric or antisymmetric, and it is always the sum of the two densities for one particle being in region A and the other in region B. This is the result to obtain, if the particles are not identical, and we are interested only in having one of them, whichever, in region A and the other in region B. Thus, the interference effects between exchange-degenerate state functions represented by the product term in Eq. (13.14) disappear when the coordinates of the particles do not overlap.[2]

13.5 Separation of Variables for Systems of Noninteracting Particles

A zero-order approximation for a system of N particles is obtained by neglecting the interaction between these particles. In this approximation, the (total) Hamiltonian $H_T^{(0)}$ of the system is the sum of the Hamiltonians H_α of the particles

$$H_T^{(0)} = \sum_{\alpha=1}^{N} H_\alpha(\alpha) \tag{13.15}$$

(the argument α between brackets here denotes collectively, as before, the coordinates of the αth particle). Let H_α be time-independent $\forall\alpha$. The energy eigenfunction $u^{(0)}$ of the system in this approximation will be the product of the energy eigenfunctions u_α of the individual particles, and the energy eigenvalue E of the system will be the sum of the energy eigenvalues E_α of these individual particles. Indeed:

$$H_T^{(0)} u^{(0)} = \sum_{\alpha=1}^{N} H_\alpha(\alpha) u^{(0)}. \tag{13.16}$$

We search for a solution in the form:

$$u^{(0)}(1,2,\ldots,N) = u_1(1)u_2(2)\ldots u_N(N), \tag{13.17}$$

[2]Notice that in calculating quantum probabilities what are added are the probability amplitudes. This gives rise to the appearance of product terms, which explains the interference phenomenon. While calculating classical probabilities, what is added are the probability densities, and no interference term appears.

and notice that operators of one particle have no effect on the operators or wave functions of any other particle, therfore

$$\sum_{\alpha=1}^{N} H_\alpha u_1 u_2 \ldots u_N = \sum_{\alpha=1}^{N} u_1 u_2 \ldots (H_\alpha u_\alpha) u_{\alpha+1} \ldots u_N = E u^{(0)}, \tag{13.18}$$

dividing by $u^{(0)}$:

$$\sum_{\alpha=1}^{N} \frac{1}{u_\alpha(\alpha)} H_\alpha(\alpha) u_\alpha(\alpha) = E. \tag{13.19}$$

The terms in the left side of this equation are independent, since each one of them depends only on the coordinates of only one particle. Therefore, the only way to have their sum constant is to have each of them constant, i.e.,

$$\begin{aligned} \frac{1}{u_\alpha} H_\alpha u_\alpha &= E_\alpha, \\ H_\alpha u_\alpha &= E_\alpha u_\alpha, \ \forall \alpha, \end{aligned} \tag{13.20}$$

hence

$$\sum_{\alpha=1}^{N} E_\alpha = E. \tag{13.21}$$

If we would not neglect the interaction between the particles, an additional term $H'(1, 2, \ldots, N)$ would be added to the right side of Eq. (13.15). This interaction term combines the coordinates of different particles preventing Eq. (13.16) to reduce to the N independent Eqs. (13.20), and the eigenstates of the total Hamiltonian H_T would not be simple products of those of H_α any more. To the total energy E an additional term arising from the interaction must be added.

13.6 *Pauli* Exclusion Principle

If the particles of the system are identical, all the H_α's are equal, and we denote any one of them by H, that is

$$H_T^{(0)} = \sum_{\alpha=1}^{N} H(\alpha) = H(1) + H(2) + \cdots + H(N), \tag{13.22}$$

and

$$u^{(0)}(1, 2, \ldots, N) = u_k(1) u_m(2) \ldots u_n(N), \tag{13.23}$$

(the product of one particle eigenfunctions of H),

$$E = E_k + E_m + \cdots + E_n, \tag{13.24}$$

where

$$H(1)u_k(1) = E_k u_k(1), \ldots \text{ and so forth.} \tag{13.25}$$

If the particles are electrons, they must be described by an antisymmetric function (see below), which can be most easily written as the determinant

$$u_A^{(0)}(1, 2, \ldots, N) = \frac{1}{\sqrt{N!}} \begin{vmatrix} u_k(1) & u_k(2) & \ldots & u_k(N) \\ u_m(1) & u_m(2) & \ldots & u_m(N) \\ \ldots & \ldots & \ldots & \ldots \\ u_n(1) & u_n(2) & \ldots & u_n(N) \end{vmatrix} \tag{13.26}$$

assuming that $(u_j(\alpha), u_l(\alpha)) = \delta_{jl},\ \forall \alpha = 1, 2, \ldots, N$.

The determinant form of $u_A^{(0)}$ shows, transparently, that it vanishes if two or more of the one particles eigenstates are the same, this means that such states can not exist, and no more than one electron can occupy a one particle eigenstate. This is the *Pauli* exclusion principle which he postulated (before the discovery of quantum theory) to explain the *Mendeleyev* periodic system of chemical elements (the *Coulomb* repulsion between the electrons in the atom which removes degeneracy of the energy levels is needed to find the mechanism of the filling of the different orbits in the atom, see [5, Ch. 26]).[3]

13.7 Types of Symmetry and Statistics

Experimental and theoretical investigations of systems of identical particles have shown as mentioned before that there are only two kinds of particles. Systems of the first kind of identical particles are described by symmetric wave functions, and systems of the second kind are described by antisymmetric ones, and there are no in nature other kinds of particles which would be described by wave functions of other symmetries.

In the zero-order approximation (neglecting the interaction between the identical particles), any number of particles of the first kind (that is described by symmetric wave functions) can occupy any one particle state, e.g., $u_k(1)u_k(2)\ldots u_k(N)$ is the symmetric function corresponding to all the N particles in the one particle state u_k. But it is meaningless to associate a particular particle with a specific state u_k or u_m, ..., etc., because rigorously

[3] A second comment of philosophical nature:
An independent science is always founded (logically) on its own postulates. The postulates of chemistry, for example, for a long time, were the phenomenological laws of constant ratios, multiple ratios, valance, equivalent weights, ...etc, which had an important role for the atomic theory for the structure of matter to prevail and win out. These postulates, at last, were unified by *Mendeleyev* in his periodic system of elements which is now fully explained by the quantum theory. Can one any more now speak of chemistry as an independent science? Or it is rather a branch (very large and very important) of physics, it is one of the physical sciences like, Astronomy, for example?

speaking, here the word particular has no meaning, since all the particles are identical. All what one can say is that there are N_k particles in the state u_k, N_m particles in the state u_m, ..., and N_n particles in the state u_n and this leads to *Bose-Einstein* statistics. A particle of this kind is called a boson.

The number of identical particles of the second kind described by antisymmetric wave functions, which can occupy a one particle state is exclusively 0 or 1, since if more than one of this kind of particles occupy a one particle state, the antisymmetric wave function of the system would vanish. Here also, it is meaningless to tell which particle in which state because the particles are identical, although the one particle states are not. All what we can say, here also, is how many particles are there in each one particle state with the restriction mentioned above that this number is either 0 or 1 only. This leads to *Fermi-Dirac* statistics, and a particle of this kind is called a fermion. In relativistic quantum field theory, it is shown that particles of integral spin ($s = 0, 1, 2, \dots$) are bosons and those of half an odd integer spin ($s = \frac{1}{2}, \frac{3}{2}, \dots$) are fermions. Examples of bosons are mesons: π, K both with $s = 0$ (spinless mesons), ρ with $s = 1$ (vector meson). Gauge particles $\gamma, W^{\pm}, Z, G$ are vector bosons with $s = 1, \dots$, etc. Examples of fermions are electrons, μ, τ leptons and baryons like protons, neutrons, Λ, Σ baryons all of which are of spin $s = \frac{1}{2}$, Δ, Ω^- baryons are of spin $s = \frac{3}{2}$.

Tightly bound aggregates of particles, like atomic nuclei can be considered as particles, if their internal structure is negligible in the problem under investigation. An interchange of two identical nuclei each is made up of a A fermions (protons+neutrons) is equivalent to A simultaneous interchanges of pairs of fermions, and each fermion interchange reverses the sign, the overall effect on the wave function is to be multiplied by a factor $(-1)^A$. Therefore, nuclei with even A like ${}_2^4He$ (α-particles), ${}_6^{12}C$ (carbon), ${}_7^{14}N$ (nitrogen) are bosons, while nuclei with odd A like ${}_1^1H$ (hydrogen), ${}_3^7Li$ (lithium) are fermions.

13.8 Helium Atom

We study here the ground state and the s-state of the first excited level of the helium atom in a crude approximation, just to give an illustration for many of the ideas explained above. As was mentioned earlier, more than one space function is needed to describe different spin states (see § 11.4.4).

In the approximation we are adopting, spin dependent forces are neglected, therefore, spin and space variables are separable, and the wave function will have the form

$$\psi(1,2) = \varphi(\vec{r}_1, \vec{r}_2)\chi(s_1, s_2). \tag{13.27}$$

χ is a linear combination of the direct products of $\chi_{z\pm}$, the spin states of an electron ($s = \frac{1}{2}$). It is convenient to take these combinations to be the simultaneous eigenstates of s^2, s_z where $\vec{S} = \vec{S}_1 + \vec{S}_2$ is the total spin of the two electrons. When the *Coulomb* repulsion between the two electrons, is taken as the perturbation, the energy eigenfunction $\phi^{(0)}$ in the zero-order approximation will be a linear combination of the products of energy eigenfunctions of a hydrogen-like atom with $Z = 2$.

13.8.1 Spin Functions

When $\vec{S}_1$ with $s_1 = \frac{1}{2}$ is added to $\vec{S}_2$ with $s_2 = \frac{1}{2}$, an $\vec{S}$ is obtained with $s = 0$ or 1 (see § 11.3). There are only four linearly independent spin functions $\uparrow\uparrow, \uparrow\downarrow, \downarrow\uparrow, \downarrow\downarrow$ for a pair of electrons. Notice that the first and the last ones are symmetric and are eigenfunctions of s_z corresponding to the eigenvalues $\pm\hbar$, therefore the only needed eigenfunction of s_z which completes the triplet ($s = 1$) is that corresponding to the eigenvalue zero of s_z. Since space rotations linearly mix the components of the spin, and in the same time result in unitary transformations of the quantum states which preserve the type of symmetry of the states, i.e., it transforms symmetric states into symmetric ones and antisymmetric states into antisymmetric ones. Therefore, under space rotations, symmetric and antisymmetric states do not mix. Therefore, $\uparrow\uparrow$ and $\downarrow\downarrow$ must be completed for the triplet by the symmetric combination

$$\frac{\uparrow\downarrow + \downarrow\uparrow}{\sqrt{2}}. \tag{13.28}$$

The other linear combination orthogonal to the symmetric one is the antisymmetric combination

$$\frac{\uparrow\downarrow - \downarrow\uparrow}{\sqrt{2}}, \tag{13.29}$$

which is reserved for the singlet.

Exercise 13.1

1. *Show that $\uparrow\downarrow \pm \downarrow\uparrow$ are mutually orthogonal.*

2. *Show by calculation that the three states of the triplet are eigenstates of s^2 corresponding to the eigenvalue $2\hbar^2$, and the singlet corresponds to the eigenvalue zero of s^2.*
Hint. $s^2 = \frac{\hbar^2}{4}(\vec{\sigma}_1 + \vec{\sigma}_2)^2$. *Prove that* $s^2 = \frac{\hbar^2}{2}(3 + \vec{\sigma}_1 \cdot \vec{\sigma}_2)$, *then find the effect of* $\vec{\sigma}_1 \cdot \vec{\sigma}_2$ *on* $\uparrow\uparrow, \downarrow\downarrow, \uparrow\downarrow, \downarrow\uparrow$ *noting that, e.g.,* $\vec{\sigma}_1 \cdot \vec{\sigma}_2 \uparrow\downarrow = (\vec{\sigma}_1 \uparrow) \cdot (\vec{\sigma}_2 \downarrow)$, *etc.*

If the triplet is denoted by ${}^3\chi$ and the singlet by ${}^1\chi$, we have:

$${}^3\chi_1 = \uparrow\uparrow, \quad {}^3\chi_0 = \frac{\uparrow\downarrow + \downarrow\uparrow}{\sqrt{2}}, \quad {}^3\chi_{-1} = \downarrow\downarrow\,. \tag{13.30}$$

It is said that the two electron spins are parallel in the triplet spin states. Also

$$^{1}\chi_0 = \frac{\uparrow\downarrow - \downarrow\uparrow}{\sqrt{2}}, \tag{13.31}$$

and it is said that the two electron spins are antiparallel in the singlet spin state.

Exercise 13.2 *Find from the tables the necessary corresponding Clebsch-Gordan coefficients, and substitute for them in the equation*

$$|jm\rangle = |m_1 m_2\rangle\langle m_1 m_2|jm\rangle \tag{13.32}$$

to realize that the triplet and singlet spin states are indeed the eigenfunctions of s^2, s_z corresponding to the mentioned eigenvalues.

13.8.2 The Space Part of the Wave Function

As we mentioned, the space part of the wave function is a linear combination of the energy eigenfunctions u_{nlm} of the hydrogen-like atom with $Z = 2$. In spectroscopic notations, the unperturbed ground state is $(1s)^2$. Both electrons are in the $1s$ state u_{100}, i.e.,

$$(1s)^2 = u_{100}(\vec{r}_1)u_{100}(\vec{r}_2), \tag{13.33}$$

which is symmetric and therefore must be multiplied by the antisymmetric singlet spin function to have an overall antisymmetric wave function for the two electrons.

In the first excited level we shall, for simplicity, consider only $(1s)(2s)$ states. Here there are only two linearly independent space functions

$$u_{100}(\vec{r}_1)u_{200}(\vec{r}_2), \quad u_{100}(\vec{r}_2)u_{200}(\vec{r}_1) \tag{13.34}$$

and the perturbation operator

$$H'(\vec{r}_1, \vec{r}_2) = \frac{e^2}{r_{12}} \tag{13.35}$$

is the *Coulomb* repulsion between the two electrons. In the above basis

$$H' = \begin{bmatrix} J & K \\ K & J \end{bmatrix} \tag{13.36}$$

with

$$\begin{aligned} J &= \int u_{100}^*(\vec{r}_1)u_{200}^*(\vec{r}_2)\frac{e^2}{r_{12}}u_{100}(\vec{r}_1)u_{200}(\vec{r}_2)d^3r_1d^3r_2, \\ K &= \int u_{100}^*(\vec{r}_1)u_{200}^*(\vec{r}_2)\frac{e^2}{r_{12}}u_{100}(\vec{r}_2)u_{200}(\vec{r}_1)d^3r_1d^3r_2. \end{aligned} \tag{13.37}$$

J is the *Coulomb* energy and K is called the exchange energy. Diagonalizing this matrix

$$\left(\begin{bmatrix} J & K \\ K & J \end{bmatrix} - W_1 I\right)\begin{pmatrix} a_1 \\ a_2 \end{pmatrix} = 0, \tag{13.38}$$

nontrivial solutions for a_1, a_2 are obtained if

$$\begin{vmatrix} J - W_1 & K \\ K & J - W_1 \end{vmatrix} = 0, \tag{13.39}$$

that is,

$$W_1 = J \pm K. \tag{13.40}$$

For $W_1 = J + K$: $a_2 = a_1$ and the space part of the wave function is

$$\frac{1}{\sqrt{2}}[u_{100}(\vec{r}_1)u_{200}(\vec{r}_2) + u_{100}(\vec{r}_2)u_{200}(\vec{r}_1)], \tag{13.41}$$

which is symmetric and therefore must be multiplied by the antisymmetric singlet spin function. This is the case of parahelium.

For $W_1 = J - K$: $a_2 = -a_1$ and the space part of the wave function is

$$\frac{1}{\sqrt{2}}[u_{100}(\vec{r}_1)u_{200}(\vec{r}_2) - u_{100}(\vec{r}_2)u_{200}(\vec{r}_1)], \tag{13.42}$$

which is antisymmetric and therefore must be multiplied by one of the symmetric spin functions of the triplet. This is the case of orthohelium. K by calculation is positive, therefore the energy in the case of parahelium is higher than that in the case of orthohelium although spin dependent interactions were neglected. The physical explanation of this is that in orthohelium, the spins of the electrons are parallel, so they must be in different space states and tend to keep away from each other. This reduces the *Coulomb* repulsion between them and hence lowers the energy (see, e.g., [4, 14]).

Solutions to Exercises

Chapter 1

Exercise 1.1

1. Prove that

$$\lim_{\varepsilon\to 0}\int_a^b f(x)\delta_\varepsilon(x-x_0)dx = \begin{cases} f(x_0), & \forall x_0 \in (a,b), \\ 0, & \forall x_0 \notin (a,b), \end{cases}$$

for any well behaved function f in a neighborhood of x_0.

Chapter 2

Exercise 2.1

$$\begin{aligned} \hbar(\omega-\omega') &= \mathscr{E}'-mc^2, \\ \hbar(\vec{k}-\vec{k}') &= \vec{p}'. \end{aligned}$$

$\rightarrow\ \hbar^2(\omega^2+\omega'^2-2\omega\omega'\cos\theta) = c^2p'^2 = \mathscr{E}'^2 - m^2c^4 = \hbar(\omega-\omega')\{\hbar(\omega-\omega') + 2mc^2\}$

$\rightarrow\ 2\hbar^2\omega\omega'(1-\cos\theta) = 2\hbar mc^2(\omega-\omega')$

$\rightarrow\ \omega'(1+2\frac{\hbar\omega}{mc^2}\sin^2\frac{\theta}{2}) = \omega,$ continue... (remember that $\omega = \frac{2\pi c}{\lambda}$).

Chapter 5

Exercise 5.1

Omit ϕ between Eqs. (5.76), (5.77) to obtain an expression for $\frac{\partial}{\partial\theta}$ in terms of L_x, L_y.

Exercise 5.3

1. Let the state Y be an eigenstate of L_z: $L_zY = \hbar mY$ and A be any operator, then

$$\langle [A, L_z]\rangle = (Y, (AL_z - L_zA)Y) = (Y, (A\hbar m - \hbar mA)Y) = 0.$$

Since $L_x = \frac{1}{i\hbar}[L_y, L_z]$ and $L_y = -\frac{1}{i\hbar}[L_x, L_z]$, therefore $\langle L_x\rangle = 0$ and $\langle L_y\rangle = 0$ in any L_z eigenstate.

Another solution:

In spherical coordinates: $L_x + iL_y = \hbar\, e^{i\phi}(\frac{\partial}{\partial\theta} + i\cot\theta\frac{\partial}{\partial\phi})$, let the L_z eigenstate be $\mathrm{f}(\theta)\, e^{im\phi}$, then

$$(L_x + iL_y)\mathrm{f}(\theta)\, e^{im\phi} = g(\theta)\, e^{i(m+1)\phi},$$

which when multiplied by $\mathrm{f}^*(\theta)\, e^{-im\phi}$ and integrated with respect to ϕ over the whole period $[0, 2\pi]$ gives the value zero for $\langle L_x\rangle + i\langle L_y\rangle$ and since $\langle L_x\rangle$ and $\langle L_y\rangle$ both are real, therefore each of them must be zero.

2. The uncertainty relation imposes on ΔL_y and ΔL_z the condition:

$$\Delta L_y \Delta L_z \geq \frac{1}{2}|\langle[L_y, L_z]\rangle| = \frac{\hbar}{2}|\langle L_x\rangle| = 0,$$

which does not prevent ΔL_y to have a finite value while $\Delta L_z = 0$ (in an L_z eigenstate).

Chapter 8

Exercise 8.3

Let $\{\vec{e}_\alpha\}$, $\{\vec{e}'_\alpha\}$ $(\alpha = 1, 2, 3)$ be two orthonormal bases of the coordinates. Any vector $\vec{A}$ can be represented in either of these two bases:

$$\vec{A} = a_\alpha \vec{e}_\alpha = a'_\alpha \vec{e}'_\alpha \quad \text{(summation convention is applied).}$$

You can easily show that $a_\alpha = \vec{e}_\alpha \cdot \vec{A}$, $a'_\alpha = \vec{e}'_\alpha \cdot \vec{A}$ and $a_\alpha = \vec{e}_\alpha \cdot \vec{e}'_\beta a'_\beta$, or in matrix form:

$$\begin{pmatrix} a_1 \\ a_2 \\ a_3 \end{pmatrix} = \Lambda \begin{pmatrix} a'_1 \\ a'_2 \\ a'_3 \end{pmatrix}, \quad \text{where} \quad \Lambda = [\vec{e}_\alpha \cdot \vec{e}'_\beta] = \begin{pmatrix} l_1 & l_2 & l_3 \\ m_1 & m_2 & m_3 \\ n_1 & n_2 & n_3 \end{pmatrix},$$

and $(l_\beta, m_\beta, n_\beta)$ are the directions cosines of $\vec{e}'_\beta$ with respect to $\{\vec{e}_\alpha\}$. Show that Λ is orthogonal $(\Lambda^{-1} = \Lambda^T)$ which means that it is unitary since the l's, m's, n's all are real. $\Lambda^{-1} = \Lambda^T$ is also obtained from Λ by the exchange of $\{\vec{e}_\alpha\}$ and $\{\vec{e}'_\alpha\}$.

Chapter 10

Exercise 10.2

$a_{mn}(0) = \langle m|a|n\rangle = \sqrt{n}\,\delta_{m,n-1}$, $a^\dagger_{mn}(0) = \sqrt{n+1}\,\delta_{m,n+1}$.

$x = \sqrt{\frac{\hbar}{2m\omega}}(a^\dagger + a) = x_0\frac{a^\dagger+a}{\sqrt{2}}$, $p = i\sqrt{\frac{m\hbar\omega}{2}}(a^\dagger - a) = im\omega x_0\frac{a^\dagger-a}{\sqrt{2}}$,

$x_{mn}(0) = \frac{x_0}{\sqrt{2}}(\sqrt{n}\,\delta_{m,n-1}+\sqrt{n+1}\,\delta_{m,n+1})$, $p_{mn}(0) = i\frac{m\omega x_0}{\sqrt{2}}(\sqrt{n+1}\,\delta_{m,n+1}- \sqrt{n}\,\delta_{m,n-1})$.

$x_{mn}(t) = \frac{x_0}{\sqrt{2}}(\sqrt{n+1}\,\delta_{m,n+1}\,e^{i\omega t} + \sqrt{n}\,\delta_{m,n-1}\,e^{-i\omega t})$,

$p_{mn}(t) = i\frac{m\omega x_0}{\sqrt{2}}(\sqrt{n+1}\,\delta_{m,n+1}\,e^{i\omega t} - \sqrt{n}\,\delta_{m,n-1}\,e^{-i\omega t})$. Find also $x^2(t)$.

Chapter 11

Exercise 11.4

2. Let z' be the axis with spherical angles θ, ϕ, then

$$\begin{aligned} S_{z'} &= \frac{\hbar}{2}(\sigma_x \sin\theta\cos\phi + \sigma_y \sin\theta\sin\phi + \sigma_z\cos\theta) \\ &= \frac{\hbar}{2}\begin{pmatrix} \cos\theta & \sin\theta\, e^{-i\phi} \\ \sin\theta\, e^{i\phi} & -\cos\theta \end{pmatrix} = \frac{\hbar}{2}\sigma_{z'}, \end{aligned}$$

where $S_{z'}$ is the spin component in the direction of z'. Diagonalize $\sigma_{z'}$ to obtain its eigenvalues (which are expected to be ± 1) and its eigenstates. The eigenvalue equation for $\sigma_{z'}$ is

$$(\sigma_{z'} - \mu I)\begin{pmatrix} a_{1,\mu} \\ a_{2,\mu} \end{pmatrix} = 0.$$

Its characteristic equation is $|\sigma_{z'} - \mu I| = \mu^2 - 1 = 0 \to \mu = \pm 1$ as expected.

Substitute for μ by ± 1 in turn in the eigenvalue equation to obtain the eigenstates $\chi_{z'_\pm}$:

$$\chi_{z'_+} = \begin{pmatrix} \cos\frac{\theta}{2} \\ \sin\frac{\theta}{2}\, e^{i\phi} \end{pmatrix}, \qquad \chi_{z'_-} = \begin{pmatrix} \sin\frac{\theta}{2} \\ -\cos\frac{\theta}{2}\, e^{i\phi} \end{pmatrix},$$

which are orthonormal.

For the x-axis ($\theta = \frac{\pi}{2}, \phi = 0$): $\chi_{x\pm} = \frac{1}{\sqrt{2}}\begin{pmatrix} 1 \\ \pm 1 \end{pmatrix} = \frac{\chi_{z_+} \pm \chi_{z_-}}{\sqrt{2}}$.

For the y-axis ($\theta = \frac{\pi}{2}, \phi = \frac{\pi}{2}$): $\chi_{y\pm} = \frac{1}{\sqrt{2}}\begin{pmatrix} 1 \\ \pm i \end{pmatrix} = \frac{\chi_{z_+} \pm i\chi_{z_-}}{\sqrt{2}}$.

For the z-axis ($\theta = 0$): $\chi_{z+} = \begin{pmatrix} 1 \\ 0 \end{pmatrix}$, $\quad \chi_{z-} = \begin{pmatrix} 0 \\ 1 \end{pmatrix}$ as expected.

3. The spherical angles of z' are θ, ϕ. That is the electron is in the spin state $\chi_{z'_+}$ of problem 2. Therefore, $\langle \vec{S} \rangle = \frac{\hbar}{2}\chi^{\dagger}_{z'_+}\vec{\sigma}\chi_{z'_+} = \frac{\hbar}{2}(\sin\theta\cos\phi, \sin\theta\sin\phi, \cos\theta)$,

therefore, $\langle \vec{S} \rangle^2 = \frac{\hbar^2}{4}$. Since $S^2 = \frac{3\hbar^2}{4}I$, therefore $\langle S^2 \rangle = \frac{3\hbar^2}{4} \neq \langle \vec{S} \rangle^2$.

Calculate $\Delta S_x, \Delta S_y$ and verify that

$$\Delta S_x \Delta S_y = \frac{\hbar^2}{4}\sqrt{\cos^2\theta + \sin^4\theta\sin^2\phi\cos^2\phi} \geq \frac{\hbar^2}{4}|\cos\theta| = \frac{1}{2}|\langle \hbar S_z \rangle|.$$

What $\langle S_{z'} \rangle$ will be equal to? Can you explain the result?

4. The electron is in the spin state χ_{z+}. Expand it in the basis $\chi_{z'_\pm}$, where z' is making an angle θ with the z-axis:

$$\chi_{z+} = \begin{pmatrix} 1 \\ 0 \end{pmatrix} = \chi_{z'_+}\cos\frac{\theta}{2} + \chi_{z'_-}\sin\frac{\theta}{2}.$$

The probability for $S_{z'}$ to be $\frac{\hbar}{2}$ is therefore $\cos^2\frac{\theta}{2}$ and to be $-\frac{\hbar}{2}$ is $\sin^2\frac{\theta}{2}$, that for S_x to be $\frac{\hbar}{2}$ is $\frac{1}{2}$ and to be $-\frac{\hbar}{2}$ is also $\frac{1}{2}$.

Bibliography

[1] L. D. Landau and E. M. Lifshitz, *Textbook on Theoretical Physics. 4th Edition: Vol. 2: The Classical Theory of Fields (Course of Theoretical Physics Series).* Elsevier Science, 2013.

[2] M. J. Lighthill, *Introduction to Fourier Analysis and Generalized Functions.* Cambridge, 1958.

[3] A. Messiah, *Quantum Mechanics.* Dover Publications, New York, 2014.

[4] A. A. Sokolov, I. M. Ternov, and V. Zhukovskii, *Quantum mechanics,* 1984.

[5] A. A. Sokolov, T. Loskutov, and I. M. Ternov, *Quantum mechanics,* 1966.

[6] M. Born, *Atomic Physics. 8th Edition.* Dover Publications, New York, 2013.

[7] L. D. Landau and E. M. Lifshitz, *Textbook on Theoretical Physics. Vol. 5, 3rd edition, Part 1: Statistical Physics.* Elsevier Science, 2013.

[8] L. D. Landau and E. M. Lifshitz, *Textbook on Theoretical Physics. Vol. 1: Mechanics.* Butterworth Heinmann, 2000.

[9] V. A. Fock, *Fundamentals of Quantum Mechanics, 3rd printing,* 1986.

[10] L. D. Landau and E. M. Lifshitz, *Quantum Mechanics Non-Relativistic Theory,* Oxford: Pergamon Press, 1991.

[11] H. Goldstein, *Classical Mechanics, 3rd Edition.* Addison-Wesley, 2001.

[12] E. Merzbacher, *Quantum Mechanics,* New York: John Wiley & Sons, 1998.

[13] P. A. M. Dirac, *The Principles of Quantum Mechanics, 4th Edition.* Clarendon Press, 1967.

[14] L. I. Schiff, *Quantum Mechanics, 3rd Edition,* New York: McGraw-Hill, 1968.

[15] P. M. Mathews and K. Venkatesan, *A Text Book of Quantum Mechanics, 2nd Edition,* Tata McGraw-Hill Education Private Limited, 2010.

Index